High-Pressure Physics

High-Pressure Physics

Editor

Khamisi Zaman

scitus
academics

High-Pressure Physics
Edited by **Khamisi Zaman**

ISBN: 978-1-68117-716-8
Library of Congress Control Number: 2016958816

Notice

Printed in United States of America on Acid Free Paper

Preface

High-pressure science has experienced a revolution in the past 15 years. Knowledge of the behavior of matter under conditions of extreme pressure is essential for describing the interior state. The development of intense new x-ray and neutron sources, improved detectors, new instrumentation, greatly increased computation power, and advanced computational algorithms have enabled researchers to determine the behavior of matter at static pressures in excess of 400 GPa. High-pressure phenomena refers to the changes in physical, chemical, and structural characteristics that matter undergoes when subjected to high pressure. Pressure thus serves as a versatile tool in materials research, and it is especially important in the investigation of the rocks and minerals that form the deep interior of the Earth and other planets. In spite of the superficial similarity between temperature and pressure, these two variables are fundamentally different in the ways they affect a material's internal energy. Temperature alterations reflect changes in the kinetic energy and thus in the thermodynamic behaviour of vibrating atoms. Increased pressure, on the other hand, changes the energy of atomic bonds by forcing atoms closer together in a smaller volume. Pressure thus serves as a powerful analysis of atomic interactions and chemical bonding. Furthermore, pressure is an important tool for synthesizing dense structures, including superhard materials, novel solidified gases and liquids, and mineral-like phases suspected to occur deep within the Earth and other planets.

High-Pressure Physics introduces the state of the art experimental and theoretical techniques in this field. It emphases on scientific challenges in condensed matter under thrilling conditions of pressure and temperature, synthesis and characterization of novel materials, and employs state-of-the-art experimental facilities and technologies.

Emphasizing applications in a range of physics disciplines—this book cuts across many areas and delivers a solid instruction in high-pressure physics.

The book, providing an overview of key research areas and applications in high-pressure physics, will be a valuable tool for advanced graduate students and young researchers.

Table of Contents

Chapter 1

Crystal Structures and Mechanical Properties of Ca_2C at High Pressure

Qun Wei [1,*], Quan Zhang [2] and Meiguang Zhang [3,*]

[1]*School of Physics and Optoelectronic Engineering, Xidian University, Xi'an 710071, China*

[2]*School of Microelectronics, Xidian University, Xi'an 710071, China*

[3]*College of Physics and Optoelectronic Technology, Baoji University of Arts and Sciences, Baoji 721016, China*

ABSTRACT

Recently, a new high-pressure semiconductor phase of Ca_2C (space group *Pnma*) was successfully synthesized, it has a low-pressure metallic phase (space group *C2/m*). In this paper, a systematic investigation of the pressure-induced phase transition of Ca_2C is studied on the basis of first-principles calculations. The calculated enthalpy reveals that the phase transition which transforms from *C2/m*-Ca_2C to *Pnma*-Ca_2C occurs at 7.8 GPa, and it is a first-order phase transition with a volume drop of 26.7%. The calculated elastic constants show that *C2/m*-Ca_2C is mechanically unstable above 6.4 GPa, indicating that the structural phase transition is due to mechanical instability. Both of the two phases exhibit the elastic anisotropy. The semiconductivity of *Pnma*-Ca_2C and the metallicity of *C2/m*-Ca_2C have been demonstrated by the electronic band structure calculations. The quasi-direct band gap of *Pnma*-Ca_2C at 0 GPa is 0.86 eV. Furthermore, the detailed analysis of the total and partial density of states is performed to show the specific contribution to the Fermi level.

Keywords: pressure-induced phase transition; first-principles calculations; Ca_2C

1. INTRODUCTION

Hitherto, the pressure-composition (*P*-*x*) phases of binary systems have gained increasing interest and been extensively researched. Among these predicted compounds, some of them have been successfully synthesized [1,2], but the others still need further experiments to confirm their theoretical predictions [3,4]. For the Ca-C system, there are many works that have been done and obtained remarkable achievements [5,6,7,8,9,10,11,12,13,14,15,16]. Gauzzi et al. [5] found the superconductivity will be enhanced in the intercalated graphite CaC_6 at high pressure. It performs the structural instability and leads to a structural transition with pressure. Nylen et al. [6] studied the structural behavior of CaC_2 at high pressure via the first-principles calculations. Their results suggest an irreversible amorphization, corroborating the structural peculiarities of acetylide carbides, which persists at high pressure conditions. Li et al. analyzed the pressure-induced superconductivity of CaC_2 [7]. They uncovered that it is calcium that contributes to the superconducting behavior, and it is capable of stabilizing carbon sp^2hybridization at a larger range of pressure. Nourbakhsh et al. [8] investigated the magnetism in CaC ionic compound and observed a perfect Fermi level spin polarization and a half-metallic behavior.

Recently, Li et al. [9] systematically explored all the stable calcium carbides at pressures from 0 to 100 GPa. This resulted in five newly predicted stable stoichiometries (Ca_5C_2, Ca_2C, Ca_3C_2, CaC and Ca_2C_3). Using in situ synchrotron powder X-ray diffraction measurements, they successfully synthesized the Ca_2C and Ca_2C_3. The Ca_2C has two phases: the semiconducting phase *Pnma*-Ca_2C at high pressure and the metallic metastable phase *C2*/*m*-Ca_2C at low pressure. The *Pnma*-Ca_2C exists in the pressure range of 7.5–100 GPa and possesses the isolated C anions. Carbon atoms polymerize to isolated dumbbells, occurring a unique metallic metastable *C2*/*m*-Ca_2C which provides an example of 2D metal. The metal calcium atom of *C2*/*m*-Ca_2C develops a negative Bader charge, confronting a more electronegative carbon atom. Due to these intriguing properties, in this paper, we will focus on the Ca_2C, presenting its structural, elastic and electronic properties, and systematically investigating the pressure-induced phase transition mechanism. The enthalpy is calculated to reveal the phase transition pressure. In addition, the elastic constants, modulus and anisotropy are calculated to study the elastic properties. Meanwhile, the electronic band structures and the total and partial density of states are analyzed.

2. RESULTS AND DISCUSSION

The 2 × 1 × 2 supercell structures of Ca_2C are illustrated in Figure 1. The black and blue spheres represent C and Ca atoms, respectively. At zero pressure, the optimized lattice parameters of *Pnma*-Ca_2C are a = 6.677 Å, b = 4.384 Å, c = 7.979 Å with two inequivalent Ca atoms occupying 4c (0.0119, 0.2500, 0.8302), 4c (0.1476, 0.2500, 0.4109) and C atoms occupying 4c (0.2521, 0.2500, 0.0918) Wyckoff positions. For *C2*/*m*-Ca_2C, the optimized lattice parameters are a = 7.166 Å, b = 3.775 Å, c = 15.490 Å, and β = 122.9°. The Wyckoff positions of *C2*/*m*-Ca_2C are Ca1: 4i (0.2715, 0.0000, −0.1120), Ca2: 4i (−0.2440, 0.0000, −0.3781) and C: 4i (−1.0479, 0.0000, −0.9705). For *Pnma*-Ca_2C (see Figure 1a), carbon atoms are isolated anions, whereas the carbon dimers are observed in *C2*/*m*-Ca_2C (Figure 1b). The interatomic distance of Ca-C for *C2*/*m*-Ca_2C is 2.44 Å in length, and the C-C bond length is 1.29 Å.

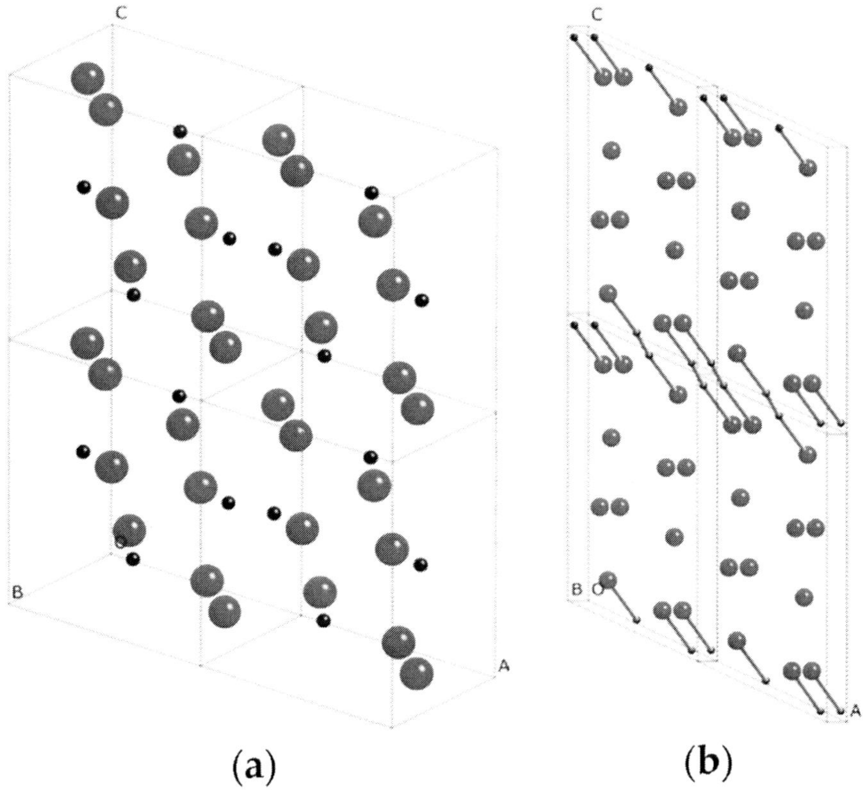

Figure 1. Crystal structures of Ca_2C. (**a**) *Pnma*-Ca_2C; (**b**) *C2*/*m*-Ca_2C. The black and blue spheres represent C and Ca atoms, respectively.

To determine the phase transition pressure of Ca_2C, the enthalpy differences between two structures are plotted as a function of pressure up to 100 GPa in Figure 2a. There is an intersection between the two enthalpy curves, indicating that the *C*2/*m*-Ca_2C phase transforms to the *Pnma*-Ca_2C phase at 7.8 GPa and the *Pnma*-Ca_2C is more stable than the *C*2/*m*-Ca_2C above this pressure point. The known transition pressure data is 7.5 GPa [9], and it is in a good agreement with our result. Meanwhile, the dependence of volume on pressure is presented in Figure 2b. The *C*2/*m*-Ca_2C is larger than the *Pnma*-Ca_2C in volume. The change of volume at 7.8 GPa shows that the phase transition is first-order with a volume drop of 26.7%. To interpret this large volume collapse, we estimated the ionic radii of the C and Ca within these two structures at 7.8 GPa through Bader charge analysis. The obtained results are listed in Table 1. The calculated charges of the two Ca_2C phase show increasing trends from *C*2/*m*-$Ca_2{}^{0.928}C^{-0.928}$ to *Pnma*-$Ca_2{}^{2.348}C^{-2.348}$ at phase transition pressure point. Compared to *C*2/*m*-Ca_2C phase (r_{Ca} = 1.871 Å, r_C = 1.534 Å), the ionic radius of Ca in the *Pnma*-Ca_2C phase is much shorter (1.485 Å), whereas the ionic radius of C (1.788 Å) in *Pnma*-Ca_2C is relatively longer. Since the contribution from Ca atom is much more than that of C atom to the volume of Ca_2C, the volume collapse from *C*2/*m* to *Pnma* phase is very large.

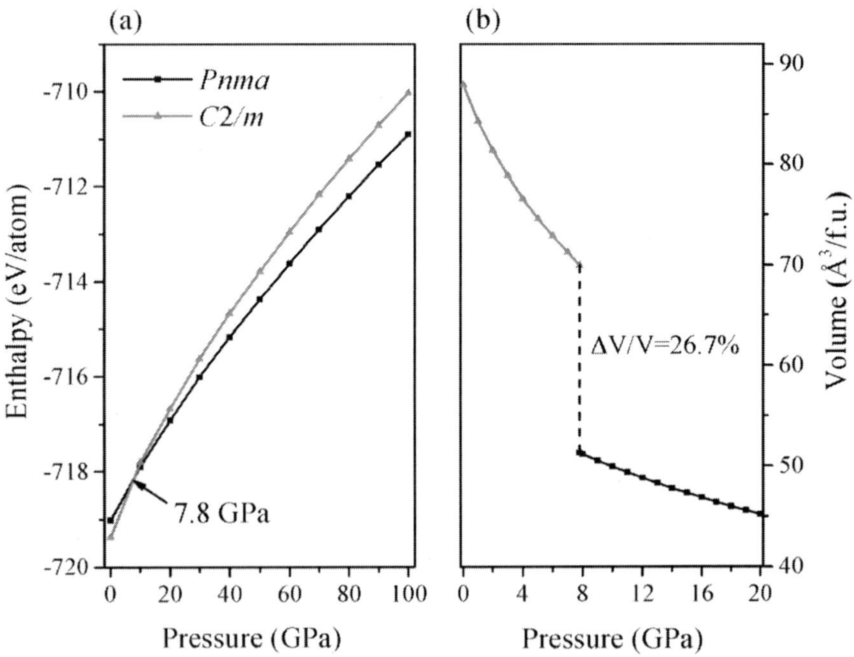

Figure 2. Enthalpy (**a**) and volume (**b**) as a function of pressure. The black and red solid lines represent *Pnma*-Ca_2C and *C*2/*m*-Ca_2C, respectively.

Table 1. Bader charge analysis for *C2/m*- and *Pnma*-Ca_2C at 7.8 GPa.

Phase	Ionic Radius (Å)		Charge Transfers (*e*)
	C	Ca	Ca → C
C2/m	1.534	1.871	0.928
Pnma	1.788	1.485	2.348

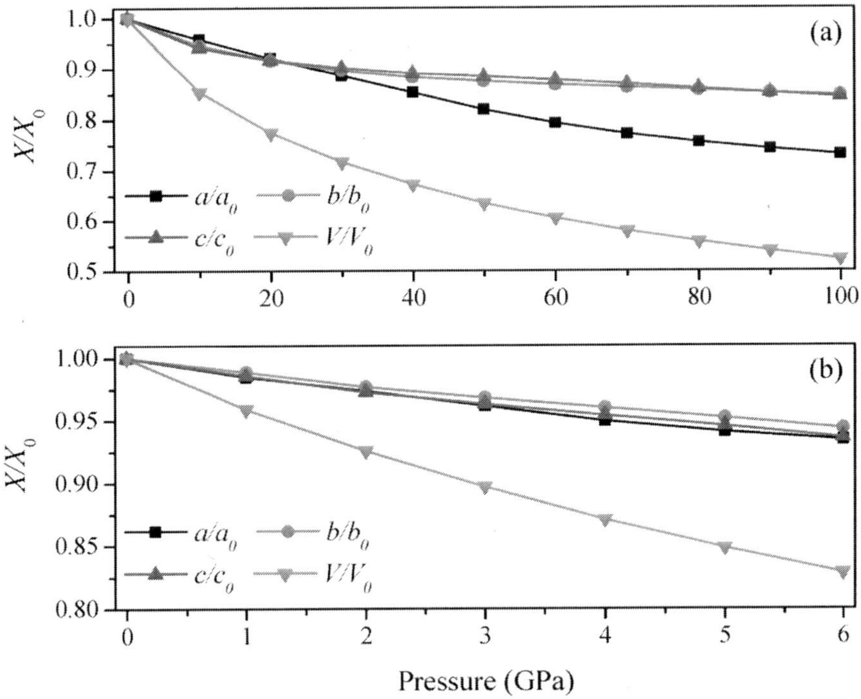

Figure 3. Lattice parameters X/X_0 as a function of pressure. (**a**)*Pnma*-Ca_2C; (**b**) *C2/m*-Ca_2C.

The lattice parameters of Ca_2C at different pressures are listed in Table 2. In Table 2, an excellent agreement with the previous theoretical and experimental values is shown [9]. The calculated lattice parameters decrease with pressure. To get more details, the variations of lattice parameters X/X_0 of the two Ca_2C phases with pressure are shown in Figure 3. For *Pnma*-Ca_2C (see Figure 3a), along the *b*- and*c*-axis, the degrees of anti-compression along these two directions are almost the same. At low pressure range ($P < 23$ GPa), the incompressibility along *a*-axis is larger than that along *b*- and *c*-axis, which is contrary to the case at high pressure range ($P > 23$ GPa). In Figure 3b, the changes of lattice parameters along the *a*-, *b*- and *c*-axis are similar for *C2/m*-Ca_2C when below 6 GPa, suggesting the same incompressibility along these three directions.

Table 2. Lattice parameters of Ca_2C at various pressures.

Phase	Pressure (GPa)	*a* (Å)		*b* (Å)		*c* (Å)		*β* (deg)		*V* (Å³)	
Pnma	0	6.677	6.689 [a]	4.384	4.389 [a]	7.979	7.981 [a]			233.58	234.32 [a]
	10	6.404	6.415 [a]	4.150	4.154 [a]	7.513	7.518 [a]			199.68	200.35 [a]
		6.449 [b]		4.157 [b]		7.523 [b]				201.7 [b]	
	30	5.919	5.929 [a]	3.931	3.933 [a]	7.195	7.204 [a]			167.41	168.04 [a]
*C*2/*m*	0	7.166		3.775		15.490		122.9		351.59	
	5	6.743	6.701 [a]	3.594	3.587 [a]	14.65	14.68 [a]	122.8	122 [a]	298.32	
	6.4	6.674		3.554		14.45		122.6		288.87	

[a] Calculated data in Ref. [9]; [b] Experimental results in Ref. [9].

The calculated elastic constants and moduli of Ca_2C at 0 GPa and high pressures are shown in Table 3. The strain-stress method was used to calculate the single crystal elastic constants. A small finite strain was applied on the optimized structure and the atomic position was fully optimized. Then, the elastic constants were obtained from the stress of the strained structure. The generalized Born's mechanical stability criteria of orthorhombic phase at 0 GPa are given by [17,18]:

$$C_{ii} > 0, i = 1 \ldots 6, \quad (1)$$

$$[C_{11} + C_{22} + C_{33} + 2(C_{12} + C_{13} + C_{23})] > 0, \quad (2)$$

$$(C_{11} + C_{22} - 2C_{12}) > 0, \quad (3)$$

$$(C_{11} + C_{33} - 2C_{13}) > 0, \quad (4)$$

$$(C_{22} + C_{33} - 2C_{23}) > 0 \quad (5)$$

The stability criteria of monoclinic phase at 0 GPa are given by [17,18]:

$$C_{ii} > 0, i = 1 \ldots 6, \quad (6)$$

$$[C_{11} + C_{22} + C_{33} + 2(C_{12} + C_{13} + C_{23})] > 0, \quad (7)$$

$$(C_{33}C_{55} - C_{35}^2) > 0, \quad (8)$$

Table 3. Calculated elastic constants C_{ij} (GPa), bulk modulus B(GPa), shear modulus G (GPa), Young's modulus E (GPa), Poisson's ratio υ, and B/G ratio of Ca_2C at various pressures.

Pressure (GPa)	*Pnma*-Ca_2C			*C2/m*-Ca_2C		
	0	50	100	0	6	6.4
C_{11}	92	212	454	32	78	71
C_{22}	87	329	551	62	86	84
C_{33}	67	331	523	56	80	86
C_{44}	40	79	114	8	13	14
C_{55}	26	97	137	19	23	20
C_{66}	38	101	135	13	17	15
C_{12}	23	158	250	11	26	28
C_{13}	35	170	248	16	27	29
C_{23}	27	185	347	7	18	23
C_{15}				−1	0.75	4
C_{25}				−1.4	−0.06	3
C_{35}				0.3	0.44	−1.31
C_{46}				0.04	2.65	2.67
B	46	203	353	24	43	45
G	30	75	122	15	21	19
E	74	200	328	37	54	50
υ	0.23	0.34	0.35	0.24	0.29	0.31
B/G	1.53	2.71	2.89	1.60	2.05	2.37

$$(C_{44}C_{66} - C_{46}^2) > 0, \tag{9}$$

$$(C_{22} + C_{33} - 2C_{23}) > 0, \tag{10}$$

$$[C_{22}(C_{33}C_{55} - C_{35}^2) + 2C_{23}C_{25}C_{35} - C_{23}^2C_{55} - C_{25}^2C_{33}] > 0, \tag{11}$$

$$\begin{aligned} &2[C_{15}C_{25}(C_{33}C_{12} - C_{13}C_{23}) + C_{15}C_{35}(C_{22}C_{13} - C_{12}C_{23}) + C_{25}C_{35}(C_{11}C_{23} - C_{12}C_{13})] \\ &-[C_{15}^2(C_{22}C_{33} - C_{23}^2) + C_{25}^2(C_{11}C_{33} - C_{13}^2) + C_{35}^2(C_{11}C_{22} - C_{12}^2)] + C_{55}g > 0, \end{aligned} \tag{12}$$

$$g = C_{11}C_{22}C_{33} - C_{11}C_{23}^2 - C_{22}C_{13}^2 - C_{33}C_{12}^2 + 2C_{12}C_{13}C_{23} \tag{13}$$

The mechanical stability in crystals under isotropic pressure is provided by Ref. [19]. This requires the symmetric matrix

$$\hat{G} = \begin{bmatrix} \tilde{C}_{11} & \tilde{C}_{12} & \tilde{C}_{13} & 2C_{14} & 2C_{15} & 2C_{16} \\ \tilde{C}_{21} & \tilde{C}_{22} & \tilde{C}_{23} & 2C_{24} & 2C_{25} & 2C_{26} \\ \tilde{C}_{31} & \tilde{C}_{32} & \tilde{C}_{33} & 2C_{34} & 2C_{35} & 2C_{36} \\ 2C_{41} & 2C_{42} & 2C_{43} & 4\tilde{C}_{44} & 4C_{45} & 4C_{46} \\ 2C_{51} & 2C_{52} & 2C_{53} & 4C_{54} & 4\tilde{C}_{55} & 4C_{56} \\ 2C_{61} & 2C_{62} & 2C_{63} & 4C_{64} & 4C_{65} & 4\tilde{C}_{66} \end{bmatrix} \tag{14}$$

has a positive determinant. In $\hat{G}$ matrix,

$$\begin{aligned} &\tilde{C}_{\alpha\alpha} = C_{\alpha\alpha} - P, \quad \alpha = 1,2,\ldots,6 \\ &\tilde{C}_{12} = C_{12} + P, \quad \tilde{C}_{13} = C_{13} + P, \quad \tilde{C}_{23} = C_{23} + P \end{aligned} \tag{15}$$

where P is the isotropic pressure.

If the elastic constants satisfy these stability criteria, it means the structure is mechanically stable. From Table 3, one can see that orthorhombic *Pnma*-Ca_2C is mechanical stable up to at least 100 GPa. For monoclinic *C2/m*-Ca_2C, the criteria C˜44C˜66−C246>0, which is similar to the Equation (9), is obeyed only up to 6.4 GPa, as seen in Figure 4, showing that it has mechanical stability below 6.4 GPa. Furthermore, the phonon spectra are presented in Figure 5 to ensure the dynamical stability. As observed, there is no imaginary frequency in the whole Brillouin zone, indicating that *Pnma*-Ca_2C is dynamically stable up to at least 100 GPa and that the *C2/m*-Ca_2C is dynamically stable below 6.4 GPa. The elastic constants as a function of pressure are displayed in Figure 6 with an approximately upward tendency. We noticed that, for *Pnma*-Ca_2C, C_{11} is larger than C_{22} or C_{33} at 0 GPa, whereas it is less thanC_{22} or C_{33} at high pressures, which is in consistent with our previous analyses on the incompressibility along the *a*-, *b*-, and *c*-axis.

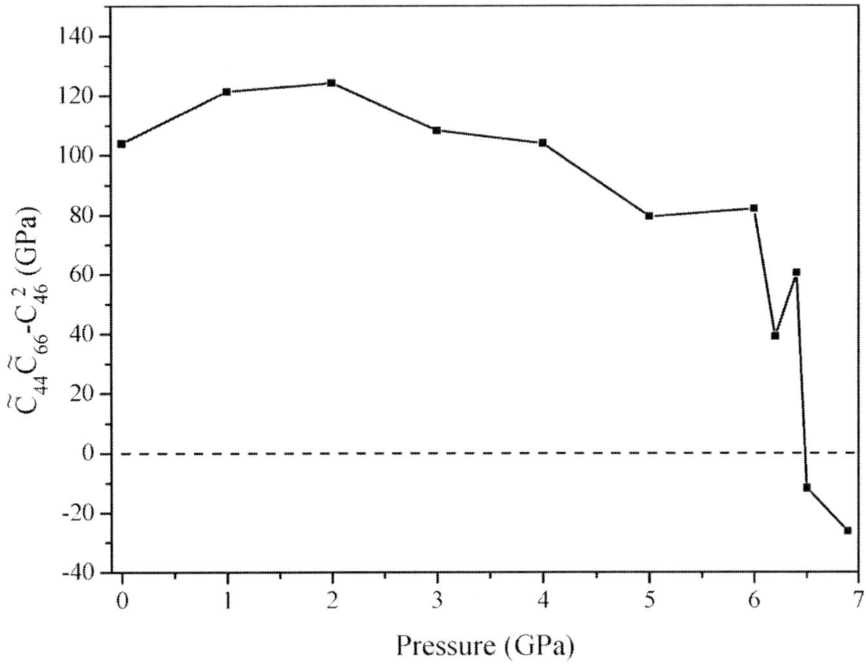

Figure 4. Calculated C˜44C˜66−C246 of *C*2/*m*-Ca_2C under different pressures.

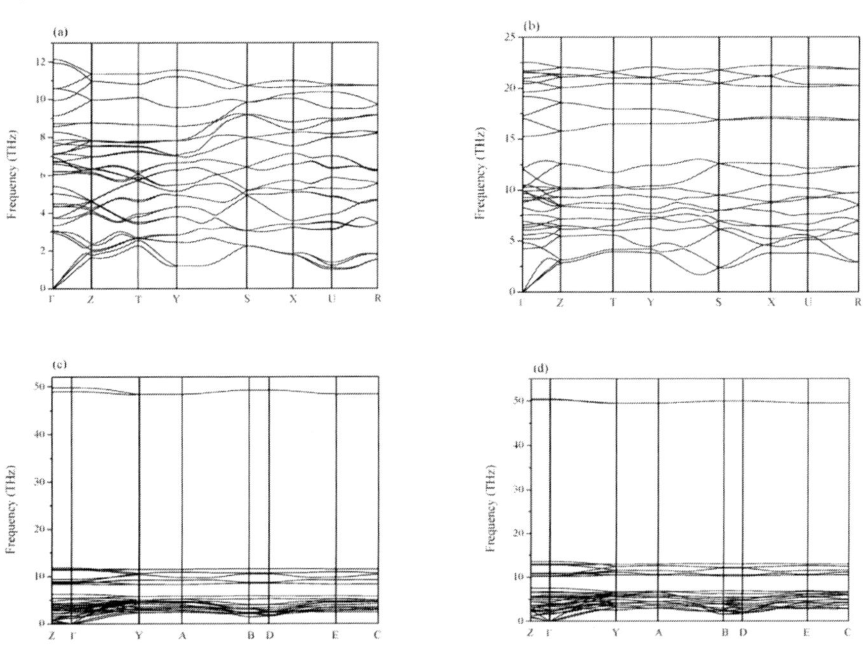

Figure 5. Phonon spectra for (**a**) *Pnma*-Ca_2C at 0 GPa; (**b**) *Pnma*-Ca_2C at 100 GPa; (**c**) *C*2/*m*-Ca_2C at 0 GPa; (**d**) *C*2/*m*-Ca_2C at 6.4 GPa.

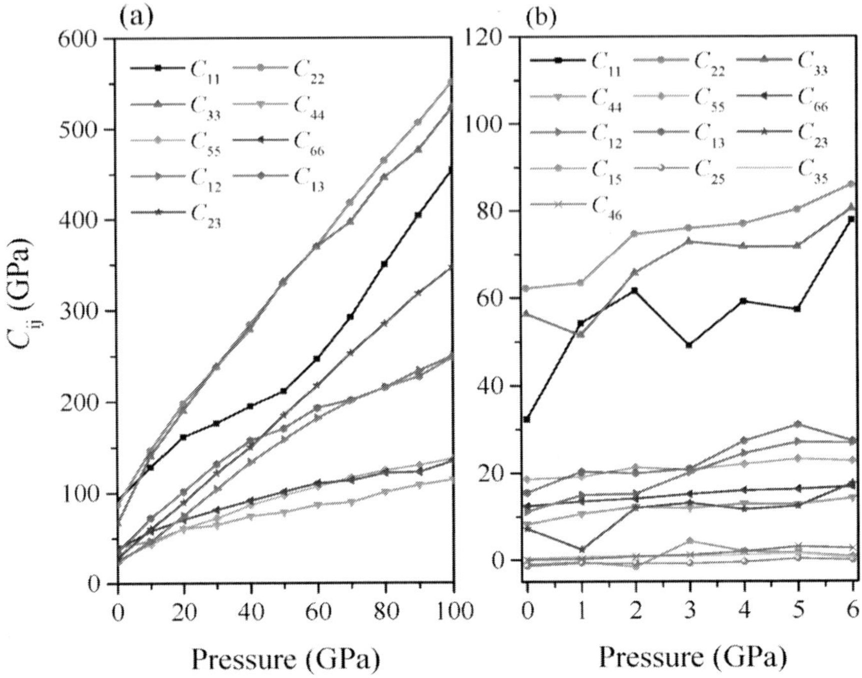

Figure 6. Elastic constants as a function of pressure. (**a**) *Pnma*-Ca_2C; (**b**) *C*2/*m*-Ca_2C.

In Table 3, the bulk modulus *B* and shear modulus *G* are calculated by Voigt-Reuss-Hill approximations [20,21,22]. The Young's modulus *E* and Poisson's ratio, υ are given by the following equations [22]:

$$E = \frac{9BG}{3B+G}, \upsilon = \frac{3B-2G}{2(3B+G)} \tag{16}$$

The *Pnma*-Ca_2C is larger than *C*2/*m*-Ca_2C in bulk modulus, shear modulus and Young's modulus at 0 GPa, as listed in Table 3. All the elastic modulus increase with pressure for *Pnma*-Ca_2C. According to Pugh [23], the brittle material has a small *B*/*G* ratio ($B/G < 1.75$), whereas, the ductile material has a larger ratio ($B/G > 1.75$). It is interesting that *Pnma*-Ca_2C and *C*2/*m*-Ca_2C show the brittle manner at 0 GPa and transform to ductile manner at 9.3 GPa and 2.0 GPa, respectively.

Calculating the elastic anisotropy of crystal is of great importance to further study the physical and chemical properties. The calculated universal elastic anisotropy index (A^U), shear anisotropic factors (A_1, A_2 and A_3) and percentage of anisotropy in compressibility and shear (A_B and A_G) are listed inTable 4. For arbitrary symmetry, the universal elastic anisotropy index A^U is obtained by [24,25]:

$$A^U = 5\frac{G_V}{G_R} + \frac{B_V}{B_R} - 6 \tag{17}$$

Table 4. Calculated universal elastic anisotropy index A^U, shear anisotropic factors A_1, A_2 and A_3, and percentage of anisotropy in compressibility and shear A_B and A_G (in %) of Ca_2C.

Phase	Pressure (GPa)	A^U	A_1	A_2	A_3	A_B	A_G
Pnma	0	0.37	1.79	1.06	1.14	0.6	3.5
	50	0.62	1.56	1.34	1.81	3.8	5.1
	100	0.15	0.95	1.44	1.06	1.3	1.2
C2/m	0	0.95	0.58	0.72	0.69	2.7	8.2
	6	0.54	0.53	0.70	0.62	0.06	5.1

When A^U is 0, it means the solid is isotropic, otherwise the solid is anisotropic. The results of *Pnma*-Ca_2C are 0.37 at 0 GPa, 0.62 at 50 GPa and 0.15 at 100 GPa, respectively. And the results of *C2/m*-$CaC_2$are 0.95 at 0 GPa and 0.54 at 6 GPa, respectively. All of them are larger than 0, indicating an elastic anisotropy. The shear anisotropic factors provide a measure of the degree of anisotropy in the bonding between atoms in different planes. The shear anisotropic factor for the {100} shear plane between the <011> and <010> directions is [26,27]:

$$A_1 = \frac{4C_{44}}{C_{11} + C_{33} - 2C_{13}} \tag{18}$$

For the {010} shear plane between the <101> and <001> directions, it is:

$$A_2 = \frac{4C_{55}}{C_{22} + C_{33} - 2C_{23}} \tag{19}$$

For the {001} shear plane between the <110> and <010> directions, it is:

$$A_3 = \frac{4C_{66}}{C_{11} + C_{22} - 2C_{12}} \tag{20}$$

The factors A_1, A_2 and A_3 are 1.0 for any isotropic crystals. As observed in Table 3, all the calculated shear anisotropic factors are not 1.0, presenting the elastic anisotropy. The percentage anisotropy in compressibility and shear are defined as [26]:

$$A_B = \frac{B_V - B_R}{B_V + B_R}, \tag{21}$$

$$A_G = \frac{G_V - G_R}{G_V + G_R}, \tag{22}$$

where B and G are the bulk and shear modulus, and the subscripts V and R represent the Voigt and Reuss bounds. The values of isotropic crystal are 0.0. In Table 3, the values of A_B and A_G suggest that these two structures of Ca_2C are anisotropic in compressibility and shear.

To intuitively illustrate the elastic anisotropy, the directional dependence of elastic anisotropy was calculated by the ELAM code [28], which shows the 2D figures of the differences in each direction. The calculated Young's modulus along different directions as well as the projections in different planes are demonstrated in Figure 7. The ratios of E_{max}/E_{min} are 1.76 (1.19) and 2.32 (1.85) for *Pnma*-Ca_2C at 0 (100) GPa and *C*2/*m*-Ca_2C at 0 (6.0) GPa, respectively, which means *C*2/*m*-Ca_2C has greater anisotropy. The anisotropy in *yz* plane is the greatest for *Pnma*-Ca_2C at both 0 and 100 GPa (see Figure 7a,b). In Figure 7c,d, the *C*2/*m*-Ca_2C also has the largest anisotropy in *yz* plane at both 0 and 6 GPa. The 2D representations of Poisson's ratio are revealed in Figure 8. All of them show the elastic anisotropy. FromFigure 8a,b, it can be found that the *Pnma*-Ca_2C has the greatest anisotropy in *yz* plane at 0 GPa and in*xz* plane at 100 GPa. However, the greatest anisotropy of *C*2/*m*-Ca_2C is in *yz* plane at both 0 and 6 GPa (see Figure 8c,d). The *C*2/*m*-Ca_2C is more anisotropic than the *Pnma*-Ca_2C in Poisson's ratio. As far as the 2D projections of shear modulus in *xy*, *yz*, and *xz* planes shown in Figure 9, both *C*2/*m*-Ca_2C and *Pnma*-Ca_2C exhibit the obvious elastic anisotropy. From Figure 9a,b, it is seen that the 2D projections of shear modulus in *xz* plane at 0 GPa and in *yz* plane at 100 GPa are almost a perfect circle, showing a slight anisotropy character in these two cases. The anisotropy of *Pnma*-Ca_2C at high pressure is smaller than that at 0 GPa. The same case occurred for *C*2/*m*-Ca_2C, as seen in Figure 9c,d. Similar to the anisotropy of Poisson's ratio, the shear modulus of *Pnma*-Ca_2C has the greatest anisotropy in *yz* plane at 0 GPa and in*xz* plane at 100 GPa, and that of *C*2/*m*-Ca_2C is the most anisotropic in *yz* plane at both 0 and 6 GPa.

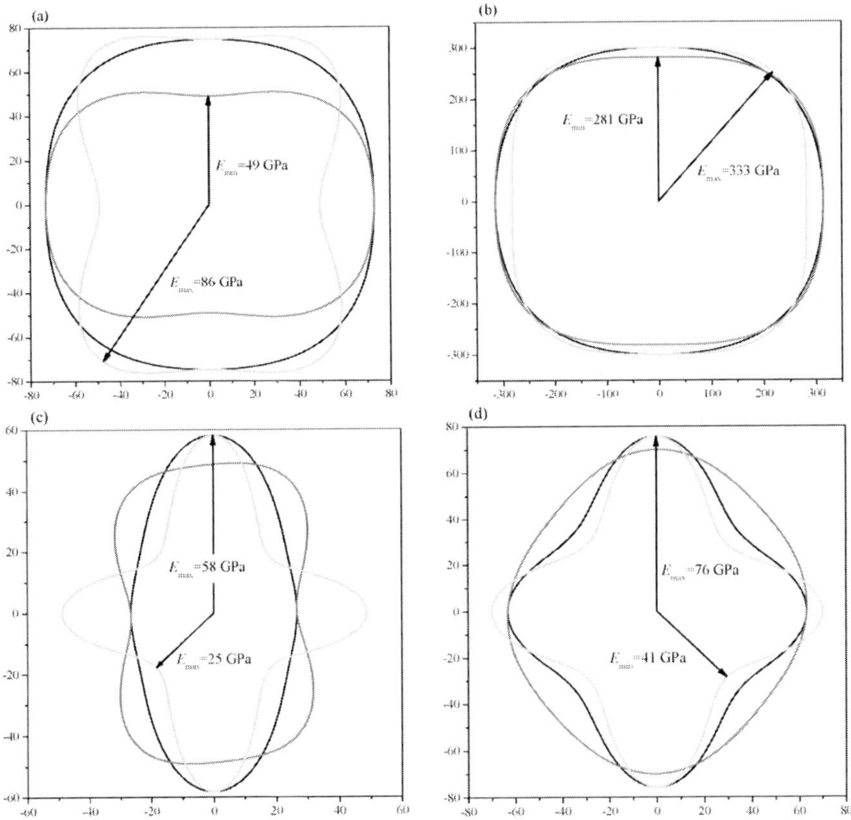

Figure 7. 2D representations of the Young's modulus. (**a**) *Pnma*-Ca_2C at 0 GPa; (**b**) *Pnma*-Ca_2C at 100 GPa; (**c**) *C*2/*m*-Ca_2C at 0 and 6 GPa; (**d**) *C*2/*m*-Ca_2C at 6 GPa. The black, red and green lines represent the*xy*, *xz* and *yz* planes, respectively.

As shown in Figure 10, a research of the electronic band structure and density of state (DOS) of Ca_2C at 0 GPa was also made. The dashed line represents the Fermi level (E_F). From Figure 10a, one can see that *Pnma*-Ca_2C is a semiconductor characterized by a quasi-direct band gap of 0.86 eV (the direct band gap at Γ point is 0.87 eV). The conduction band minimum (CBM) is just at Γ point, and the valence band maximum (VBM) locates at (0, 0, 0.378) along the Γ-Z direction. The calculated band gap of *Pnma*-Ca_2C at 14 GPa is direct band gap with 0.65 eV, which is close to the previous value of 0.64 eV [9]. It is known that the calculated band gap with DFT is usually underestimated by 30%–50%, so the ideal band gap is larger than this calculated result. The DOS of *Pnma*-Ca_2C near Fermi level is mainly originated from the contributions of C-*p* orbital electrons. In Figure 10b, the calculated electronic band structure crosses the Fermi level along many directions in the Brillouin zone, showing the metallic character. And the DOS near Fermi level is mainly characterized by the Ca-*d* orbital electrons.

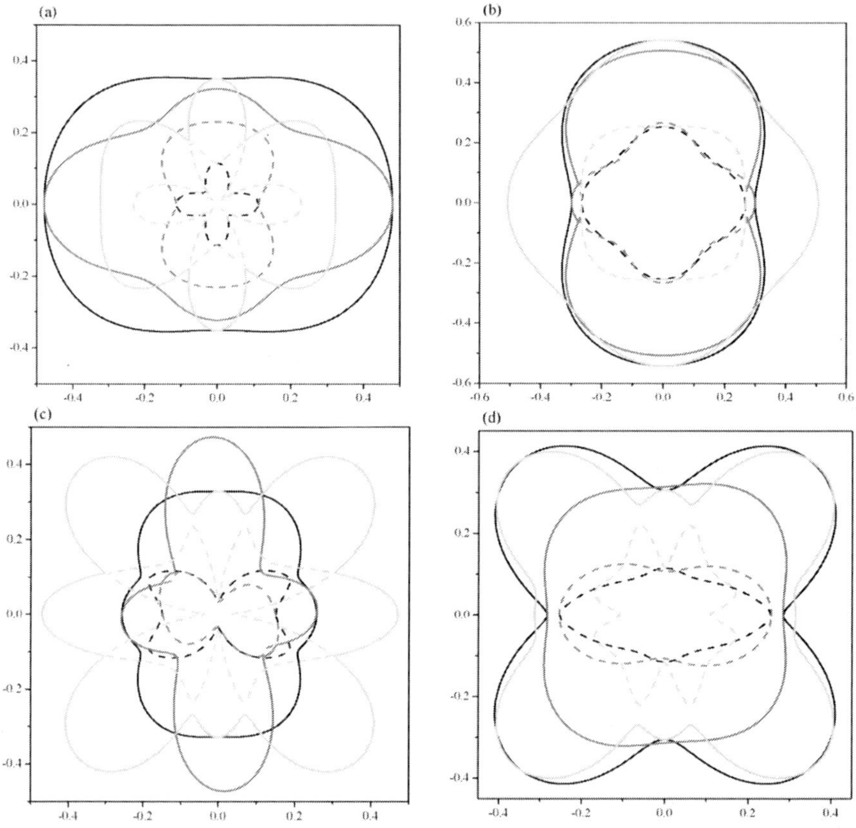

Figure 8. 2D representations of Poisson's ratio. (**a**) *Pnma*-Ca_2C at 0 GPa; (**b**) *Pnma*-Ca_2C at 100 GPa; (**c**) *C2/m*-Ca_2C at 0 GPa; (**d**) *C2/m*-Ca_2C at 6 GPa. The solid and dash lines represent the maximal and minimal positive values, respectively. The black, red and green lines represent the *xy*, *xz* and *yz* planes, respectively.

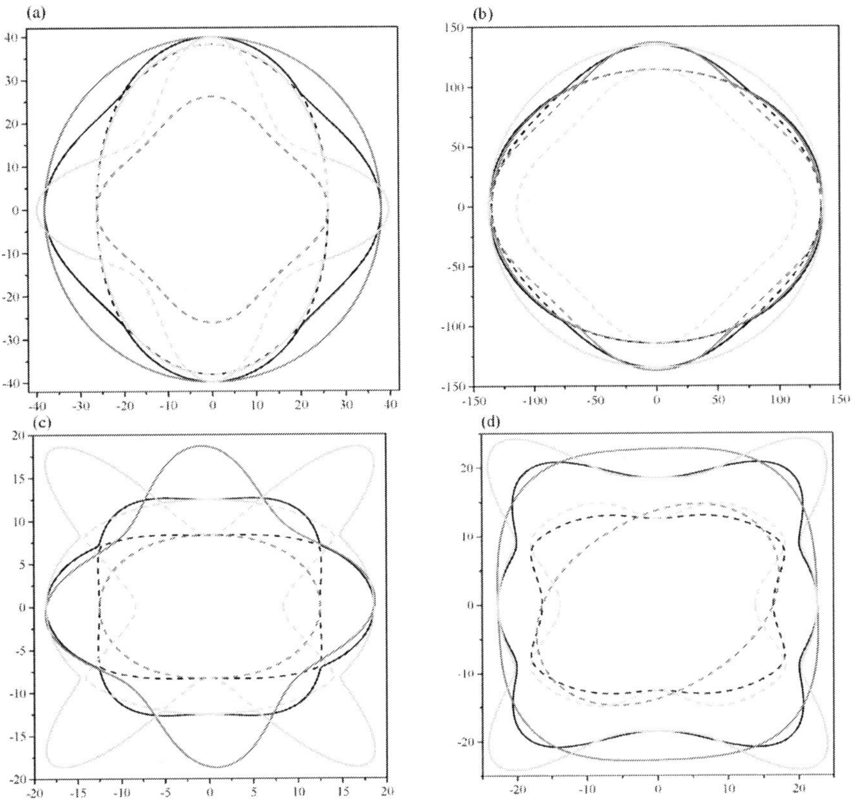

Figure 9. 2D representations of shear modulus. (**a**) *Pnma*-Ca_2C at 0 GPa; (**b**) *Pnma*-Ca_2C at 100 GPa; (**c**) *C*2/*m*-Ca_2C at 0 GPa; (**d**) *C*2/*m*-Ca_2C at 6 GPa. The solid and dash lines represent the maximal and minimal positive values, respectively. The black, red and green lines represent the *xy*, *xz* and *yz* planes, respectively.

3. COMPUTATIONAL METHODS

Our calculations are performed via the generalized gradient approximation (GGA) parameterized by Perdew-Burke-Eruzerhof (PBE) [29] in the Cambridge Serial Total Energy Package (CASTEP) code [30], which is based on the density functional theory (DFT) [31,32]. For the two Ca_2C phases, the ultrasoft pseudo-potential [33] which describes the interactions between the ionic core and valence electrons is used with the cutoff energy of 420 eV. The k-points of Pnma-Ca_2C (7 × 11 × 6) and *C*2/*m*-Ca_2C (6 × 9 × 3) in the

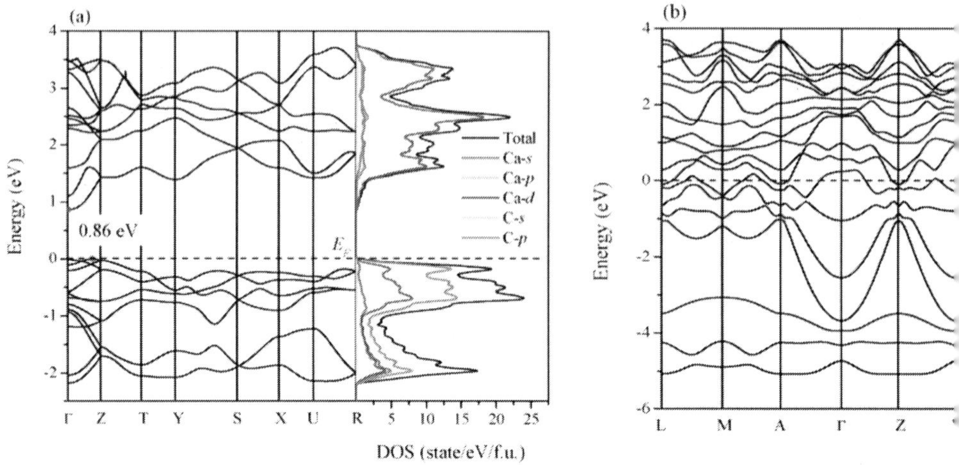

Figure 10. Electronic band structure and density of state of *Pnma*-Ca_2C (**a**) and *C*2/*m*-Ca_2C (**b**) at 0 GPa.

first irreducible Brillouin zone are generated using Monkhorst-Pack mesh scheme [34]. Furthermore, the Broyden-Fletcher-Goldfarb-Shanno (BFGS) minimization scheme [35] is used in geometry optimization. The convergence is within 1 meV/atom in the total energy convergence tests for all calculation parameters. The self-consistent convergence of the total energy is 5×10^{-6} eV/atom, the maximum force on the atom is 0.01 eV/Å, the maximum stress is 0.02 GPa and the maximum ionic displacement is 5×10^{-4} Å.

4. CONCLUSIONS

A systematic analysis of the pressure-induced phase transition of Ca_2C is made by first-principles calculations. The enthalpy and dependence of volume on pressure of Ca_2C are performed. We found that there is a phase transition which occurs at 7.8 GPa transforming from *C*2/*m*-Ca_2C to *Pnma*-Ca_2C with a volume drop of 26.7%. The *Pnma*-Ca_2C is larger than *C*2/*m*-Ca_2C in the calculated bulk modulus, shear modulus, Young's modulus and Poisson's ratio at 0 GPa. Both of them exhibit the elastic anisotropy. The low-pressure phase *C*2/*m*-Ca_2C, which is mechanically stable up to 6.4 GPa, has the greater anisotropy over the *Pnma*-Ca_2C. The electronic band structures reveal the semiconductivity of*Pnma*-Ca_2C and the metallicity of *C*2/*m*-Ca_2C. The quasi-direct band gap of *Pnma*-Ca_2C at 0 GPa is 0.86 eV.

Furthermore, the total and partial density of states is provided to study the specific contribution to Fermi level.

ACKNOWLEDGMENTS

This work was financially supported by the Natural Science Foundation of China (No. 11204007), Natural Science Basic Research plan in Shaanxi Province of China (Nos. 2016JM1026 and 2016JM1016), and Education Committee Natural Science Foundation in Shaanxi Province of China (Grant No. 16JK1049). X. Dai (Xidian University) is thanked for allowing to use the Materials Studio, including CASTEP code.

AUTHOR CONTRIBUTIONS

Qun Wei and Meiguang Zhang designed the project; Quan Zhang performed the calculations, Qun Wei and Quan Zhang prepared the manuscript, Meiguang Zhang revised the paper and coordinated this work, all authors discussed the results and commented on the manuscript.

CONFLICTS OF INTEREST

The authors declare no conflict of interest.

REFERENCES

1. Zhang, W.; Oganov, A.R.; Goncharov, A.F.; Zhu, Q.; Boulfelfel, S.E.; Lyakhov, A.O.; Stavrou, E.; Somayazulu, M.; Prakapenka, V.B.; Konôpková, Z. Unexpected stable stoichiometries of Sodium Chlorides. *Science* **2013**, *342*, 1502–1505.
2. Kurakevych, O.O.; Strobel, T.A.; Kim, D.Y.; Cody, G.D. Synthesis of Mg_2C: A Magnesium Methanide. *Angew. Chem. Int. Ed.* **2013**, *52*, 8930–8933.
3. Fan, Q.; Chai, C.; Wei, Q.; Yang, Y.; Yang, Q.; Chen, P.; Xing, M.; Zhang, J.; Yao, R. Prediction of novel phase of silicon and Si-Ge alloys. *J. Solid State Chem.* **2016**, *233*, 471–483.
4. Wei, Q.; Yan, H.; Zhu, X.; Lin, Z.; Yao, R. Theoretical investigations on the elastic and thermodynamic properties of Rhenium Phosphide. *Z. Naturforsch. A* **2016**, *71*, 1–8.

5. Gauzzi, A.; Takashima, S.; Takeshita, N.; Terakura, C.; Takagi, H.; Emery, N.; Hérold, C.; Lagrange, P.; Loupias, G. Enhancement of superconductivity and evidence of structural instability in intercalated Graphite CaC_6 under high pressure. *Phys. Rev. Lett.* **2007**, *98*, 067002.

6. Nylén, J.; Konar, S.; Lazor, P.; Benson, D.; Häussermann, U. Structural behavior of the acetylide carbides Li_2C_2 and CaC_2 at high pressure. *J. Chem. Phys.* **2012**, *137*, 224507.

7. Li, Y.L.; Luo, W.; Zeng, Z.; Lin, H.Q.; Mao, H.K.; Ahuja, R. Pressure-induced superconductivity in CaC_2. *Proc. Natl. Acad. Sci. USA* **2013**, *110*, 9289–9294.

8. Nourbakhsh, Z.; Hashemifar, S.J.; Akbarzadeh, H. First-principles insights into *p* magnetism in CaC and CaN ionic compounds. *J. Magn. Magn. Mater.* **2013**, *341*, 56–59.

9. Li, Y.L.; Wang, S.N.; Oganov, A.R.; Gou, H.; Smith, J.S.; Strobel, T.A. Investigation of exotic stable calcium carbides using theory and experiment. *Nat. Commun.* **2015**, *6*, 6974.

10. Dong, S.; Ding, H.; Zhou, B.; Duan, C.; Wu, P.; Zhao, H. First-principles and Monte Carlo studies on the magnetic stability of half-metallic zinc-blende CaC and similar compounds. *J. Magn. Magn. Mater.* **2015**, *378*, 469–477.

11. Debessai, M.; Hamlin, J.J.; Schilling, J.S.; Rosenmann, D.; Hinks, D.G.; Claus, H. Superconductivity for CaC_6 to 32 GPa hydrostatic pressure. *Phys. Rev. B* **2010**, *82*, 132502.

12. Mazin, I.I.; Boeri, L.; Dolgov, O.V.; Golubov, A.A.; Bachelet, G.B.; Giantomassi, M.; Andersen, O.K. Unresolved problems in superconductivity of CaC_6. *Physics C* **2007**, *460–462*, 116–120.

13. Zhang, L.; Xie, Y.; Cui, T.; Li, Y.; He, Z.; Ma, Y.; Zou, G. Pressure-induced enhancement of electron-phonon coupling in superconducting CaC_6 from first principles. *Phys. Rev. B* **2006**, *74*, 184519.

14. Moradi, M.; Mohammadi, A.; Afshari, M.; Soltani, Z. The half-metallicity of zinc-blende CaC/GaAs (001) heterojunction: A density functional theory study. *J. Magn. Magn. Mater.* **2013**,*332*, 81–84.

15. Nourbakhsh, Z.; Hashemifar, S.J.; Akbarzadeh, H. Ab initio study of different structures of CaC: Magnetism, bonding, and lattice dynamics. *J. Alloys Compd.* **2013**, *579*, 360–364.

16. Gao, G.Y.; Yao, K. Surface *sp* half-metallicity of zinc-blende calcium monocarbide. *J. Appl. Phys.***2009**, *106*, 053703.

17. Wu, Z.-J.; Zhao, E.-J.; Xiang, H.-P.; Hao, X.-F.; Liu, X.-J.; Meng, J. Crystal structures and elastic properties of superhard IrN_2 and IrN_3 from first principles. *Phys. Rev. B* **2007**, *76*, 054115.

18. Fan, Q.Y.; Wei, Q.; Yan, H.Y.; Zhang, M.G.; Zhang, D.Y.; Zhang, J.Q. A New Potential Superhard Phase of OsN_2. *Acta Phys. Pol. A* **2014**, *126*, 740–746.

19. Sin'Ko, G.V.; Smirnov, N.A. Ab initio calculations of elastic constants and thermodynamic properties of bcc, fcc, and hcp Al crystals under pressure. *J. Phys. Condens. Matter* **2002**, *14*, 6989–7005.

20. Voigt, W. *Lehrburch der Kristallphysik*; Teubner: Leipzig, Germany, 1928.

21. Reuss, A. Berechnung der Fließgrenze von Mischkristallen auf Grund der Plastizitätsbedingung für Einkristalle. *J. Appl. Math. Mech.* **1929**, *9*, 49–58.

22. Hill, R. The Elastic Behaviour of a Crystalline Aggregate. *Phys. Soc. Lond. Sect. A* **1952**, *65*, 349–354.

23. Pugh, S.F. Relations between the elastic moduli and the plastic properties of polycrystalline pure metals. *Lond. Edinb. Dublin Philos. Mag. J. Sci.* **1954**, *45*, 823–843.

24. Ranganathan, S.I.; Ostoja-Starzewski, M. Universal Elastic Anisotropy Index. *Phys. Rev. Lett.* **2008**,*101*, 055504.

25. Feng, J.; Xiao, B.; Zhou, R.; Pan, W.; Clarke, D.R. Anisotropic elastic and thermal properties of the double perovskite slab-rock salt layer $Ln_2SrAl_2O_7$ (Ln = La, Nd, Sm, Eu, Gd or Dy) natural superlattice structure. *Acta Mater.* **2012**, *60*, 3380–3392.

26. Ravindran, P.; Fast, L.; Korzhavyi, P.A.; Johansson, B.; Wills, J.; Eriksson, O. Density functional theory for calculation of elastic properties of orthorhombic crystals: Application to $TiSi_2$. *J. Appl. Phys.* **1998**, *84*, 4891–4904.

27. Wei, Q.; Zhang, M.; Yan, H.; Lin, Z.; Zhu, X. Structural, electronic and mechanical properties of Imma-carbon. *Europhys. Lett.* **2014**, *107*, 27007.

28. Marmiera, A.; Lethbridge, Z.A.D.; Walton, R.I.; Smith, C.W.; Parker, S.C.; Evans, K.E. ElAM: A computer program for the analysis and representation of anisotropic elastic properties. *Comput. Phys. Commun.* **2010**, *181*, 2102–2115.

29. Perdew, J.P.; Burke, K.; Ernzerhof, M. Generalized Gradient Approximation Made Simple. *Phys. Rev. Lett.* **1996**, *77*, 3865–3868.

30. Clark, S.J.; Segall, M.D.; Pickard, C.J.; Hasnip, P.J.; Probert, M.I.J.; Refson, K.; Payne, M.C. First principles methods using CASTEP. *Z. Kristallogr.* **2005**, *220*, 567–570.

31. Hohenberg, P.; Kohn, W. Inhomogeneous Electron Gas. *Phys. Rev.* **1964**, *136*, B864–B871.

32. Kohn, W.; Sham, L.J. Self-Consistent Equations Including Exchange and Correlation Effects. *Phys. Rev.* **1965**, *140*, A1133–A1138.

33. Vanderbilt, D. Soft self-consistent pseudopotentials in a generalized eigenvalue formalism. *Phys. Rev. B* **1990**, *41*, 7892–7895.

34. Monkhorst, H.J.; Pack, J.D. Special points for Brillouin-zone integrations. *Phys. Rev. B* **1976**, *13*, 5188–5192.

35. Pfrommer, B.G.; Côté, M.; Louie, S.G.; Cohen, M.L. Relaxation of Crystals with the Quasi-Newton Method. *J. Comput. Phys.* **1997**, *131*, 233–240.

Chapter 2

High-Pressure Orthorhombic Ferromagnesite as a Potential Deep-Mantle Carbon Carrier

Jin Liu, Jung-Fu Lin & Vitali B. Prakapenka

Department of Geological Sciences, Jackson School of Geosciences, The University of Texas at Austin, Austin, TX 78712, USA

ABSTRACT

Knowledge of the physical and chemical properties of candidate deep-carbon carriers such as ferromagnesite [(Mg,Fe)CO_3] at high pressure and temperature of the deep mantle is necessary for our understanding of deep-carbon storage as well as the global carbon cycle of the planet. Previous studies have reported very different scenarios for the (Mg,Fe)CO_3 system at deep-mantle conditions including the chemical dissociation to (Mg,Fe)O+CO_2, the occurrence of the tetrahedrally-coordinated carbonates based on CO_4 structural units, and various high-pressure phase transitions. Here we have studied the phase stability and compressional behavior of (Mg,Fe)CO_3 carbonates up to relevant lower-mantle conditions of approximately 120 GPa and 2400 K. Our experimental results show that the rhombohedral siderite (Phase I) transforms to an orthorhombic phase (Phase II with *Pmm*2 space group) at approximately 50 GPa and 1400 K. The structural transition is likely driven by the spin transition of iron accompanied by a volume collapse in the Fe-rich (Mg,Fe)CO_3 phases; the spin transition stabilizes the high-pressure phase II at much lower pressure conditions than its Mg-rich counterpart. It is conceivable that the low-spin ferromagnesite phase II becomes a major deep-carbon carrier at the deeper parts of the lower mantle below 1900 km in depth.

INTRODUCTION

The existence of carbon-bearing solids, fluids, and melts in the Earth's deep interior can affect a series of geophysical and geochemical properties of our planet1′2′3′4′5. Laboratory studies of carbon-bearing minerals at high pressures and temperatures (*P-T*) can thus provide crucial constraints on the role and behavior of the deep-carbon phases in the Earth's mantle as well as the mantle's role in the global carbon cycle6′7. Among all carbonates that can be potentially subducted into the Earth's deep mantle through plate subductions, iron-bearing magnesite [($Mg,Fe)CO_3$] has been proposed to be a major deep-carbon host because of its existence in subducting plates as well as its wide *P-T*stability field8′9′10′11′12. $(Mg,Fe)CO_3$ forms a continuous solid solution between magnesite [$MgCO_3$] and siderite [$FeCO_3$] in the rhombohedral$R\bar{3}c$ structure, in which the Mg-rich compositions are called ferromagnesite and the Fe-rich part is named magnesiosiderite.

A number of previous studies have reported different behaviors of the $(Mg,Fe)CO_3$ system at relevant *P-T* conditions of the mantle10′13′14′15′16′17′18; various high-pressure $(Mg,Fe)CO_3$ polymorphic phase transitions have been reported to occur at high *P-T* including the formation of phases based on tetrahedral CO_4 units (similar to the SiO_4units of silicate minerals)14′15, and the chemical decomposition of $(Mg,Fe)CO_3$ into (Mg,Fe)O and CO_2 (refs. 9,18). It has been experimentally observed that the rhombohedral structure of the end-member magnesite is stable up to approximately 100 GPa and 2000 K, but then transforms into an orthorhombic structure, named magnesite II, at approximately 115 GPa and 2100 K (ref. 10). Theoretical calculations13′14′15 and other experimental studies16, on the other hand, have shown that magnesite transforms into a high-pressure phase built with $(CO_4)^{4-}$ tetrahedral groups energetically favored in the monoclinic structure. At high *P-T* conditions, $(Mg,Fe)CO_3$ and a mixture of oxides ((Mg,Fe)O and CO_2) have been suggested to transform into a number of Fe^{3+}-bearing high-pressure carbonate, magnetite, and nano-diamonds via intra-crystalline reactions between Fe^{2+} and $(CO_3)^{2-}$ (ref. 16). A new high *P-T* phase with the $Fe_4(CO_4)_3$ chemical formula and all iron as $Fe^{3+}$was reported to occur through high-pressure reactions between FeO and saturated CO_2 at approximately 40 GPa and 1500 K (ref. 17). After temperature quenching to room temperature, this new phase undergoes a structural transformation in compression17. These experimental results suggest that the oxygen fugacity in the $(Mg,Fe)CO_3$system may strongly affect the reaction products at high *P-T*conditions18′19. In the Earth's mantle, the oxygen fugacity (fO_2) may be controlled by the carbon-oxygen-hydrogen-sulfur and ferric-ferrous iron equilibria, resulting in a general decrease in fO_2 with depth through the reduction of volatile species, such that the ferric iron content can be used as an indicator for the redox state of the Earth's interior (see ref. 20 for a review). However, the self-disproportionation model of Fe^{2+} into Fe^{3+} and Fe^{0} has been presented as a mechanism acting in the lower mantle where the formation of Al-bearing silicate perovskite is associated with the disproportionation of Fe^{2+} into Fe^{3+} in perovskite and metallic Fe (a charge balanced reaction). Thus far, the exact crystal structure of the high-pressure phase in the (Mg,Fe)CO_3system and the C-O coordination geometry remain highly debated10′13′14′15′16. Since the lower-mantle magnesite is expected to contain a certain amount of siderite in the solid solution21, studying the phase stability, crystal structure, and thermal equation of state (EoS) of ferromagnesite, especially through the recently discovered spin transition of iron22′23′24, at relevant *P-T* conditions of the deep mantle is of particular importance to our understanding of the deep-carbon storage6′25.

An iso-symmetric electronic spin transition of iron in rhombohedral (Mg,Fe)CO_3 has been reported to occur at approximately 45 GPa and high temperatures11′12′26′27. Evaluation of the pressure-volume relationship of siderite across the spin transition showed that the spin transition is associated with a volume collapse of approximately 10%, such that the low-spin siderite exhibits a smaller unit cell volume than the endmember $MgCO_3$ counterpart11′28. That is, the low-spin siderite has an effective ionic radius of the Fe^{2+} smaller than that of the $Mg^{2+}$and is much denser than the rhombohedral $MgCO_3$ and the extrapolated high-spin siderite11′29′30. Since the iron ions occupy the isolated octahedral sites in the rhombohedral lattice of siderite, the Fe^{2+}-Fe^{2+} exchange interactions between neighboring Fe^{2+} likely have a negligible effect on the transition pressure from the high-spin state to the low-spin state12. The weak iron-iron exchange interactions as well as drastic reduction of the effective Fe^{2+} ionic radius in rhombohedral (Mg,Fe)CO_3 raise the possibility that the low-spin siderite likely undergoes a structural transition at a relatively lower pressure than that of magnesite11′30. To decipher the phase stability of the (Mg,Fe)CO_3system in the lower mantle, here we have investigated the crystal structures of siderite and magnesiosiderite [$(Fe_{0.65}Mg_{0.35})CO_3$] samples at high *P-T* conditions using synchrotron X-ray diffraction (XRD) in a laser-heated diamond anvil cell (DAC). Together with electron microprobe analyses of the quenched samples, our results show a structural phase transition from the rhombohedral (space group: $R\bar{3}c$) phase (Phase I) to the orthorhombic phase (Phase II) at above approximately 50 GPa and 1400 K. The *P-T* range for the structural transition in the Fe-rich parts of the (Mg,Fe)CO_3 system can be understood in terms of the volume collapse due to the spin transition of iron at high *P-T*. Here we apply these results to decipher physical and chemical behaviors of candidate deep-mantle carbonates in the lower mantle.

EXPERIMENTAL RESULTS AND DISCUSSIONS

High-Pressure Phase Transition in Siderite

XRD patterns were collected up to 120 GPa and 2400 K for siderite (Figs. 1 and 2, and Table S1). At ambient temperature, rhombohedral siderite undergoes a volume collapse at approximately 42 GPa that can be associated with an iso-symmetric spin transition, consistent with previous results11′12 (Fig. S1). However, upon laser heating to above 1400 K at approximately 50 GPa, new diffraction peaks appear, which are incompatible with the rhombohedral structure and are systematically observed at high *P*-*T* (Fig. 1*B* and Fig. S2). These peaks were also observed in the temperature-quenched samples at high pressures. Analyses of these new diffraction peaks show that they do not belong to any previously proposed high-pressure structural models of $(Fe,Mg)CO_3$including a pyroxene-like structure13′14, a monoclinic *C*2/*m* structure15, and a monoclinic $P2_1/c$ structure16. The diffraction patterns also differ from that for the $Fe_4(CO_4)_3$ phase which was reported to be temperature unquenchable at high pressures17. Furthermore, these peaks cannot be simply assigned to any of the previously proposed structures for products of siderite dissociation, such as FeO, Fe_2O_3, Fe_3O_4, Fe_4O_5, CO_2, C (nano-diamond or graphite), or Fe_3C, at high *P*-*T*and variable redox conditions15′16′17′18′19′31.

Experimental observations of the phases were collected from three different samples in this study, and are shown as black symbols (+, * and ×) (Table S1). Pluses (+): siderite I; plus/crosses (*): two-phase co-existing region (siderite I and II); crosses (×): siderite II; dash-dotted curves: transition boundaries between phase I and mixed I+II phases or phase II and mixed I+II phases. Spin crossover phase diagram of iron in siderite I below 1200 K from Liu et al.27 is plotted in colored area for comparison with the observations of the structural phases in this study. HS: high-spin state; LS: low-spin state. The color bar at the right represents the fraction of the LS state27. Dashed curves: the LS fraction of iron in siderite I extrapolated from 1200 K to 2700 K (blue: the initial occurrence of the LS state; red: the last occurrence of the HS state, ref. 27). Grey solid curve: expected lower-mantle geotherm41. The observation of the mixed-phase region may be due to *P*-*T* uncertainty as well as the sluggishness (the kinetic barrier) of the phase transformation.

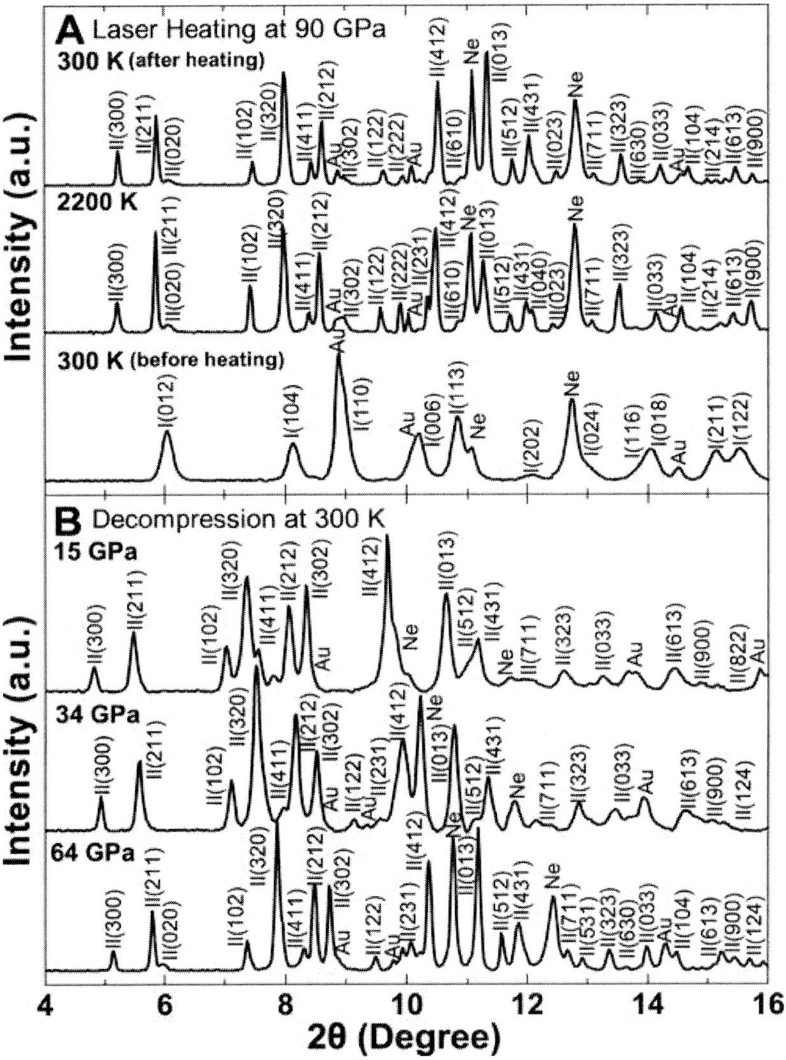

Figure 1: Representative X-ray diffraction patterns of siderite I and II phases [$FeCO_3$] at high *P-T*. (*A*) $FeCO_3$ heated up to 2200 K at 90 GPa. (*B*) Decompression of siderite II at room temperature. Gold (Au) was used as the primary pressure calibrant, while neon (Ne) was used as the thermal insulator, pressure medium as well as the secondary pressure calibrant49. Miller indices (hkl) of siderite I and II phases are labeled as I(hkl) and II(hkl), respectively. The wavelength of the monochromatic X-ray beam was 0.3344 Å.

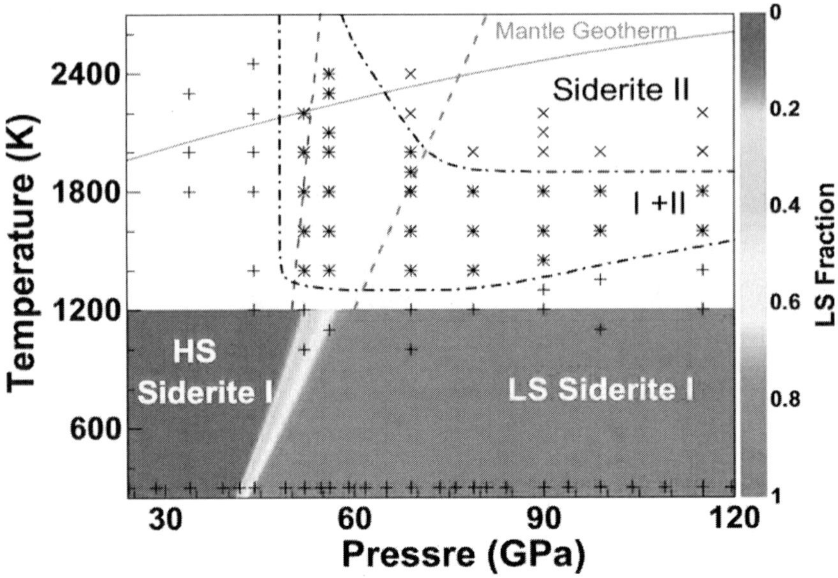

Figure 2: Schematic phase relations of siderite [$FeCO_3$] at high *P-T*relevant to the lower mantle conditions.

The SEM-EDX mapping of the recovered sample from 90 GPa and 2200 K exhibited homogeneous distributions of Fe, C, and O in the laser-heated area, showing no compositional dissociation and diffusion during laser heating nor in *P-T* quenching (Fig. S3). Further EDX analyses of the O/Fe ratios of the recovered sample showed a consistent O/Fe ratio of three throughout the heated area, indicating that the valence state of ferrous iron should have remained unchanged during laser heating and *P-T* quenching (Table S2). These results indicate that the high-pressure phase should have the same chemical composition as the starting siderite sample [$FeCO_3$] and that dissociation or valence state change of iron in siderite did not occur in our sample at the *P-T* conditions of our study. Hereafter, this high-pressure polymorph is referred to as siderite II (or Phase II), and the rhombohedral siderite is referred to as siderite I (Phase I; or siderite for simplicity).

Siderite II was initially observed upon laser heating at approximately 1400 K and 50 GPa where it co-existed with siderite I over a certain *P-T*range (Fig. 2). At *P-T* conditions higher than approximately 70 GPa and 2200 K, siderite I totally disappeared and siderite II became a single phase. X-ray diffraction patterns of the temperature-quenched siderite II were collected at 300 K in compression between 50 and 120 GPa and in decompression down to 15 GPa; below 15 GPa, siderite II became amorphous. All diffraction peaks of the siderite II can be well indexed with an orthorhombic unit cell (Fig. 1, Fig. S2, and Table S3). Based on the principles of reflection conditions for the indexed Miller indices of the diffraction peaks32, four potential space groups are found to be reasonable starting models for the orthorhombic structure: *P*222, $P222_1$,*Pmm*2, and *Pmmm*. Using the GSAS software package33, representative LeBail refinements of the XRD patterns at 90 GPa and 300 K showed that the *Pmm*2 space group for the siderite II structural model gave the best fit with the smaller residues to the experimental spectra (Fig. 3). The *Pmm*2 structural model for siderite II contains 12 formula units in the primitive cell (Z); at a representative pressure of 90 GPa, its unit cell parameters are *a* = 10.9902 (±0.0028) Å, *b* = 6.3405 (±0.0021) Å, and *c* = 5.2726 (±0.0009) Å (Table S3). We note that previously proposed monoclinic and orthorhombic structural models of the high-pressure magnesite phase contain 12 and 6 formula units in the primitive cell, respectively10′14′15′16. We had also checked previously reported high-pressure phases of other carbonates and did not find any of them consistent with this siderite II structural model (see Table S4). It is worth noting that our observed X-ray diffraction patterns are consistent with that for the high-pressure phase of magnesite by Isshiki et al.10, although our proposed orthorhombic structural model has a larger unit cell than that proposed by Isshiki et al.10 The differences here were due to the use of high-quality diffraction peaks, including 4 diffraction peaks with *d*-spacing greater than 3.0 Å. The high-quality X-ray diffraction patterns with extended *d*-spacing range (30 peaks in total) permit us to resolve the crystal structure of the siderite II phase reliably. On the other hand, Boulard et al.16 proposed the monoclinic high-pressure phase of magnesite and magnesiosiderite as a result of chemical dissociation in laser-heated carbonates. These experiments were conducted without any pressure-transmitting medium in the sample chamber16, which may contribute to the differences in the observed phases.

It should be noted that our diffraction results do not permit the use of the Rietveld full-profile refinement to refine the carbon atom positions as well as the C-O bonding geometry and characters, because the relatively light carbon atoms do not contribute to X-ray diffraction signals significantly enough for such an analysis34. It has been reported that high-pressure phases of carbonates can contain tetrahedrally-coordinated carbon $(CO_4)^{4-}$ units with the sp^3 bonding characters15′16′35or the coexistence of both trigonally and tetrahedrally coordinated units36, instead of the trigonally-coordinated CO_3^{2-} units alone in the siderite I. The C-O bonding characters of our observed siderite II at high pressures thus remain to be further investigated using other more sensitive methods.

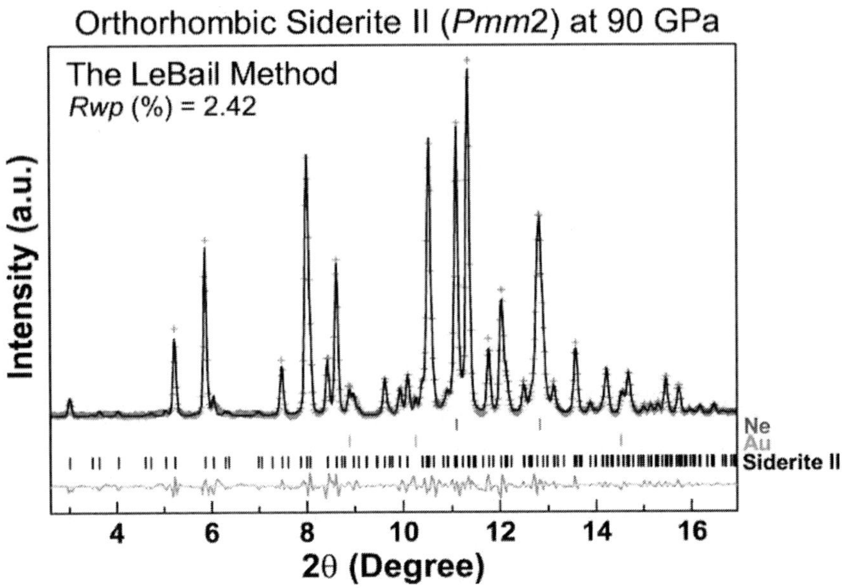

Figure 3: Representative LeBail fit of an X-ray diffraction spectrum of siderite II at 90 GPa and room temperature.

The sample was temperature-quenched to room temperature from 2200 K at 90 GPa. Pluses: measured powder diffraction pattern after background subtraction; black solid curve: refined profile; grey solid curve: residual between the observation and the refinement; vertical ticks: Ne (blue), Au (orange), and siderite II (black).

Pressure-Volume (*P-V*) Relations in Siderite I and II Phases

Here we have used the *Pmm*2 structural model of siderite II to re-index previous diffraction results10'16 and to discuss the *P-V* relations of siderite I and II phases in the (Mg,Fe)CO_3 system at high pressures. For re-indexing the literature data, our high-pressure orthorhombic structural model and the dichotomy method were used to obtain the lattice parameter by minimizing the d_{Obs}-d_{Calc} values in our calculations. Previously proposed monoclinic or orthorhombic structural model was not used here. The *Pmm*2 structural model can be used to index previous high-pressure X-ray diffraction patterns in the (Mg,Fe)CO_3 system including magnesite II (ref. 10) and the high-pressure phase of magnesiosiderite [($Mg_{0.25}Fe_{0.75}$)CO_3] (ref. 16). Using the orthorhombic structure model of siderite II, the uncertainty between the observed and calculated *d* spacings can be fairly improved by re-indexing the experimental data for magnesite II at 119 GPa and 300 K from Ref. 10 and

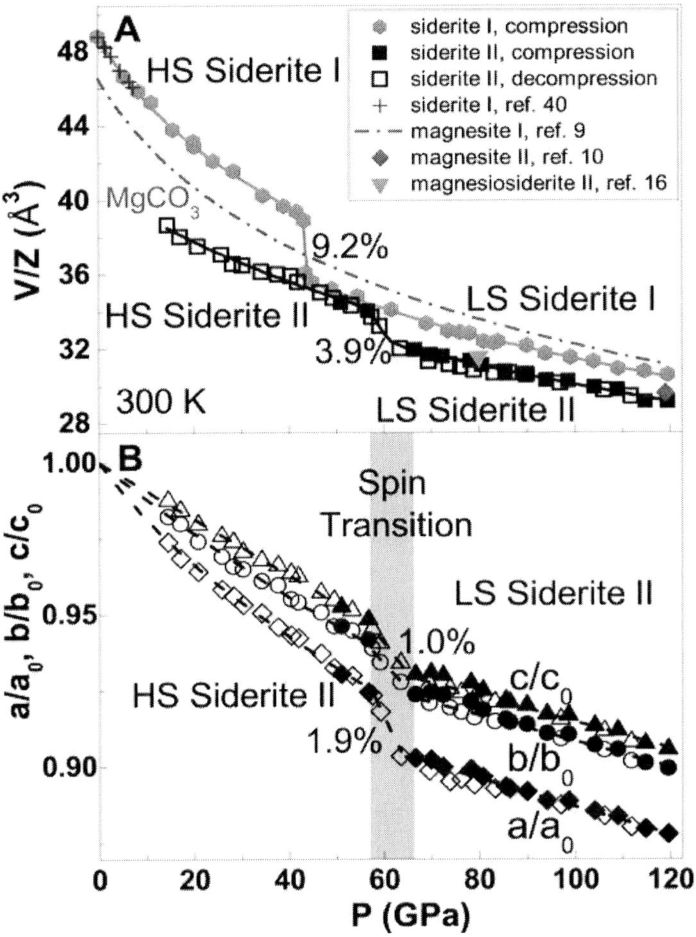

Figure 4: Comparison of the pressure-volume relations in the $(Mg,Fe)CO_3$ phases. (*A*) Unit cell volume of siderite I and II phases as a function of pressure at ambient temperature. The vertical axis is plotted as the unit cell volume per formula unit (*V/Z*). The number of molecules per unit cell (*Z*) is 6 for siderite I and 12 for siderite II. HS: high-spin state; LS: low-spin state. Solid curves: modeled BM EoS fits of the experimental results. The volume collapse of 9.2 (±0.5)% and 3.9 (±0.4)% for siderite I and II can be associated with their respective spin transition at high pressures. Solid diamond: *V/Z* of magnesite II re-calculated from Isshiki et al.10 (Table S5); Solid triangle: *V/Z* of the high-pressure phase of magnesiosiderite $[(Mg_{0.25}Fe_{0.75})CO_3]$ re-calculated from Boulard et al.16 (Table S6). (*B*) Lattice parameters of the siderite II as a function of pressure at 300 K. The lattice collapse in siderite II is 1.9%, 1.0%, and 1.0% for a/a_0, b/b_0, and c/c_0, respectively, at approximately 60 GPa. Filled symbols: compression; open symbols: decompression.

the high-pressure phase of magnesiosiderite at 80 GPa and 300 K from Ref. 16 (Tables S5 and S6). Despite the variation in the $FeCO_3$ content in these previous studies, these re-indexed unit cell parameters at high pressures are generally consistent with our experimental *P-V* results for the siderite II structural model (Fig. 4).

Examinations of the *P-V* curve of siderite I show a sharp volume reduction of 9.2 (±0.5)% at approximately 42 GPa and 300 K, that can be associated with the previously reported spin transition in the $(Mg,Fe)CO_3$ system11'12 (Fig. 4). Across the spin transition, the volume of the low-spin siderite I becomes smaller than the end-member magnesite via Fe-O bond distance shrinkage in the octahedral site37, indicating that the effective ionic radius of Fe^{2+} in the low-spin configuration is smaller than that of Mg^{2+} (refs. 11,29). The volume difference between the low-spin siderite I and magnesite I is ~3.6% at 45 GPa and ~2.1% at 120 GPa, indicating that the low-spin siderite I is more incompressible than the end-member magnesite. Such a volume reduction related to Fe spin transition increases with increasing the molar concentration of $FeCO_3$ in the $(Mg,Fe)CO_3$ system12'27.

Our *P-V* results for siderite II also show an abrupt change of 3.9 (±0.4)% in the unit cell volume at approximately 60 GPa (Fig. 4*A*). Since there's no evidence of any further structural changes in our diffraction patterns, one can thus relate this volume change to an iso-symmetric electronic transition of Fe in siderite II from the low-spin state to the high-spin state in decompression. This spin transition for siderite II occurs at 60 GPa which is approximately 18 GPa higher than that in siderite I at 42 GPa. We note that the pressure range of the spin transition in siderite II was also where the structural transition from siderite I to siderite II was observed. The low-spin siderite II is approximately 4.5% smaller in the molar volume than the low-spin siderite I, while the high-spin siderite II is approximately 11.4% smaller than the high-spin siderite I. Furthermore, the unit cell volume of the low-spin siderite II is approximately 2% smaller than that of magnesite II at 119 GPa and 300 K and approximately 1% smaller than that of the magnesiosiderite II $[(Mg_{0.25}Fe_{0.75})CO_3]$ at 80 GPa and 300 K (Fig. 4*A*). Compared with the 9.2% volume reduction across the spin transition in siderite I, the smaller volume reduction of 3.9% in siderite II is consistent with the denser structure of the high-pressure siderite II phase. Fitting the *P-V* results to the third-order Birch-Murnaghan equation of state (EoS)38 shows that the low-spin siderite II exhibits a higher incompressibility than the high-spin siderite II as well as the low-spin siderite I (Tables S7 and S8). Specifically, the low-spin siderite II exhibits an isothermal bulk modulus (K_{0T}) of 251 (±17) GPa with a fixed pressure derivative of the K_{0T} (K_{0T}') at 4, while the high-spin siderite II has a K_{0T} of 222 (±13) GPa with a fixed K_{0T}' of 4.

The lattice parameters of siderite II also exhibit abrupt changes across the pressure-induced spin transition (Fig. 4*B*); the axial length of the *a* axis is reduced by 1.9% which is almost twice as much as that in the *b* and *c* axes. In contrast, the axial lengths of the *a* and *c* axes in siderite I are reduced by approximately 2.0% and 3.0%, respectively, across the spin crossover[12]. The difference in the changes of the unit cell parameters across the spin transition in siderite I and II likely indicates relatively denser packing structure and less compressible Fe-O bonding characters in the high-pressure phase II (See Table S8 for the EoS parameters of the siderite I and II).

Phase Diagram of the (Mg,Fe)CO_3 Carbonates at High Pressure and Temperature

Together with previous studies[10,16,27,39,40], here we use our experimental results on $FeCO_3$ and $(Fe_{0.65}Mg_{0.35})CO_3$ to decipher the compositional and spin transition effects on the *P-T* phase diagram of the (Mg,Fe)CO_3 system (Fig. 5). Analyses of the *P-T* conditions for the occurrence of phase II in the (Mg,Fe)CO_3 system show that siderite II started to occur at 50 GPa and 1400 K, magnesiosiderite II [$(Fe_{0.65}Mg_{0.35})CO_3$] at 60 GPa and 1500 K, and magnesite II at 85 GPa and 2400 K. This indicates a strong effect of Fe-Mg substitution on the transitional pressure from the rhombohedral phase I to the orthorhombic phase II and that the high-pressure phase II in the $FeCO_3$-rich part of the system occurs at lower *P-T* conditions than that in the $MgCO_3$-rich counterpart (Fig. 5). Previous high *P-T* studies on the system have shown that the spin transition in phase I occurred sharply at approximately 45 GPa and 300 K (refs. 11,12) and that the spin crossover broadened and shifted toward higher pressures at elevated temperatures up to 1200 K (ref. 27). Because the FeO_6 octahedra in the system are well isolated from each other by the rigid CO_3 units, the spin transition pressure is shown to be not affected by the amount of $FeCO_3$ in the system[27]; there has been no observable compositional effect on the spin transition pressure in the $MgCO_3$-$FeCO_3$ system[11,12,27]. Due to the iron substitution, however, the $FeCO_3$-rich part of the system in the low-spin state has a smaller unit cell volume and a higher density than the $MgCO_3$-rich high-spin counterpart, effectively reducing the pressure (energy) needed to transform to the high-pressure phase II by as much as approximately 35–40 GPa between the $FeCO_3$ and $MgCO_3$ end-members. Considering the structural transition pressures and the occurrence of the spin transition in both phase I and II, it is evident that the combination of the $FeCO_3$ substitution and the spin transition of iron in the $MgCO_3$-$FeCO_3$ system promotes the occurrences of the high-pressure low-spin phase II at much lower *P-T* conditions than the $MgCO_3$-rich counterpart[30].

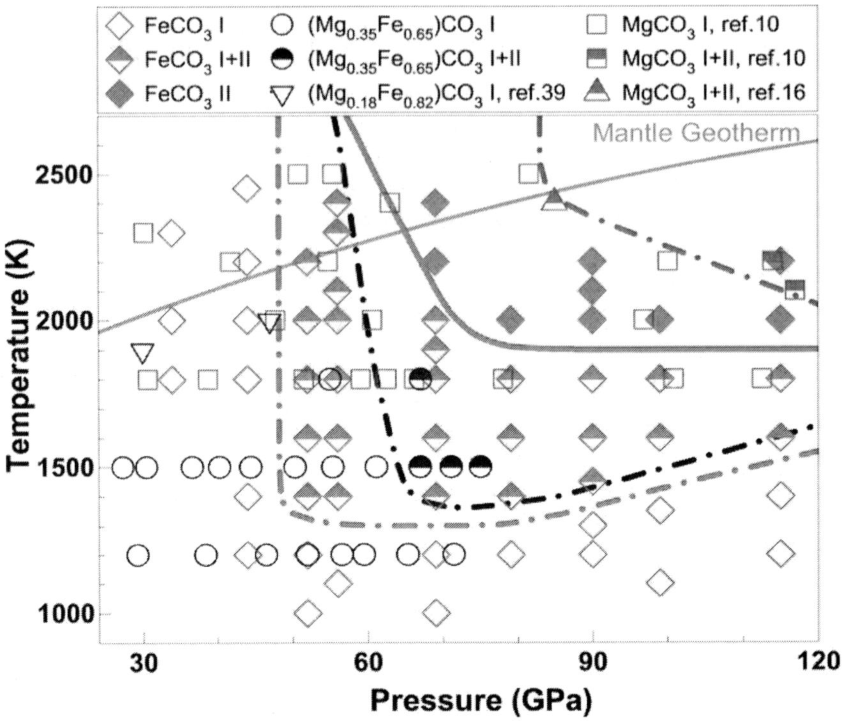

Figure 5: Experimentally observed phases in the $(Mg,Fe)CO_3$system at high *P-T*. Open, half-filled, and solid diamonds: phase I, I+II, and II, respectively, for siderite [$FeCO_3$] (this study); open and half-filled squares: phase I and I+II, respectively, for magnesite [$MgCO_3$] (ref. 10); half-filled triangle: phase I and II for magnesite16; open and half-filled circles: phase I and I+II, respectively, for $(Mg_{0.35}Fe_{0.65})CO_3$ (this study); open triangles: phase I for magnesiosiderite [$(Mg_{0.09}Fe_{0.82}Ca_{0.09})CO_3$] (ref. 39). Red, black, and blue curves: phase transformation boundary between phase I and II for siderite, magnesiosiderite, and magnesite, respectively. Gray solid curve: expected lower-mantle geotherm41.

Implications for the Deep-Mantle Carbon Storage

Spin transitions of iron in candidate mantle minerals such as perovskite [$(Mg,Fe)SiO_3$] and ferropericlase [(Mg,Fe)O] have been observed to affect their physical and chemical properties including density, sound velocities, iron partitioning coefficient, as well as transport properties22′23′24. Our results here show that the spin transition in $(Mg,Fe)CO_3$ can affect the phase diagram of the system by stabilizing the high-pressure phase II in the orthorhombic structure to lower *P-T*conditions (Figs. 2 and 5). Previous studies have indicated that magnesite likely contains approximately 15 at%

iron in the Earth's mantle[7,21]. Along an expected lower-mantle geotherm[41], the high-spin ferromagnesite containing 15% iron [$(Mg_{0.85}Fe_{0.15})CO_3$] would start to transform to the low-spin state at approximately 55 GPa and 2200 K (approximately 1400 km depth) and eventually to the ferromagnesite II phase at 80 GPa and 2400 K (approximately 1900 km depth) (Figs. 2 and5). That is, the ferromagnesite II phase in the lower part of the lower mantle likely assumes the low-spin state. Since the low-spin iron ion is energetically more stable at high pressures than its high-spin counterpart, iron ions would partition favorably into the low-spin phase with respect to the high-spin phases in order to minimize the Gibbs free energy of the system in the lower mantle[42]. A recent experimental study shows that the partition coefficient of iron between lower-mantle silicate perovskite (Pv) and ferropericlase (Fp) [$K_D = (Fe/Mg)_{Pv}/(Fe/Mg)_{Fp}$] decreases from 0.85 to 0.52 at approximately 40–47 GPa and 2000 K, which corresponds to the expected pressure-temperature conditions for the spin transition zone of ferropericlase in the lower mantle[22]. That is, the low-spin ferropericlase becomes enriched in Fe^{2+} as a result of the spin transition[23,42]. Considering the volume reduction across the spin transition in ferromagnesite, it is conceivable that iron would preferentially partition into the low-spin ferromagnesite II with respect to other surrounding high-spin phases in the lower mantle.

Depending on the oxygen fugacity and pressure-temperature (*P-T*) conditions, carbon can exist in various forms, including carbides, diamond, graphite, hydrocarbons, CO_2, and carbonates, in the Earth's interior[43]. Subducted slabs have been proposed to play a certain role in transporting a certain amount of carbon into the deep lower mantle[4,7]. Among the transported carbon-bearing materials, ferromagnesite likely can survive the extreme *P-T* conditions of the subducting slabs in the mantle and becomes a stable carbon-bearing phase in the lower mantle[8,10,13]. Such a transport mechanism for delivering carbon into the Earth's lower mantle is also supported by the observations of the ferromagnesian carbonate inclusions in eclogite xenoliths from the deep mantle[44] as well as magnesite-bearing inclusions in natural deep-mantle diamonds[45,46]; though, several previous studies have suggested that polymorphs of $(Mg,Fe)CO_3$ can be replaced by the occurrence of the reduced forms of carbon such as diamond and iron carbides depending on redox conditions of the mantle regions[3,7,47]. Although the oxygen fugacity of the subducting slab materials may be not compatible with the stability of carbonates or carbonate-rich liquid[48], the observation of carbonate inclusions in diamonds potentially brought up to the Earth's surface from the deep mantle indicates that carbonates can exist in the mantle at least locally. Future studies are needed to further understand the role of oxygen fugacity on the stability of the high-pressure ferromagnesite at relevant conditions of the lower mantle. Our results here show that ferromagnesite inside the relative cold and oxidizing slabs likely undergoes an electronic spin transition to the low-spin ferromagnesite at the mid-lower mantle conditions and then becomes stable in the orthorhombic *Pmm2* structure in the low-spin state toward the lower part of the lower mantle. It is thus conceivable that the low-spin orthorhombic ferromagnesite is a major carbon host in the deep lower mantle.

METHODS

Natural single-crystal specimens of siderite (no. NMNH R11313) from the Tsumeb Mine (Namibia) were obtained from the Department of Mineral Sciences, Smithsonian Institution40. Based on the wavelength dispersive spectrometer analyses (WDS) of an electron microprobe (JEOL JXA-8200), the sample has a composition of $FeCO_3$ with a very minor content of $MnCO_3$ (<0.2 mol%); for simplicity, the composition of this sample is referred to as $FeCO_3$ by neglecting the very minor manganese content in the formula. Synchrotron X-ray diffraction patterns showed the lattice parameters of the sample to be a = 4.6909 (±0.0005) Å and c= 15.3687 (±0.0049) Å under ambient conditions. The magnesiosiderite sample was obtained from the Vargas Gem and Mineral Collection at the University of Texas at Austin (collection number: V3817), and has a chemical composition of $(Fe_{0.65}Mg_{0.33}Mn_{0.02})CO_3$; for simplicity, this sample is presented as $(Mg_{0.35}Fe_{0.65})CO_3$, with the minor Mn content counted toward the total Mg content12'27. Synchrotron X-ray diffraction patterns showed the lattice parameters of the sample to be a = 4.6753(±0.0012) Å and c = 15.2794 (±0.0030) Å under ambient conditions.

Each sample was ground to micro-sized powder and then mixed with 3 wt% of micro-sized Au powder as the pressure calibrant49; the use of the inert Au also helped avoid potential chemical reactions with the iron-bearing sample27. The mixture was then slightly pressed to form a platelet approximately 10 μm thick. A small platelet with a diameter of approximately 60 μm was loaded into the sample chamber of a symmetric DAC having a pair of 200 μm flat anvils or 150–300 μm beveled anvils. Neon or dried NaCl layers were also loaded with the sample in the sample chamber and used as the thermal insulator and pressure medium. To avoid any potential air and moisture contamination in the sample chamber, the whole DAC was evacuated at 10^{-2} mbar for 30 minutes before the sample chamber was closed in a vacuum using the high-pressure gas loading system in the Mineral Physics Laboratory of the University of Texas at Austin. High-pressure synchrotron X-ray diffraction experiments were conducted at the 13IDD beamline of the GSECARS of the Advanced Photon Source (APS), Argonne National Laboratory (ANL) using a focused monochromatic X-ray beam with a wavelength (λ) of 0.3344 Å and a beamsize of 2 μm in diameter. For the laser heating, two infrared laser beams were focused down to 25 μm diameter on both sides of the sample, and were co-axially aligned with the incoming X-ray beam using the X-ray induced luminescence on NaCl or the sample50. Temperatures of the laser-heated sample were measured using thermal radiation spectra fitted to the blackbody radiation function50. We note that the siderite sample absorbed the infrared laser very efficiently and evenly on both sides, resulting in temperature uncertainties typically less than 50 K at high pressures. Pressure was measured through the thermal equation of state of Au at high temperature49. The lattice parameters were determined from the diffraction patterns processed by using the LeBail method with the GSAS software package33. Electron microprobe analyses of the quenched samples were conducted using the Environmental Scanning Electron Microscope (FEI Quanta 650 ESEM) and the energy dispersive spectroscopy (Bruker XFlash Detector 5010) with the electron beam resolution of 2.5 nm at 30 kV at the Texas Materials Institute of the University of Texas at Austin. The FEI Quanta 650 ESEM analyses did not require the use of surface conductive coating (such as gold or carbon), which helped preserve original characteristics of the sample especially for carbon analyses. Note that this ESEM does not have capability to do quantitative analysis on carbon concentrations, instead of qualitative determination.

ACKNOWLEDGEMENTS

We acknowledge X. Wu, A. Nayak, J. Zhu, D. Fan, C. Prescher, and K. Zhuravlev for experimental assistance, and thank P. Dera, Z. Mao, and J. Yang for constructive discussions. We thank Laura Dafov for assisting with editing the manuscript. J.F. Lin acknowledges supports from the U.S. National Science Foundation (EAR-0838221), Energy Frontier Research Centers (EFree), the Extreme Physics and Chemistry Directorate of the Deep Carbon Observatory (DCO), and Center for High Pressure Science and Technology Advanced Research (HPSTAR). Synchrotron works of the study were performed at the GSECARS of the APS, ANL. APS is supported by 263 DOE-BES, under Contract No. DE-AC02-06CH11357.

CONTRIBUTIONS

J.F.L. & J.L. conceived & designed the research; J.L., J.F.L., V.B.P. conducted the experiments; J.L. & J.F.L. wrote the paper. All authors reviewed the manuscript.

COMPETING INTERESTS

The authors declare no competing financial interests.

REFERENCES

1. Manning, C. E., Shock, E. L. & Sverjensky, D. A. The chemistry of carbon in aqueous fluids at crustal and upper-mantle conditions: experimental and theoretical constraints. *Rev. Mineral. Geochem.* **75**, 109–148, 10.2138/rmg.2013.75.5 (2013).
2. Dasgupta, R. & Hirschmann, M. M. Melting in the Earth's deep upper mantle caused by carbon dioxide. *Nature* **440**, 659–662 (2006).
3. Rohrbach, A. & Schmidt, M. W. Redox freezing and melting in the Earth's deep mantle resulting from carbon-iron redox coupling. *Nature* **472**, 209–212, 10.1038/nature09899 (2011).
4. Walter, M. J. *et al.* Deep mantle cycling of oceanic crust: evidence from diamonds and their mineral inclusions. *Science* **334**, 54–57, 10.1126/science.1209300 (2011).
5. Jana, D. & Walker, D. The impact of carbon on element distribution during core formation. *Geochim. Cosmochim. Acta* **61**, 2759–2763, 10.1016/S0016-7037(97)00091-4 (1997).
6. Hazen, R. M. & Schiffries, C. M. Why deep carbon? *Rev. Mineral. Geochem.* **75**, 1–6, 10.2138/rmg.2013.75.1 (2013).
7. Dasgupta, R. & Hirschmann, M. M. The deep carbon cycle and melting in Earth's interior. *Earth Planet. Sci. Lett.* **298**, 1–13, 10.1016/j.epsl.2010.06.039 (2010).
8. Biellmann, C., Gillet, P., Guyot, F. o., Peyronneau, J. & Reynard, B. Experimental evidence for carbonate stability in the Earth's lower mantle. *Earth Planet. Sci. Lett.* **118**, 31–41, 10.1016/0012-821x(93)90157-5 (1993).
9. Fiquet, G. *et al.* Structural refinements of magnesite at very high pressure. *Am. Mineral.* **87**, 1261–1265 (2002).
10. Isshiki, M. *et al.* Stability of magnesite and its high-pressure form in the lowermost mantle. *Nature* **427**, 60–63 (2004).
11. Lavina, B. *et al.* Siderite at lower mantle conditions and the effects of the pressure-induced spin-pairing transition. *Geophys. Res. Lett.* **36**, L23306, 10.1029/2009gl039652 (2009).
12. Lin, J.-F., Liu, J., Jacobs, C. & Prakapenka, V. B. Vibrational and elastic properties of ferromagnesite across the electronic spin-pairing transition of iron. *Am. Mineral.* **97**, 583–591, 10.2138/am.2012.3961 (2012).

13. Panero, W. R. & Kabbes, J. E. Mantle-wide sequestration of carbon in silicates and the structure of magnesite II. *Geophys. Res. Lett.***35**, L14307, 10.1029/2008GL034442 (2008).
14. Skorodumova, N. V., Belonoshko, A. B., Huang, L., Ahuja, R. &Johansson, B. Stability of the $MgCO_3$ structures under lower mantle conditions. *Am. Mineral.* **90**, 1008–1011, 10.2138/am.2005.1685 (2005).
15. Oganov, A. R., Ono, S., Ma, Y., Glass, C. W. & Garcia, A. Novel high-pressure structures of $MgCO_3$, $CaCO_3$ and CO_2 and their role in Earth's lower mantle. *Earth Planet. Sci. Lett.* **273**, 38–47, 10.1016/j.epsl.2008.06.005 (2008).
16. Boulard, E. *et al.* New host for carbon in the deep Earth. *Proc. Natl. Acad. Sci. USA* **108**, 5184–5187 (2011).
17. Boulard, E. *et al.* Experimental investigation of the stability of Fe-rich carbonates in the lower mantle. *J. Geophys. Res.* **117**, B02208, 10.1029/2011jb008733 (2012).
18. Gallagher, P. K. & Warne, S. S. J. Thermomagnetometry and thermal decomposition of siderite. *Thermochim. Acta* **43**, 253–267, 10.1016/0040-6031(81)85183-0 (1981).
19. Tao, R., Fei, Y. & Zhang, L. Experimental determination of siderite stability at high pressure. *Am. Mineral.* **98**, 1565–1572, 10.2138/am.2013.4351 (2013).
20. Frost, D. J. & McCammon, C. A. The redox state of Earth's mantle.*Annu. Rev. Earth Planet. Sci.* **36**, 389–420, 10.1146/annurev.earth.36.031207.124322 (2008).
21. Dasgupta, R., Hirschmann, M. M. & Withers, A. C. Deep global cycling of carbon constrained by the solidus of anhydrous, carbonated eclogite under upper mantle conditions. *Earth Planet. Sci. Lett.* **227**, 73–85, 10.1016/j.epsl.2004.08.004 (2004).
22. Irifune, T. *et al.* Iron partitioning and density changes of pyrolite in Earth's lower mantle. *Science* **327**, 193–195 (2010).
23. Lin, J.-F., Speziale, S., Mao, Z. & Marquardt, H. Effects of the electronic spin transitions of iron in lower mantle minerals: implications for deep mantle geophysics and geochemistry. *Rev. Geophys.* **51**, 244–275, 10.1002/rog.20010 (2013).
24. Mao, Z., Lin, J.-F., Liu, J. & Prakapenka, V. B. Thermal equation of state of lower-mantle ferropericlase across the spin crossover.*Geophys. Res. Lett.* **38**, L23308, 10.1029/2011gl049915 (2011).
25. Oganov, A. R., Hemley, R. J., Hazen, R. M. & Jones, A. P. Structure, bonding, and mineralogy of carbon at extreme conditions. *Rev. Mineral. Geochem.* **75**, 47–77, 10.2138/rmg.2013.75.3 (2013).

26. Mattila, A. *et al.* Pressure induced magnetic transition in siderite $FeCO_3$ studied by x-ray emission spectroscopy. *J. Phys. Condens. Matter* **19**, 386206 (2007).
27. Liu, J., Lin, J.-F., Mao, Z. & Prakapenka, V. B. Thermal equation of state and spin transition of magnesiosiderite at high pressure and temperature. *Am. Mineral.* **99**, 84–93, 10.2138/am.2014.4553 (2014).
28. Shi, H., Luo, W., Johansson, B. & Ahuja, R. First-principles calculations of the electronic structure and pressure-induced magnetic transition in siderite $FeCO_3$. *Phys. Rev. B* **78**, 155119 (2008).
29. Shannon, R. D. Revised effective ionic radii and systematic studies of interatomic distances in halides and chalcogenides. *Acta Crystallogr. Sect. A* **32**, 751–767, 10.1107/S0567739476001551 (1976).
30. Hazen, R. M., Downs, R. T. & Prewitt, C. T. Principles of comparative crystal chemistry. *Rev. Mineral. Geochem.* **41**, 1–33, 10.2138/rmg.2000.41.1 (2000).
31. Lavina, B. *et al.* Discovery of the recoverable high-pressure iron oxide Fe_4O_5. *Proc. Natl. Acad. Sci. USA* **108**, 17281–17285, 10.1073/pnas.1107573108 (2011).
32. Looijenga-Vosa, A. & Buerger, M. J. in *International Tables for Crystallography* Vol A. (ed Theo Hahn) Ch. 3.1, 46–51 (Springer, 2006).
33. Toby, B. EXPGUI, a graphical user interface for GSAS. *J. Appl. Cryst.* **34**, 210–213, 10.1107/S0021889801002242 (2001).
34. McCusker, L. B., Von Dreele, R. B., Cox, D. E., Louer, D. & Scardi, P.Rietveld refinement guidelines. *J. Appl. Cryst.* **32**, 36–50, 10.1107/S0021889898009856 (1999).
35. Arapan, S., Souza de Almeida, J. & Ahuja, R. Formation of sp^3hybridized bonds and stability of $CaCO_3$ at very high pressure.*Phys. Rev. Lett.* **98**, 268501 (2007).
36. Merlini, M. *et al.* Structures of dolomite at ultrahigh pressure and their influence on the deep carbon cycle. *Proc. Natl. Acad. Sci. USA* **109**, 13509–13514, 10.1073/pnas.1201336109 (2012).
37. Lavina, B. *et al.* Structure of siderite $FeCO_3$ to 56 GPa and hysteresis of its spin-pairing transition. *Phys. Rev. B* **82**, 064110 (2010).
38. Birch, F. Finite strain isotherm and velocities for single-crystal and polycrystalline NaCl at high pressures and 300 K. *J. Geophys. Res.* **83**, 1257–1268, 10.1029/JB083iB03p01257 (1978).
39. Santillan, J. & Williams, Q. A high-pressure infrared and X-ray study of $FeCO_3$ and $MnCO_3$: comparison with $CaMg(CO_3)_2$-dolomite. *Phys. Earth Planet. Inter.* **143–144**, 291–304, 10.1016/j.pepi.2003.06.007 (2004).

40. Zhang, J., Martinez, I., Guyot, F. & Reeder, R. J. Effects of Mg-Fe^{2+}substitution in calcite-structure carbonates: thermoelastic properties. *Am. Mineral.* **83**, 280–287 (1998).
41. Katsura, T., Yoneda, A., Yamazaki, D., Yoshino, T. & Ito, E.Adiabatic temperature profile in the mantle. *Phys. Earth Planet. Inter.* **183**, 212–218, 10.1016/j.pepi.2010.07.001 (2010).
42. Badro, J., Fiquet, G. & Guyot, F. in *Earth's Deep Mantle: Structure, Composition, and Evolution* Vol. 160 *Geophysical Monograph Series*(eds Robert D. van der Hilst, Jay D. Bass, Jan Matas, & Jeannot Trampert) 241–260 (AGU., 2005).
43. Palyanov, Y. N. *et al.* Mantle–slab interaction and redox mechanism of diamond formation. *Proc. Natl. Acad. Sci. USA* **110**, 20408–20413, 10.1073/pnas.1313340110 (2013).
44. Liou, J. G., Zhang, R. Y., Ernst, W. G., Rumble, D. & Maruyama, S.High-pressure minerals from deeply subducted metamorphic rocks. *Rev. Mineral. Geochem.* **37**, 33–96 (1998).
45. Wang, A., Pasteris, J. D., Meyer, H. O. A. & Dele-Duboi, M. L.Magnesite-bearing inclusion assemblage in natural diamond.*Earth Planet. Sci. Lett.* **141**, 293–306, 10.1016/0012-821X(96)00053-2 (1996).
46. Berg, G. W. Evidence for carbonate in the mantle. *Nature* **324**, 50–51 (1986).
47. Dasgupta, R. Ingassing, Storage, and Outgassing of Terrestrial Carbon through Geologic Time. *Rev. Mineral. Geochem.* **75**, 183–229, 10.2138/rmg.2013.75.7 (2013).
48. Stagno, V., Ojwang, D. O., McCammon, C. A. & Frost, D. J. The oxidation state of the mantle and the extraction of carbon from Earth's interior. *Nature* **493**, 84–88, 10.1038/nature11679 (2013).
49. Fei, Y. *et al.* Toward an internally consistent pressure scale. *Proc. Natl. Acad. Sci. USA* **104**, 9182–9186, 10.1073/pnas.0609013104 (2007).
50. Prakapenka, V. B. *et al.* Advanced flat top laser heating system for high pressure research at GSECARS: application to the melting behavior of germanium. *High Pressure Res.* **28**, 225–235, 10.1080/08957950802050718 (2008).

Chapter 3

Lattice Dynamics Study of Phonon Instability and Thermal Properties of Type-I Clathrate K_8Si_{46} under High Pressure

Wei Zhang [1,4], Zhao Yi Zeng [2,*], Ni Na Ge [3] and Zhi Guo Li [4]

[1]*School of Science, Southwest University of Science and Technology, Mianyang 610064, Sichuan, China*

[2]*College of Physics and Electronic Engineering, Chongqing Normal University, Chongqing 400047, China*

[3]*State Key Laboratory Cultivation Base for Nonmetal Composites and Functional Materials, Southwest University of Science and Technology, Mianyang 610064, Sichuan, China*

[4]*Laboratory for Shock Wave and Detonation Physics Research, Institute of Fluid Physics, Chinese Academy of Engineering Physics, Mianyang 621900, Sichuan, China*

ABSTRACT

For a further understanding of the phase transitions mechanism in type-I silicon clathrates K_8Si_{46}, ab initio self-consistent electronic calculations combined with linear-response method have been performed to investigate the vibrational properties of alkali metal K atoms encapsulated type-I silicon-clathrate under pressure within the framework of density functional perturbation theory. Our lattice dynamics simulation results showed that the pressure induced phase transition of $K_8Si_{46}$was believed to be driven by the phonon instability of the calthrate lattice. Analysis of the evolution of the partial phonon density of state with pressure, a legible dynamic picture for both guest K atoms and host lattice, was given. In addition, based on phonon calculations and combined with quasi-harmonic approximation, the specific heat of K_8Si_{46} was derived, which agreed very well with experimental results. Also, other important thermal properties including the thermal expansion coefficients and Grüneisen parameters of K_8Si_{46} under different temperature and pressure were also predicted.

Keywords: phonon spectrum; clathrate compounds; lattice dynamics; high pressure

1. INTRODUCTION

Nanostructured type I clathrates are composed of two 20-atoms (small) and six 24-atoms (large) cages of group-IV elements which can host different kinds of atoms. The framework atoms admit partial substitution by other atomic species. Such good tailorability enables these clathrate compounds to have extensive potential applications in areas of superconductivity [1], wide-band-gap semiconductors [2], optoelectronics [3], magnetism [4], thermoelectric [5] and photovoltaics [6], etc. As illustrated in Figure 1, there are three distinct Wyckoff symmetry sites, i.e., 6*c*, 16*i* and 24*k* for the framework atoms. Encaptured metal atoms located at the center of small and large cages correspond to 2*a* sites and 6*d* sites, respectively. Since type-I Ba_8Si_{46} had been realized in a multi-anvil press by the group of Yamanaka in 2000 [7], high pressure technique has opened up a new door to synthesize these kind of clathrates compounds and provided new ideas to explore the interaction mechanism between guest atoms and the host lattice. During the high pressure experiments in studying the stability and compressibility of these silicon clathrates, some clathrates doped with large guest atoms such as K_8Si_{46} [8,9], Ba_8Si_{46}[10,11,12,13,14,15], $I_8Si_{44}I_2$ [16,17], Rb_6Si_{46} [18] exhibit an abrupt pressure induced volume collapse transition while the overall crystal symmetry is preserved. Various mechanisms (e.g., the change of hybridization between guest atoms and frame cages [12] and an electronic topological transition [14,15]) for this high pressure phase transition have been proposed to explain this phenomenon for Ba_8Si_{46}. In our recent work [19], detailed investigations have been performed to study the effect of forming lattice vacancies on the mechanical and electronic properties of Ba_8Si_{46} under high pressure. The results indicate that the compressbility of Ba_8Si_{46} is governed by the Si framework and the pressure induced escape of host Si atoms would cause the sudden volume collapse during the compress, which is consistent with the explanation from Iitaka et al. according to their atomistic model [20]. A more coherent picture of the phase transition is given by the newest experimental study of the clathrate collapse in mixed $Ba_8(Si,Ge)_{46}$ clathrates which suggests that the volume collapse is a second order transition through a Landau modeling ,with a symmetry-breaking mechanism related to the Ba displacement initiated either by vacancy creation or by local distortion of the framework structure [21].

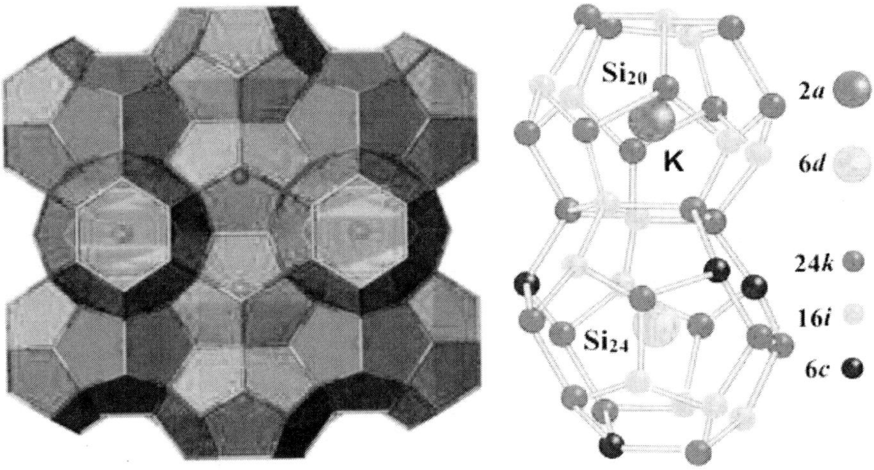

Figure 1. The sketch map of ball and stick representation of the type I clathrate K_8Si_{46} with an illustration of Wyckoff sites of the structure.

In the case of K_8Si_{46}, a noticeable change in the compressibility had been found at around 15 GPa followed by an abrupt change of volume at around 20 GPa. Above 32 GPa, the sample transformed into an amorphous phase. Earlier ab initio phonon band structure calculations using finite displacement method [22] had found that vibrational frequencies of the K atoms in the large cavities became imaginary at a pressure of 16 GPa which suggested a phase transition triggered by positional disordered guest atoms [8]. However, it is known that the supercell should be constructed in a phonon calculation within this method [22], which is obviously unbearable for an originally large primitive cell of K_8Si_{46}. Therefore, in their calculations, just primitive cell was supposed to be used. Although the primitive K_8Si_{46} cell under ground state has a large lattice constant of somewhat more than 10 Å, the magnitude of the force constants beyond that distance is generally expected to be negligible. In a covalent semiconductor and complex structured system, the required cutoff radius should be larger than 10 Å as a rule, thus such a treatment may cause unreliable results when the K_8Si_{46} cell is compressed under high pressure (the lattice constant of K_8Si_{46} under 15 GPa is about 9.68 Å). For a confirmation of this point, we have also performed the phonon calculation within finite displacement method using a unit cell of K_8Si_{46} and indeed found that a large number of imaginary frequencies appeared when the pressure was applied up to 15 GPa. So, in the present work, we employ the density functional perturbation theory (DFPT) [23,24] which avoids the use of a supercell and allows calculation of the dynamical matrix exactly at an arbitrary ***q*** vector to study the stability of K_8Si_{46} under pressure. For a perfect crystal, the dynamical instability of the phonon is found to be originated from the framework of K_8Si_{46} which is quite different to the results reported by J. S. Tse et al. [8]. From the obtained partial phonon density of the state, a clearer dynamic picture of guest K atoms and host lattice under high pressure is given. Besides, based on the phonon calculation, we derived the thermal properties of K_8Si_{46} from quasi-harmonic approximation which were believed to offer a useful reference to design and synthetize a new ternary type-I Si clathrate based on K_8Si_{46} with enhanced thermoelectric performance.

2. COMPUTATIONAL METHODS

Phonon dispersion spectrum was calculated by using the linear-response method [25,26] within the density functional perturbation theory (DFPT) [23,24]. The full phonon dispersion curves were obtained through Fourier interpolation. Norm-conserving pseudopotentials generated using the kinetic energy optimization scheme developed by Lin et al. [27] and Lee [28] were employed to describe the ion-electron interactions with an energy cutoff of 800 eV to expand the valence electronic wave functions. Monkhorst-Pack *k*-points mesh of 6 × 6 × 6 had been chosen. The electronic exchange-correlation interactions were treated within the local density approximation (LDA) [29]. During the pseudopotentials calculations, pseudo atomic calculations were performed for K (3*s*, 3*p*, 4*s*) and Si (3*s*, 3*p*). By employing the Parrinello–Rahman method [30,31], hydrostatic pressure load on crystal was realized within the variable cell approach. Both the cell parameters and the atomic internal coordinates were fully relaxed at each target external pressure by applying Broyden, Fletcher, Goldfarb and Shanno (BFGS) scheme [32]. All these total energy electronic structure calculations were carried out by Cambridge Serial Total Energy Package (CASTEP) code [33,34].

3. RESULTS AND DISCUSSION

3.1. Phonon Spectra

Within the framework of DFPT method, phonon frequencies are computed as second-order derivatives of the total energy with respect to a given perturbation in the form of atomic displacements. The force constants matrix can be obtained by differentiating the Hellmann-Feynman forces on atoms with respect to ionic coordinates. Based on the interatomic force constants, we can obtain the phonon spectra by using Fourier interpolation with specific treatment of the long-range dipole-dipole interaction [35]. The phonon band structure calculations were performed up to 40 GPa with an interval of 10 GPa. A more careful calculation at 39 GPa was performed for determining the exact pressure value when the phonon became instable.

The obtained ground state phonon dispersion relations in K_8Si_{46} crystal are shown in Figure 2a. Clear flat vibrational bands can be easily recognized in the low frequency region around 100 cm^{-1}. By analysis of partial phonon density of state (PPDOS) as shown in Figure 3a, we find that these low frequencies are corresponding to a sharp peak located at about 98.8 cm^{-1} which are originated from the localized vibration of 6*d* sites K atoms at Si_{24} cages. Another somewhat weaker but still clear peak at 128.2 cm^{-1}also from K(6*d*) can be attributed to the asymmetry of the large Si_{24} cage which takes ellipsoidal shape. A similar phenomenon had also been observed in another type-I silicon-clathrate Na_8Si_{46} reported by Li et al. [36]. Moreover, a strong and broad peak centered at 172.1 cm^{-1} is found to be contributed by the mixing vibrations of K atoms at Si_{20} cages and the framework Si atoms indicates the intense interaction hybridization between them, this is because the size of Si_{20} cage is much smaller than Si_{24} cage which yields a shorter interatomic distance of K-Si. Experimentally, the measured Raman spectrum of K_8Si_{46}showed noticeable peaks at 94, 119 and 177 cm^{-1} related to the vibration of K atoms [9], which is consistent with our results. By inelastic neutron scattering, Mélinon et al. [37] and Reny et al. [38] obtained very similar vibrational spectrum of K_8Si_{46} and found two identified peaks centered at about (100 cm^{-1}, 170 cm^{-1}) and (98 ± 2 cm^{-1}, 161 ± 5 cm^{-1}) respectively. Our calculated results perfectly reproduced the main characteristics of experimental observation. Under a pressure of 30 GPa, due to the shrinkage of the silicon cages under compress, the PPDOS shows strong mixing of K and Si vibrations. The frequency of K atoms' motion at both 2*a* and 6*d* sites became higher, which indicates a further localization of these atoms, as shown in Figure 3b. However, it is noted that the appearance of massive low frequencies that originate from the framework atoms suggests a collective "soften" of Si-Si bonds which will finally make the host lattice unstable. As the pressure is applied at a value of 40 GPa, it can be seen clearly from Figure 2c that the frequencies around the M symmetry point become imaginary, which clearly shows the instability of clathrate framework. As illustrated in in Figure 3c, the PPDOS under 40 GPa also shows a dramatic reduction of frequencies from host lattice as presented in the phonon spectrum. Our calculation results show that a mechanical instability of the silicon framework is believed to be responsible for the pressure-induced volume collapse at about 20 GPa and subsequent amorphization at 32 GPa of K_8Si_{46} observed experimentally.

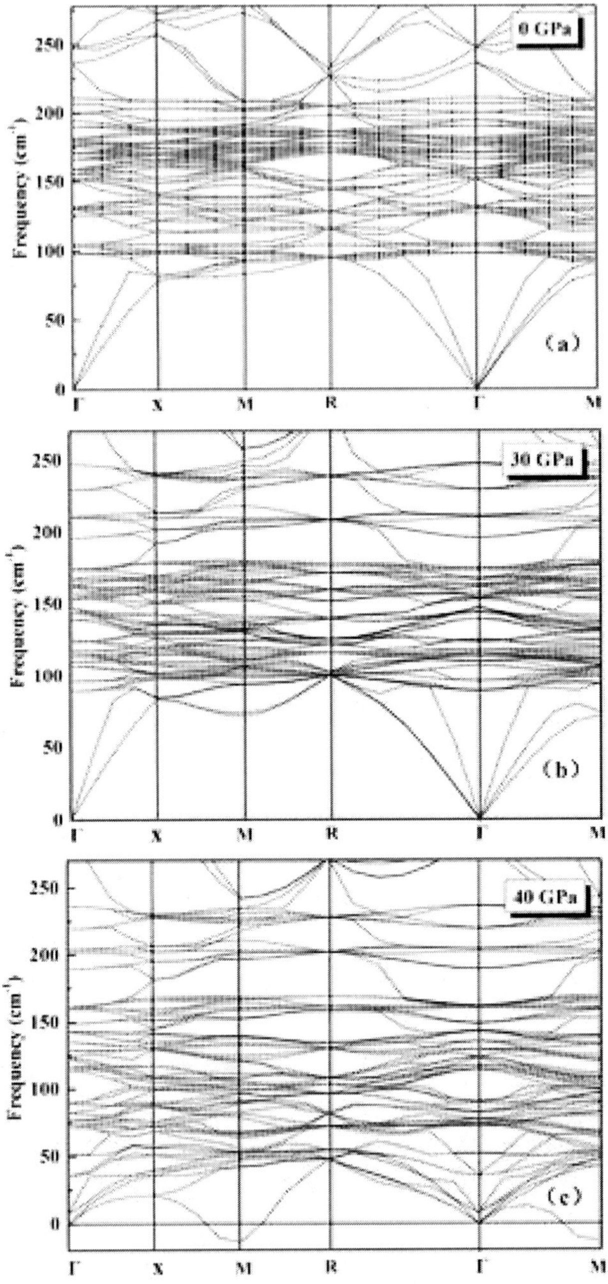

Figure 2. The phonon dispersion relation of K_8Si_{46} at (**a**) 0 GPa; (**b**) 30 GPa and (**c**) 40 GPa.

However, it is noted that the unstable pressure given in the present work is obviously larger than the experimental observation. This is because our calculations are performed using a perfect K_8Si_{46} crystal from the view of lattice dynamics without consideration of other possible transition mechanisms (e.g., vacancies formation [20], local symmetry-breaking [21] or an electronic topological transition [14,15], etc.) associated with the pressure collapse of the clathrate structure. If multi-mechanisms are involved in phase transition, the value of transition pressure would be affected a lot. For instance, the lattice vacancies are actually very likely to be produced in these type-I clathrates, especially in 6*c* sites. The experimental observed Cs_8Sn_{44} was formed from the missing two Sn atoms in the 6*c* sites [39]. Besides, theoretical calculations by Iitaka et al. also showed that 6*c* sites lattice vacancies formation under high pressure was indeed energetically preferable. By the model of partially occupied Si sites they explained the transition pressure and change of Raman spectra of both K_8Si_{46} and Ba_8Si_{46} [20]. Moreover, our recent work showed that 6*c* sites lattice vacancies increased the compressibility of clathrate greatly while guest atoms vacancies hardly had any influence on this property [19].

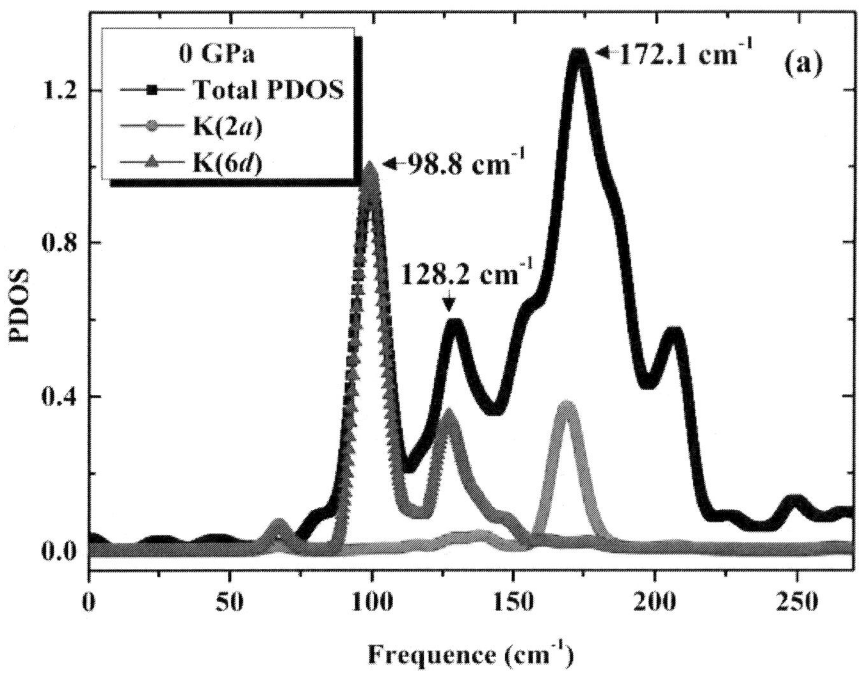

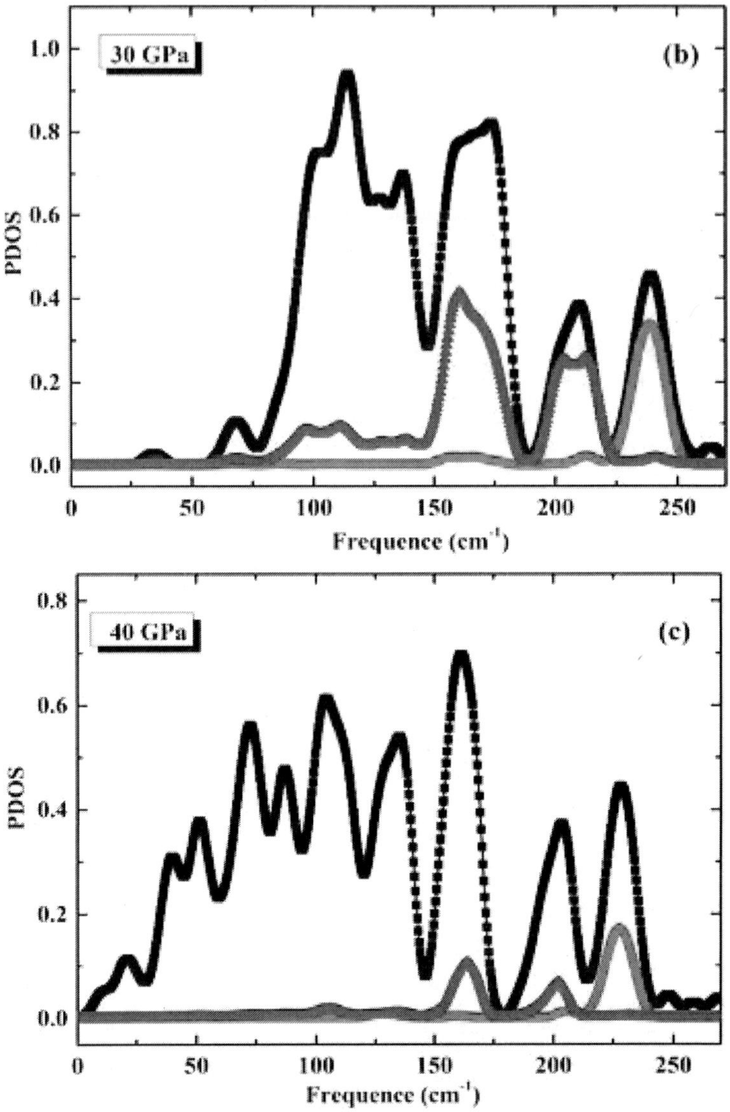

Figure 3. The partial phonon density of state of K_8Si_{46} at (**a**) 0 GPa; (**b**) 30 GPa and (**c**) 40 GPa.

Experimentally, by performing high quality in situ high-pressure angle-dispersive X-ray powder diffraction measurements, Li et al. found a highly disordered Si framework from analysis of the obtained anomalously large Si thermal parameters [13]. Also, present lattice dynamics calculation for K_8Si_{46} shows that the Si-Si bond "softens" under high pressure which again provides theoretical possibility for the formation of lattice vacancies. If one considers this mechanism, the volume collapse pressure of K_8Si_{46} is believed to be reduced. Thus, in view of these results, guest K atoms displacement induced phonon instability from earlier ab initio phonon band structure calculations [8] is indeed more likely to be caused by the disadvantages of the finite displacement method in treating these large cell calthrate compound, especially under high pressure. Our calculated results based on DFTP method obviously gives a more convincing and clear physical picture for the instability of K_8Si_{46} under high pressure.

3.2. Thermal Properties from Quasi-Harmonic Approximation

The results of calculated phonon spectra and phonon density of state can be used to compute the thermodynamic properties using the quasi-harmonic approximation (QHA) [40]. In the QHA, the phonon Helmholtz free energy is given by:

$$F^*_{vib}(T) = k_B T \int_0^{\infty} [\frac{1}{2}\hbar\omega + k_B T \ln(1 - e^{-\frac{\hbar\omega}{k_B T}})] f(\omega) d\omega \quad (1)$$

where k_B is the Boltzmann constant. $\hbar$ is Planck's constant and $f(\omega)$ is the phonon density of states (PDOS). Through a series calculation of PDOS of K_8Si_{46} with different volumes, the volume dependence of Helmholtz free energy $F_{vib}(V,T)$ can be obtained. Then the vibrational contribution to the entropy, the specific heat at constant volume and isothermal bulk modulus can be derived by:

$$S_{vib}(T) = (\frac{\partial F_{vib}}{\partial T})_V \quad (2)$$

$$C_v(T) = -T(\frac{\partial^2 F_{vib}}{\partial T^2})_V \quad (3)$$

$$B_T = -V(\frac{\partial P}{\partial V})T = V(\frac{\partial^2 F_{vib}}{\partial V^2})_T \quad (4)$$

The Grüneisen parameters can be computed by the volume derivative of (−TS):

$$\gamma = -\frac{V}{C_V T}(\frac{\partial(-TS)}{\partial V})_T \quad (5)$$

Then, the volume coefficient of thermal expansion and constant pressure heat capacity (C_p) follows:

$$\alpha_V = -\frac{1}{V}\left(\frac{\partial V}{\partial T}\right)_P = \frac{\gamma C_V}{VB_T} \tag{6}$$

$$C_P = C_V(1+\gamma\alpha_V T) \tag{7}$$

From QHA calculation, zero-point energy F_{vib} (T = 0) of K_8Si_{46} compound is determined to be 2.688 eV. Moreover, the calculated variation of volume thermal expansion coefficient α_V with temperature under different pressures are illustrated in Figure 4 from which we can find that α_V increases rapidly with temperature below about 200 K and pressure imposes a strong restraint on the lattice expansion. At room temperature and zero pressure, the α_V is predicted as 6.26 × 10^{-5} K^{-1}, corresponding to a linear thermal expansion coefficient α_L as 2.09 × 10^{-5} K^{-1} which is very close to the experimental value of Na_8Si_{46} (about 2.0 × 10^{-5} K^{-1}) given by Qiu et al. [41]. In addition, another important thermodynamic quantity of Grüneisen parameters which is difficult to determine experimentally can also be predicted by QHA method. The obtained temperature dependencies of Grüneisen parameters under pressure of 0, 10, 20 and 30 GPa are presented in Figure 5. It can be found that the Grüneisen parameters become almost constant in relation to the varied temperature under high pressure. At ambient conditions, the Grüneisen parameter is found to be 2.47. For congener compound Na_8Si_{46} under same condition, this value was reported as 2.68 [41].

These similarities in thermdynamic properties of type I clathrates doped with same group elements have also been reported by many experimental works, for example, the thermal expansion coefficients of $Ba(Sr)_8Ga_{16}Ge_{30}$ and $Sr_8Ga_{16}Ge_{30}$ [42] were just found to be almost identical to each other, and so were $Rb_8Sn_{44}\square_2$ and $Cs_8Sn_{44}\square_2$ (□ means lattice vacancy) [43]. Consequently, in general, the thermal expansion coefficient of intermetallic clathrates is assumed to mainly depend on the bonding of the framework atoms. The nature of the guest atoms just make a small contribution due to their weaker ionic bonding to the host structure. However, in the case of Ba_8Si_{46}, which has been attracting extensive attention due to the discovery of its superconductivity, the measured α_L at room temperature was found to be only 1.2 × 10^{-5} K^{-1}. Besides, a drop of α_L of Ba_8Si_{46} occurred at the superconducting transition temperature and the frequency-dependent Grüneisen parameter of Ba_8Si_{46} indicated strong anharmonicity of the lattice vibrations for low energy mode with a value of up to 8.6 while higher energy modes are much less anharmonic with a value somewhat below 2 [44]. These novel results of Ba_8Si_{46} are quite different

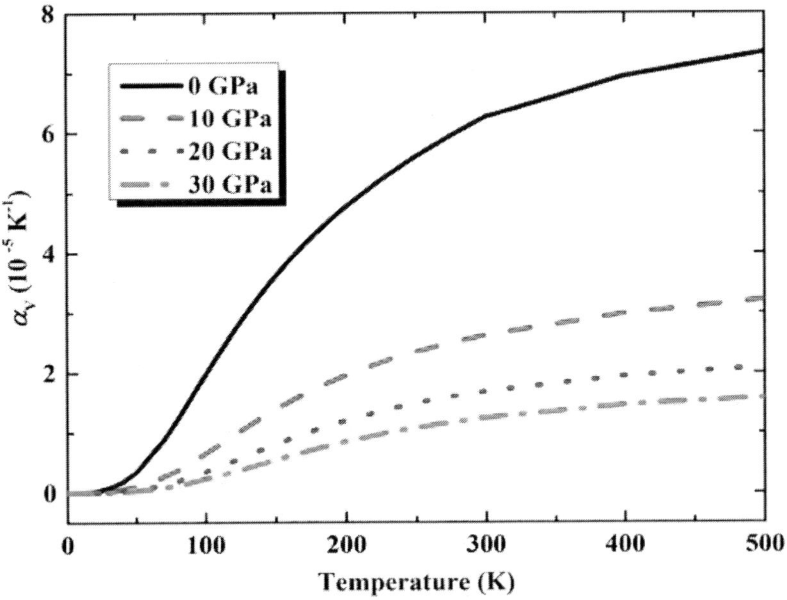

Figure 4. The calculated variation of volume coefficient of thermal expansion of K_8Si_{46} with temperature under 0, 10, 20 and 30 GPa.

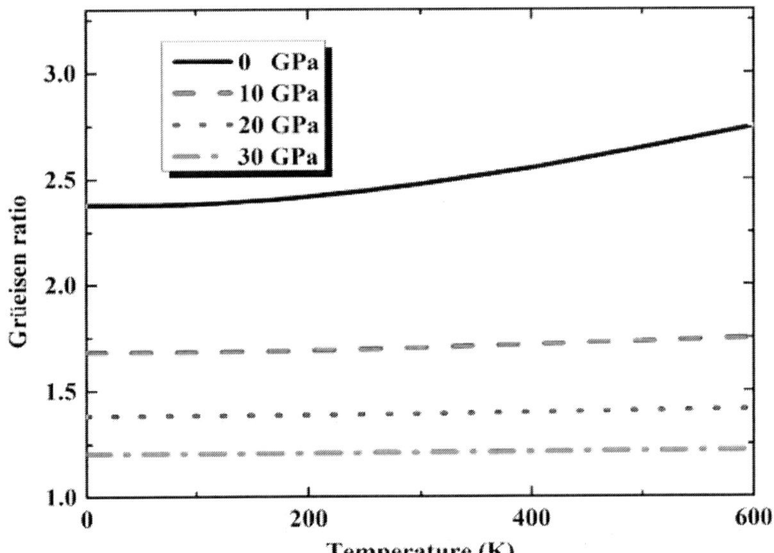

Figure 5. The calculated variation of Grüneisen parameters of $K_8Si_{46}$with temperature under 0, 10, 20 and 30 GPa.

from those of $Na(K)_8Si_{46}$ which reveals the non-negligible role of different hybridization between the guest and host atoms in studying the vibrational properties for doped clathrate compound. Moreover, the vacancies which are very likely to be formed in clathrate compounds can also significantly influence the system vibrational properties. Their presence can decrease the average bonding strength among host atoms and induce the displacement of guest atoms from their ideal positions. Thus, the anharmonicity of system is enhanced which would yeild an increment of the thermal expansion coefficient and Grüneisen parameters at zero pressure that had been identified by the experimental work for type-I Ge-based clathrates [45]. However, this effect became more unintelligible under high pressure because the formation of vacancies can also increase the compressbility of the clathrate which would lead the volume collapse under compress. So, further theoretical simulation with considering the formation of vacancies in hosts framework is expected to explore the effects of these vacancies on the vibrational properties of clathrate system under high pressure.

Because the experimental specific heat was conducted at constant pressure namely C_p, in Figure 6, the resulting dependence of C_p on temperature calculated from QHA method at zero pressure is illustrated, the experimental data up to room temperature from Stefanoski have also been plotted together for a comparison [46]. It can be found that they agree with each other excellently (for instance, the specific heat at room temperature determined by present work is 1151.5 $J{\cdot}mol^{-1}{\cdot}K^{-1}$, experimental measured value is 1159.6 $J{\cdot}mol^{-1}{\cdot}K^{-1}$) which indicates the validity of present lattice dynamic simulation within DFTP. At low temperatures, the temperature dependence of specific heat presents a similar behavior to thermal expansion coefficient. When the temperature is higher than about 400 K, specific heat gradually approaches the Dulong-Petit limit, i.e., $3nN_Ak_B$ (about 1397.2 $J{\cdot}mol^{-1}{\cdot}K^{-1}$) which is followed by all solid at high temperatures. In our former study [47], we found the specific heat of Na_8Si_{46} predicted by quasi-harmonic Debye model was underestimated obviously at low temperature which revealed the limitation of the Debye model in dealing with these doped clathrate compounds. The experimental heat capacities of Na_8Si_{46} were finally reproduced by treating the special “rattle” modes of captured Na atoms in cages as Einstein oscillators. In the present work, a full phonon calculation within DFPT method can give a rather exact description of vibrational properties of these clathrate compounds which avoids a discriminatory treatment of the host lattice and encapsulated atoms. The specific heat of K_8Si_{46} under pressure of 30 GPa is also presented in Figure 6, from which it can be found that pressure can decrease the specific heat considerably due to the suppression of lattice vibration.

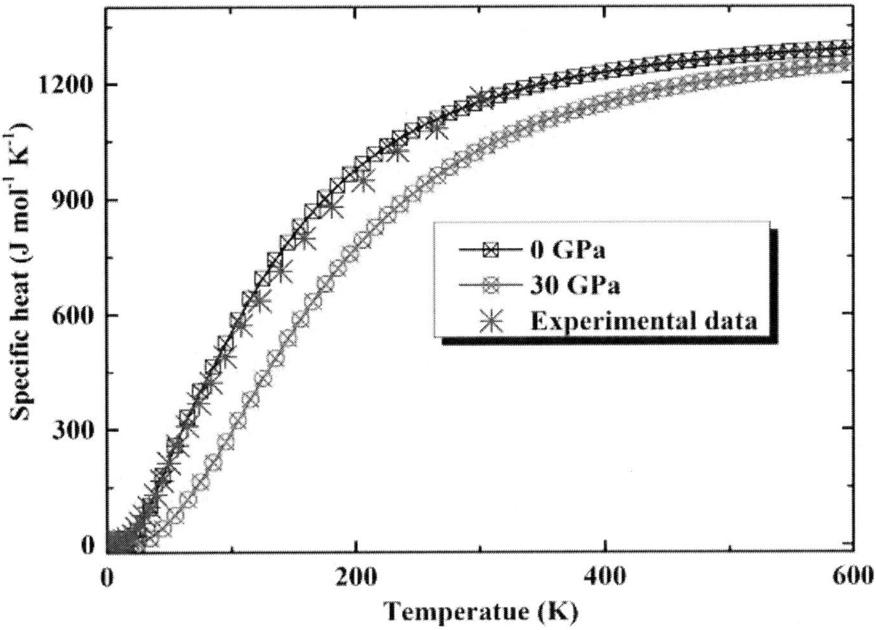

Figure 6. The value of C_P of K_8Si_{46} as a function of temperature at 0 and 30 GPa, in comparison with the experimental data [41].

The comparison of the calculated specific heat to that predicted by the Debye model leads to the concept of the temperature dependent Debye temperature. The obtained Debye temperature at the high temperature limit is 550 K, which is consistent with the reported experimental result of 577 K [48]. In addition, the thermal conductivity κ_L contributed by the lattice can be estimated by the Debye equation [49]:

$$\kappa_L = \lambda_V^m C/3 \quad (8)$$

where C is the volumetric heat capacity, λ is the mean free path of phonons, assumed as the average distance between the guest atoms, V_m is the velocity of sound which can be derived from Debye temperature [50]. In this way, the lattice thermal conductivity is given as 1.64 W m^{-1}·K^{-1} which is comparable to other reported room temperature thermal conductivities of type-I silicon based clathrate compound [48].

4. CONCLUSIONS

In summary, we have given a detailed study of the vibrational and thermal properties of type-I calthrate K_8Si_{46}. By employing the linear-response method within the density functional perturbation theory (DFPT), we obtained the phonon dispersion relation in K_8Si_{46} crystal as well as phonon density of states which contained detail information about the vibrational properties of both guest atoms and framework lattice. Under a pressure of 40 GPa, the frequencies from the motions of framework silicon atoms become imaginary around the M symmetry point suggests the instability of K_8Si_{46} crystal. The determined phonon instable pressure by lattice dynamic simulation is larger than the transition pressure observed experimentally. However, actual phase transition may be driven by multi-mechanisms which are expected to reduce the phonon instable pressure. By using quasi-harmonic approximation, the thermal properties of K_8Si_{46} including heat specific, thermal expansion coefficient and Grüneisen parameters are predicted. From a comparison of Na_8Si_{46}, we find that the thermal properties of these type-I calthrates are very insensitive to the different nature of guest atoms in the same group of the periodic table.

ACKNOWLEDGMENTS

This work was supported by the National Natural Science Foundation of China under Grant No.11347134, 11504304, 11347134, 11347019 and 11304408.

AUTHOR CONTRIBUTIONS

Wei Zhang and Zhao Yi Zeng conceived and designed the theoretical simulation; Ni Na Ge Performed partial calculations; Zhi Guo Li analyzed the data and contributed computational resources; Wei Zhang and Zhao Yi Zeng wrote the paper.

CONFLICTS OF INTEREST

The authors declare no conflict of interest. The founding sponsors had no role in the design of the study; in the collection, analyses, or interpretation of data; in the writing of the manuscript, and in the decision to publish the results.

REFERENCES

1. Herrmann, R.F.W.; Tanigaki, K.; Kuroshima, S.; Suematsu, H. Superconductivity in silicon based barium-inclusion clathrates. *Chem. Phys. Lett.* **1998**, *283*, 29–32.

2. Nolas, G.S.; Cohn, J.L.; Slack, G.A.; Schujman, S.B. Semiconducting Ge clathrates: Promising candidates for thermoelectric applications. *Appl. Phys. Lett.* **1998**, *73*, 178–180.

3. Menon, M.; Richter, E.; Subbaswamy, K. Structural and vibrational properties of Si clathrates in a generalized tight-binding molecular-dynamics scheme. *Phys. Rev. B* **1997**, *56*, 12290–12295.

4. Kawaguchi, T.; Tanigaki, K.; Yakusawa, M. Silicon clathrate with an electron system. *Phys. Rev. Lett.* **2000**, *85*, 3189–3192.

5. Martin, J.; Wang, H.; Nolas, G.S. Optimization of the thermoelectric properties of $Ba_8Ga_{16}Ge_{30}$.*Appl. Phys. Lett.* **2008**, *92*, 222110.

6. Martinez, A.D.; Krishna, L.; Baranowski, L.L.; Lusk, M.T.; Toberer, E.S.; Tamboli, A.C. Synthesis of Group IV clathrates for photovoltaics. *IEEE J. Photovolt.* **2013**, *3*, 1305–1310.

7. Yamanaka, S.; Enishi, E.; Fukuoka, H.; Yasukawa, M. High-pressure synthesis of a new silicon clathrate superconductor Ba_8Si_{46}. *Inorg. Chem.* **2000**, *39*, 56–58.

8. Tse, J.S.; Desgreniers, S.; Li, Z.Q.; Ferguson, M.R.; Kawazoe, Y. Structural Stability and Phase Transitions in K_8Si_{46} Clathrate under High Pressure. *Phys. Rev. Lett.* **2002**, *89*, 195507.

9. Kume, T.; Koda, T.; Sasaki, S.; Shimizu, H.; John, S.T. High-pressure Raman study of the potassium-doped silicon clathrate K_8Si_{46}. *Phys. Rev. B* **2004**, *70*, 052101.

10. San-Miguel, A.; Mélinon, P.; Connétable, D.; Blasé, X.; Tournus, F.; Reny, E.; Itié, J.P. Pressure stability and low compressibility of intercalated cagelike materials: The case of silicon clathrates.*Phys. Rev. B* **2002**, *65*, 054109.

11. Kume, T.; Fukuoka, H.; Koda, T.; Sasaki, S.; Shimizu, H.; Yamanaka, S. High-pressure Raman study of Ba doped silicon clathrate. *Phys. Rev. Lett.* **2003**, *90*, 155503.

12. San Miguel, A.; Merlen, A.; Toulemonde, P.; Kume, T.; Le Floch, S.; Aouizerat, A.; Itié, J.P. Pressure-induced homothetic volume collapse in silicon clathrates. *Europhys. Lett.* **2005**, *69*, 556.

13. Yang, L.; Ma, Y.M.; Iitaka, T.; Tse, J.S.; Stahl, K.; Ohishi, Y.; Jiang, J.Z. Pressure-induced phase transformations in the Ba_8Si_{46} clathrate. *Phys. Rev. B* **2006**, *74*, 245209.

14. Tse, J.S.; Flacau, R.; Desgreniers, S.; Iitaka, T.; Jiang, J.Z. Electron density topology of high-pressure Ba_8Si_{46} from a combined Rietveld and maximum-entropy analysis. *Phys. Rev. B* **2007**, *76*, 17.

15. Tse, J.S.; Yang, L.; Zhang, S.J.; Jin, C.Q.; Sahle, C.J.; Sternemann, C.; Nyrow, A.; Giordano, V.; Jiang, J.Z.; Yamanaka, S.; Desgreniers, S. Pressure-induced electron topological transitions in Ba-doped Si clathrate. *Phys. Rev. B* **2011**, *84*, 184105.

16. San-Miguel, A.; Toulemonde, P. High-pressure properties of group IV clathrates. *High. Press. Res.***2005**, *25*, 159–185.

17. Shimizu, H.; Kume, T.; Kuroda, T.; Sasaki, S.; Fukuoka, H.; Yamanaka, S. High-pressure Raman study of the iodine-doped silicon clathrate $I_8Si_{44}I_2$. *Phys. Rev. B* **2003**, *68*, 212102.

18. Machon, D.; Toulemonde, P.; McMillan, P.F.; Amboage, M.; Munoz, A.; Rodríguez-Hernández, P.; San Miguel, A. High-pressure phase transformations, pressure-induced amorphization, and polyamorphic transition of the clathrate $Rb_{6.15}Si_{46}$. *Phys. Rev. B* **2009**, *79*, 184101.

19. Zhang, W.; Ge, N.N.; Zou, Y.T.; Zeng, Z.Y.; Cai, L.C. Influence of missing guest and host atoms on the mechanical and electronic properties of type-I clathrate compound Ba_8Si_{46}. *J. Alloys Compd.***2015**, *653*, 77–87.

20. Iitaka, T. Pressure-induced isostructural phase transition of metal-doped silicon clathrates. *Phys. Rev. B* **2007**, *75*, 012106:1–012106:4.

21. Blancon, J.C.; Machon, D.; Pischedda, V.; Debord, R.; Toulemonde, P.; Le Floch, S.; San-Miguel, A. Revisiting pressure-induced phase transition in silicon clathrates using Ge substitution. *Phys. Rev. B* **2016**, *93*, 134103.

22. Parlinski, K.; Li, Z.Q.; Kawazoe, Y. First-principles determination of the soft mode in cubic ZrO_2.*Phys. Rev. Lett.* **1997**, *78*, 4063.

23. Baroni, S.; Giannonzzi, P.; Testa, A. Green's-function approach to linear response in solids. *Phys. Rev. Lett.* **1987**, *58*, 1861.

24. Giannozzi, P.; Gironcoli, S.; Pavone, P.; Baroni, S. Ab initio calculation of phonon dispersions in semiconductors. *Phys. Rev. B* **1991**, *43*, 7231.

25. Gonze, X.; Allan, D.C.; Teter, M.P. Dielectric tensor, effective charges, and phonons in neutron-diffraction studies of salicylic acid and α-quartz by variational density-functional perturbation theory. *Phys. Rev. Lett.* **1994**, *68*, 3603–3606.

26. Gonze, X. First-principles responses of solids to atomic displacements and homogeneous electric fields: Implementation of a conjugate-gradient algorithm. *Phys. Rev. B* **1997**, *55*, 10337.

27. Lin, J.S.; Qteish, A.; Payne, M.C.; Heine, V.; Lin, J.S. Optimized and transferable nonlocal separable ab initio pseudopotentials. *Phys. Rev. B* **1993**, *47*, 4174–4180.

28. Lee, M.H. Advanced pseudopotentials for large scale electronic structure calculations: With application to a study of weakly ordered material-γ-Al_2O_3. Ph.D. Thesis, Cambridge University, Cambridge, UK, 1995.

29. Vosko, S.H.; Wilk, L.; Nusair, M. Accurate spin-dependent electron liquid correlation energies for local spin density calculations: A critical analysis. *Can. J. Phys.* **1980**, *58*, 2100–2105.

30. Parrinello, M.; Rahman, A. Crystal structure and pair potentials: A molecular-dynamics study.*Phys. Rev. Lett.* **1980**, *45*, 1196–1199.

31. Parrinello, M.; Rahman, A. Polymorphic transitions in single crystals: A new molecular dynamics method. *J. Appl. Phys.* **1981**, *52*, 7182–7190.

32. Fischer, T.H.; Almlof, J. General methods for geometry and wave function optimization. *J. Phys. Chem.* **1992**, *96*, 9768–9774.

33. Payne, M.C.; Teter, M.P.; Allen, D.C.; Arias, T.A.; Joannopoulos, J.D. Iterative minimization techniques for ab initio total-energy calculations: Molecular dynamics and conjugate gradients.*Rev. Mod. Phys.* **1992**, *64*, 1045–1097.

34. Milman, V.; Winkler, B.; White, J.A.; Packard, C.J.; Payne, M.C.; Akhmatskaya, E.V.; Nobes, R.H. Electronic structure, properties, and phase stability of inorganic crystals: A pseudopotential plane-wave study. *Int. J. Quantum Chem.* **2000**, *77*, 895–910.

35. Gonze, X.; Lee, C. Dynamical matrices, Born effective charges, dielectric permittivity tensors, and interatomic force constants from density-functional perturbation theory. *Phys. Rev. B* **1997**, *55*, 10355.

36. Tse, J.S.; Li, Z.Q.; Uehara, K. Phonon band structures and resonant scattering in Na_8Si_{46} and Cs_8Sn_{46} calthrate. *Europhys. Lett.* **2001**, *56*, 261.

37. Mélinon, P.; Kéghélian, P.; Perez, A.; Champagnon, B.; Guyot, Y.; Saviot, L.; Dianoux, A.J. Phonon density of states of silicon clathrates: Characteristic width narrowing effect with respect to the diamond phase. *Phys. Rev. B* **1999**, *59*, 10099.

38. Reny, E.; San-Miguel, A.; Guyot, Y.; Masenelli, B.; Mélinon, P.; Saviot, L.; Borowski, M. Vibrational modes in silicon clathrate compounds: A key to understanding superconductivity. *Phys. Rev. B***2002**, *66*, 014532.

39. Cohn, J.L.; Nolas, G.S.; Fessatidis, V.; Metcalf, T.H.; Slack, G.A. Glasslike heat conduction in high-mobility crystalline semiconductors. *Phys. Rev. Lett.* **1999**, *82*, 779–782.

40. Otero-de-la-Roza, A.; Abbasi-Pérez, D.; Luaña, V. Gibbs2: A new version of the quasiharmonic model code. II. Models for solid-state thermodynamics, features and implementation. *Comput. Phys. Commun.* **2011**, *182*, 2232–2248.

41. Qiu, L.; White, M.A.; Li, Z.; John, S.T.; Ratcliffe, C.I.; Tulk, C.A.; Sankey, O.F. Thermal and lattice dynamical properties of Na_8Si_{46} clathrate. *Phys. Rev. B* **2001**, *64*, 024303.

42. Okamoto, N.L.; Nakano, T.; Tanaka, K.; Inui, H. Mechanical and thermal properties of single crystals of the type-I clathrate compounds $Ba_8Ga_{16}Ge_{30}$ and $Sr_8Ga_{16}Ge_{30}$. *J. Appl. Phys.* **2008**, *104*, 013529.

43. Kaltzoglou, A.; Fässler, T.; Christensen, M.; Johnsen, S.; Iversen, B.B.; Presniakov, I.; Sobolev, A.; Shevelkov, A. Effects of the order-disorder phase transition on the physical properties of $A_8Sn_{44}\square2$ (A = Rb, Cs). *J. Mater. Chem.* **2008**, *18*, 5630–5637.

44. Lortz, R.; Viennois, R.; Petrovic, A.; Wang, Y.; Toulemonde, P.; Meingast, C.; Koza, M.M.; Mutka, H.; Bossak, A.; San Miguel, A. Phonon density of states, anharmonicity, electron-phonon coupling, and possible multigap superconductivity in the clathrate superconductors Ba_8Si_{46} and $Ba_{24}Si_{100}$: Factors behind large difference in T_c. *Phys. Rev. B* **2008**, *77*, 224507.

45. Falmbigl, M.; Rogl, G.; Rogl, P.; Kriegisch, M.; Müller, H.; Bauer, E.; Reinecker, M.; Schranz, W. Thermal expansion of thermoelectric type-I-clathrates. *J. Appl. Phys.* **2010**, *108*, 043529.

46. Stefanoski, S. Synthesis and Physical Properties of Group 14 Intermetallic Clathrates. Ph.D Theses, University of South Florida, Gainesville, FL, USA, 2012.

47. Zhang, W.; Chen, Q.Y.; Li, B.; Zeng, Z.Y.; Cai, L.C. First-principles calculations for thermodynamic properties of type-I silicon clathrate intercalated by sodium atoms. *Mod. Phys. Lett. B* **2015**, *29*, 1550166.

48. Nolas, G.S. *The Physics and Chemistry of Inorganic Clathrates*; Springer Science+Business Media: Dordrecht, The Netherlands, 2014.

49. White, A.M. *Properties of Materials*; Oxford University Press: Oxford, UK, 1999.

50. Anderson, O.L. A simplified method for calculating the Debye temperature from elastic constants.*J. Phys. Chem. Solids* **1963**, *24*, 909–917.

Chapter 4

High-Pressure Raman Study on Anglesite*

Pei-Lun Lee[1], Eugene Huang[2], Shu-Cheng Yu[3], Yen-Hua Chen[3]

[1]General Education Center, National Chia-Yi University, Chiayi, Taiwan;

[2]Department of Creative Fashion Design and Management, Chung-Chou University of Science and Technology, Yenlin, Taiwan; [3]Department of Earth Science, National Cheng-Kung University, Tainan City, Taiwan.

ABSTRACT

High-pressure Raman spectroscopic experiments of anglesite, $PbSO_4$, were carried out in a diamond anvil cell up to about 35 GPa at room temperature. Two stages of changes in the pattern of wavenumber-versus-pressure plot were observed in most of the vibrational modes. The first stage of change is the kink in the slope of wavenumber-versus-pressure plot at about 14 GPa, indicating a change in the force constant of modes. Further compression up to 21 GPa causes the splitting in the S-O stretching modes, representing the distortion of SO_4 tetrahedron. The behavior of transition observed in $PbSO_4$ is similar to that of $BaSO_4$ as well as those found in the solid solution series of $BaSO_4$-$SrSO_4$. Transitions taking place at higher pressure for anglesite is due to the relatively greater value of electronegativity for Pb^{2+}, as compared with M^{2+} in other complex compounds of similar structure.

Keywords: Anglesite; High-Pressure; Raman Spectroscopy; Electronegativity

1. INTRODUCTION

High-pressure research in Earth sciences has been focused on silicate minerals that are directly relevant to the Earth's interior and much less attention has been paid to other rock-forming minerals. However, recent investigations on some sulfates [1-3] and hydroxides [4,5] had provided interesting observations for the understanding of the behaviors of OH groups in hydrous minerals as well as the structural changes on sulfates under highpressure.

High-pressure study on sulfates is important because sulfates have analogous structure as olivine but with much greater compressibility. It is comprehensible that highpressure study on the distortion of SO_4 tetrahedron and the lattice stability of sulfates may have direct implications on the seemly stiffer iso-structural silicates.

The minerals with a structure of barite-type, such as barite ($BaSO_4$), anglesite ($PbSO_4$), and celestite ($SrSO_4$), are some of the most widely distributed sulfates in the world. The structure of barite was reported as orthorhombic system with space group Pbnm [6]. In our previous investigation, we found that $BaSO_4$ shows anisotropic compressibility along three crystallographic axes with c being the most compressible axis [2]. A phase transformation at 10 GPa for barite was also reported [3]. Anglesite, $PbSO_4$, has the same structure as barite and forms complete solid solution with barite in the $BaSO_4$- $PbSO_4$ series [7,8]. Previous investigation on another Barelated solid solution, $BaSO_4$-$SrSO_4$ series, reveals that similar phase transitions occur in members of this series and the pressures at which transitions take place increases with the decrease in the size of M^{2+} cations [9,10].

Since the two M^{2+} cations, Ba^{2+} and Pb^{2+}, have very different chemical and physical behaviors, such as electronegativity and electron configuration etc., it is interesting to conduct high-pressure investigation on anglesite for the understanding of the effect of the cation exchange in the lattice dynamics of the analogous $BaSO_4$ structure.

2. EXPERIMENTAL METHODS

Raman scattering method was applied to the study of the behaviors of anglesite at high-pressure. The experimental details are described as follows.

Crystals of synthetic anglesite [11] were ground to fine powder and used as the starting material. The sample was first examined by the Raman spectroscopy for its identity [12]. A disk of T301 stainless steel with an original thickness of 250 μm was used as gasket in the high-pressure work. The gasket was compressed between the diamond anvils to make an indentation on which a hole of 150 μm was drilled for the sample chamber. After the chamber was loaded with sample, it was then filled with a 4:1 methanol-ethanol mixture as a pressure-transmitting medium and then gradually compressed to high-pressure. Ruby fluorescence method was used for pressure measurement [13] after each pressure increments. In each pressure measurement, we measured more than three ruby grains to estimate the pressure distribution of the compressed sample.

A Renishaw 2000 type spectroscopy was used in the Raman spectroscopic study. An Ar-ion laser beam (514.5 nm) was used as an excitation source. The laser light (about 5 μm in diameter) was focused on the sample in the diamond cell and back-scattered Raman signals were collected by the CCD (charge-coupled device) detector. The power for the laser source used was 100 to 500 mW and recording time for each Raman spectrum was 20 to 60 seconds. Pressure was recorded before and after each Raman spectrum was collected. Both loading and unloading processes were conducted. The uncertainties in pressure and wavenumber readings are 0.5 GPa and 1 cm^{-1}, respectively.

3. RESULTS

Some of the representative Raman spectra of anglesite taken during the loading and unloading processes of the experiment are shown in **Figure 1**. The Raman ν_1 (nondegenerate symmetric stretching mode at 978 cm^{-1}), ν_2 (doubly degenerate symmetric bending mode at 439, 450 cm^{-1}), ν_3 (triply degenerate asymmetric stretching mode at 1064, 1157 cm^{-1}) and ν_4 (triply degenerate asymmetric bending mode at 606, 616, 642 cm^{-1}) of anglesite are observed at the ambient conditions in this study. This is in excellent agreement with that reported in the literature. [10,11] Griffith [12] reported that anglesite exhibited more than nine relatively intense Raman bands at the ambient conditions (ν_3 mode (1057, 1068, 1162 cm^{-1})). However, several Raman modes are not observed in this measurement. The reason for this is probably due to a considerable mixing of the translational and rotational bands which were not easily resolved in the anglesite crystal [14].

The frequencies of Raman modes are found to increase, but their intensity becomes weaker with increasing pressure. Most Raman bands of anglesite remain quite sharp at all pressures. It is found that the ν_1 mode splits into two recognizable peaks when the pressure exceeds 21 GPa, and the intensity of the new peak becomes stronger with increasing pressure. The ν_{3b} mode also splits into two peaks when the pressure exceeds 21 GPa, but it is too weak to be defined. No sign of splitting is detected for the ν_2 and ν_4 mode throughout the entire pressure range up to 35 GPa.

The variations of the frequency with respect to pressure for all the observable Raman bands of anglesite are plotted in **Figure 2**. Interesting phenomena are observed

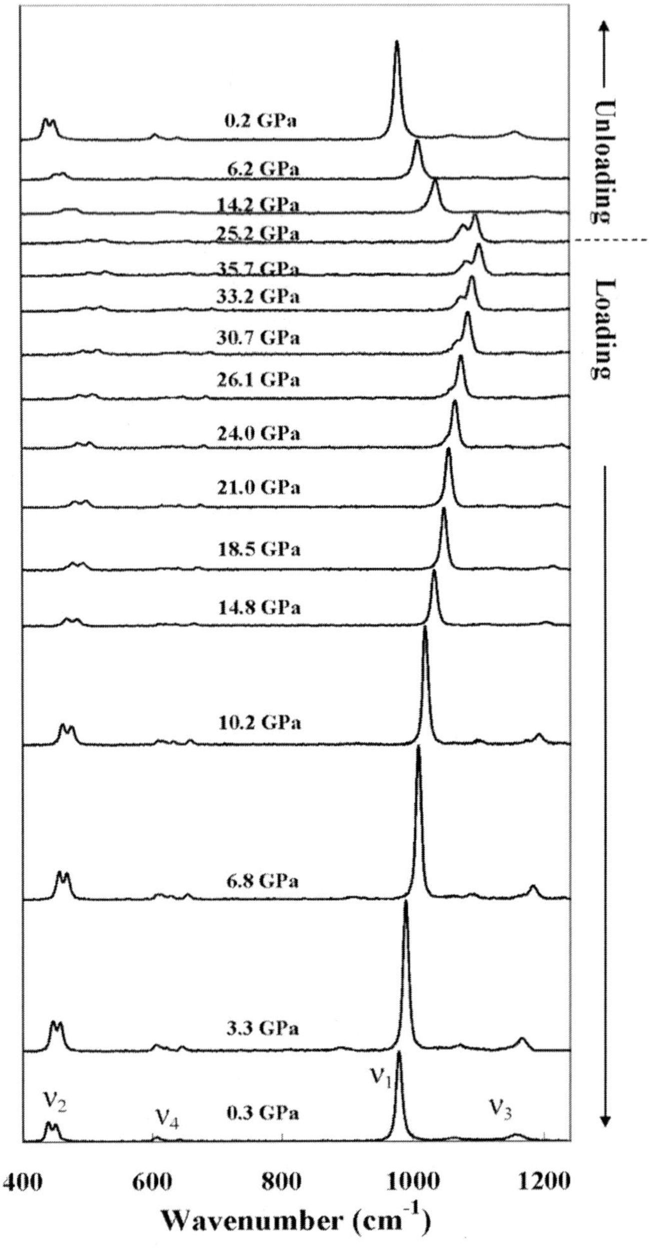

Figure 1. A series of Raman spectra of anglesite shows the effect of pressure on the four Raman modes. The ν_1 mode is the most intense Raman mode of anglesite. The anglesite phase was recovered during the unloading process.

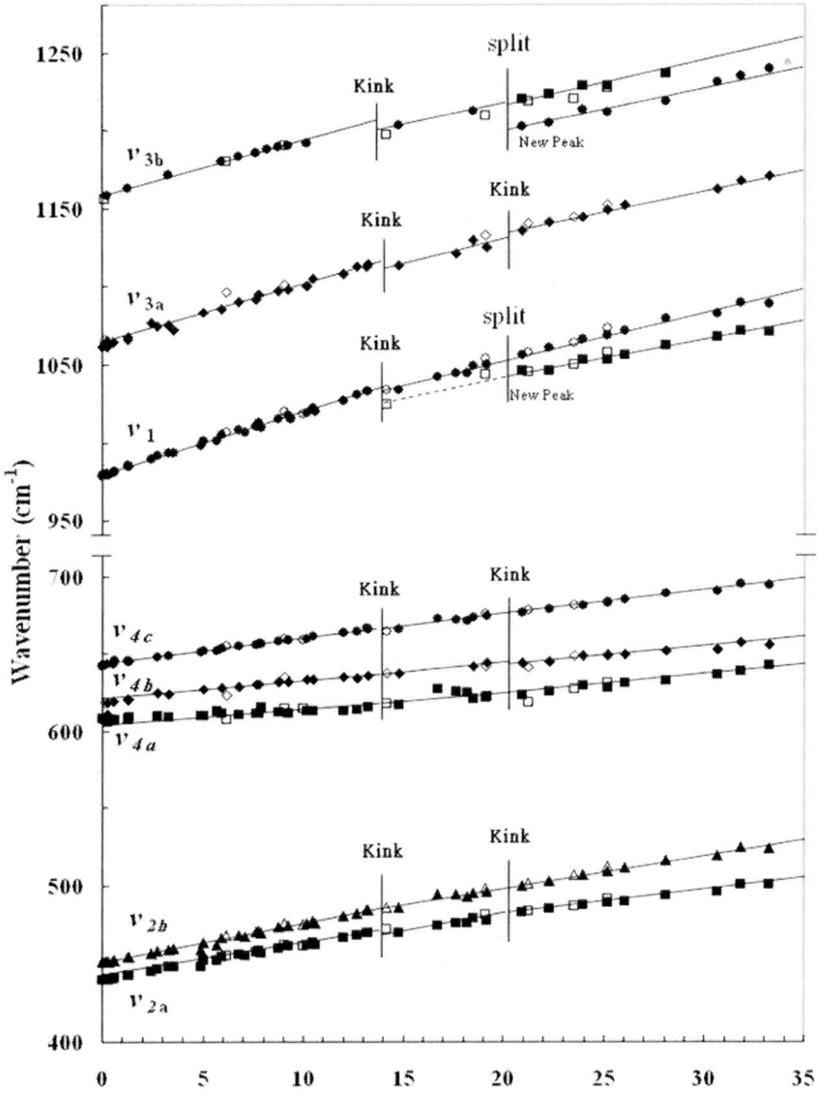

Figure 2. A plot shows the shifts of the wavenumber of each Raman mode of anglesite as a function of pressure. Data for loading process are shown as solid symbols and those for unloading process as open symbols. Two stages of change in the slope of each mode as well as the change in the number of modes can be observed. The "kink" at which a discontinuity in the slope of wavenumber vs. pressure plot is seen for each mode at 14 GPa. The solid lines are the eyeball linear fit of the data (note the difference in the scale on the vertical-axis of the pattern).

as well as the splitting of stretching modes of S-O. In **Figure 2**, it is noted that all Raman bands are seen to vary linearly with pressure up to about 14 GPa. With further compression, apparent kinks or discontinuities in the wavenumber-versus-pressure plot are detected for almost all modes.

The changes in the slope (dv/dP) of each Raman mode of anglesite with pressure are listed in **Table 1**. The S-O stretching modes, v_3 and v_1, show two discontinuities during compression at 14 and 21 GPa. Above 21 GPa, in addition to the kink in the slope (dv/dP), new mode also appears in both of v_3 and v_1. The kinks in the slope of v_2 modes and v_4 modes, though to a much less magnitude, also can be detected at 14 GPa and 21 GPa (**Figure 2**). The appearance of the kinks for all modes at 14 GPa and 21 GPa is interpreted as the changes in the force constants of the S-O vibration bands while the splitting of v_1 and v_{3b} above 21 GPa represents the onset of the distortion of SO_4tetrahedra. The additional vibrational modes that show up at high-pressure persist during the unloading process until about 10 GPa where all modes are converted to the original phase.

4. DISCUSSION

In the structure of anglesite, each sulfur atom is surrounded by four oxygen atoms and forms an SO_4 tetrahedron while each lead atom coordinates with 12 oxygens. Among the 12 oxygen atoms, 11 atoms shared with the nearest six neighboring SO_4 tetrahedra and the twelfth oxygen associates with another SO_4 tetrahedron. In other words, each Pb^{2+} is coordinated with 12 oxygen ions to form seven different SO_4 groups [6]. The Raman modes observed in this study are all assigned to the stretching and bending of the S-O modes. These S-O modes are nevertheless affected by the relative changes of Pb-O distance during the compression process. When anglesite is compressed, the change in the bond distance between Pb-O would response more to the compression force than that between S-O in the SO_4 tetrahedron due to the fact that PbO_{12}polyhedra are more compressible than SO_4 tetrahedron. Therefore, the bonding distances between Pb and O tend to decrease more than those between S and O during initial stage of compression process. However, when PbO_{12} polyhedra are compressed further, the Pb-O distances gradually decrease and their bonding strengths increase in the meantime. As a consequence, the force constants of various band of S-O in SO_4 tetrahedra will also change since their bonding oxygens are sharing with Pb in PbO_{12} polyhedra. The discontinuity in the slope dv/dP for various Raman modes at 14 GPa signifies that the PbO_{12} polyhedra are compressed to the extend that their average Pb-O distance reaches a critical state as to cause a sudden change in the bonding forces of various S-O bonds. The change in the force constant among various S-O bands in SO_4 tetrahedra result in the deviation of compression data from the normal compression curve above 14 GPa.

Table 1. The variations of the slope (dv/dP, cm^{-1}/GPa) of each Raman mode in anglesite.

slope (R^2)		Pressure (GPa)		
		14	21	
ν_1	new			2.36 (0.938)
	1a	4.00 (0.997)	3.53 (0.978)	3.18 (0.992)
ν_2	2a	2.26 (0.995)	1.88 (0.909)	1.59 (0.990)
	2b	2.53 (0.991)	2.15 (0.771)	2.10 (0.990)
ν_3	3a	3.83 (0.964)	3.55 (0.805)	3.09 (0.990)
	new			2.56 (0.822)
	3b	3.52 (0.993)	2.47 (0.999)	2.23 (0.915)
ν_4	4a	0.55 (0.775)	0.89 (0.821)	1.33 (0.950)
	4b	1.45 (0.896)	1.48 (0.974)	1.00 (0.952)
	4c	1.73 (0.998)	1.63 (0.822)	1.56 (0.984)

Althought there are changes in the force constant in various S-O bonds, the SO_4 tetrahedra in the lattice remain undistorted until they are compressed up to 21 GPa where additional modes appear in the stretching bands (**Figure 2**), accompanying the slight but noticeable change in the slope of all S-O modes (**Table 1**). These phenomena are attributed to the distortion of SO_4 tetrahedra as their bonding oxygens are further bonded with Pb of PbO_{12} polyhedra as more and more pressure is applied. The distortion of SO_4 tetrahedra persists up to 35 GPa in this experiment. The distortion of SO_4 tetrahedra as well as the change in the force constants of various S-O bonds is, nevertheless, reversible that the wavenumbers of all vibration modes follow the trend of compression process during the unloading process.

Barite, celestite and anglesite all have the same structure with different M^{2+} cations (**Table 2**). It is interesting to note that similar changes in the vibrational modes are observed among these compounds under high-pressure [3, 9; also see **Table 2**]. The variations of the frequency with respect to pressure for Raman ν_1 mode of barite, celestite and anglesite are plotted in **Figure 3**. From **Figure 3**, we know the pressures at which the kinks in slope as well as the split of SO_4tetrahedra take place are different in those minerals. The pressure at which the first kink takes place is 8, 10 and 14 GPa for barite, celestite and anglesite, respectively. The pressure at which the second kink takes place is 17, 18 and 21 GPa for barite, celestite and anglesite, respectively.

As we noted in **Table 2** that these compounds have similar M-O bonding lengths and S-O bonding force constants and, therefore, it is not surprising that they behave similarly under high pressure. Chen et al. [10] reported that 2 stages of changes in Raman modes occur at high pressure in the $BaSO_4$-$SrSO_4$ solid solution series. The compound with larger cation tends to undergo structural modifications at a pressure lower than that with smaller cation. Since the size of Ba^{2+} is larger than Sr^{2+} and Pb^{2+}, the transitions take place in $BaSO_4$ are at pressures lower than those in $SrSO_4$ and $PbSO_4$. Howeverthe fact that the pressures at which transitions occur in $SrSO_4$ are lower than those in $PbSO_4$ is not accounted for by the effect of cation size since the size of Pb^{2+} is larger than Sr^{2+}that the transitions in $PbSO_4$ should be expected to take place at pressures lower than those in $SrSO_4$. Apparently, factors that cause the transitions in $PbSO_4$ at pressures higher than those in $SrSO_4$ other than the difference in size are needed.

Table 2. Some characterized parameters of barite, celestite and anglesite.

M = Ba, Sr or Pb		Barite ($BaSO_4$)	Celestite ($SrSO_4$)	Anglesite ($PbSO_4$)
Metal ion-oxygen polyhedra (Average M-O length)		2.951$^{\#}$ 2.952*	2.827$^{\#}$ 2.831*	2.864$^{\#}$ 2.870*
S-O Bending force constants (H, md/A)*		0.46	0.47	0.42
S-O Stretching force constants (K, md/A)*		6.27	6.34	5.98
M^{2+} radius (A)		1.6	1.44	1.49
Electronegativity of M^{2+}		0.9	1.0	1.8
Pressure of Raman ν_1 mode slope change	1st	8	10	14
	2nd	17	18	21

$^{\#}$Jacobsen et al. (1998) [16]. *Miyake et al. (1978) [15].

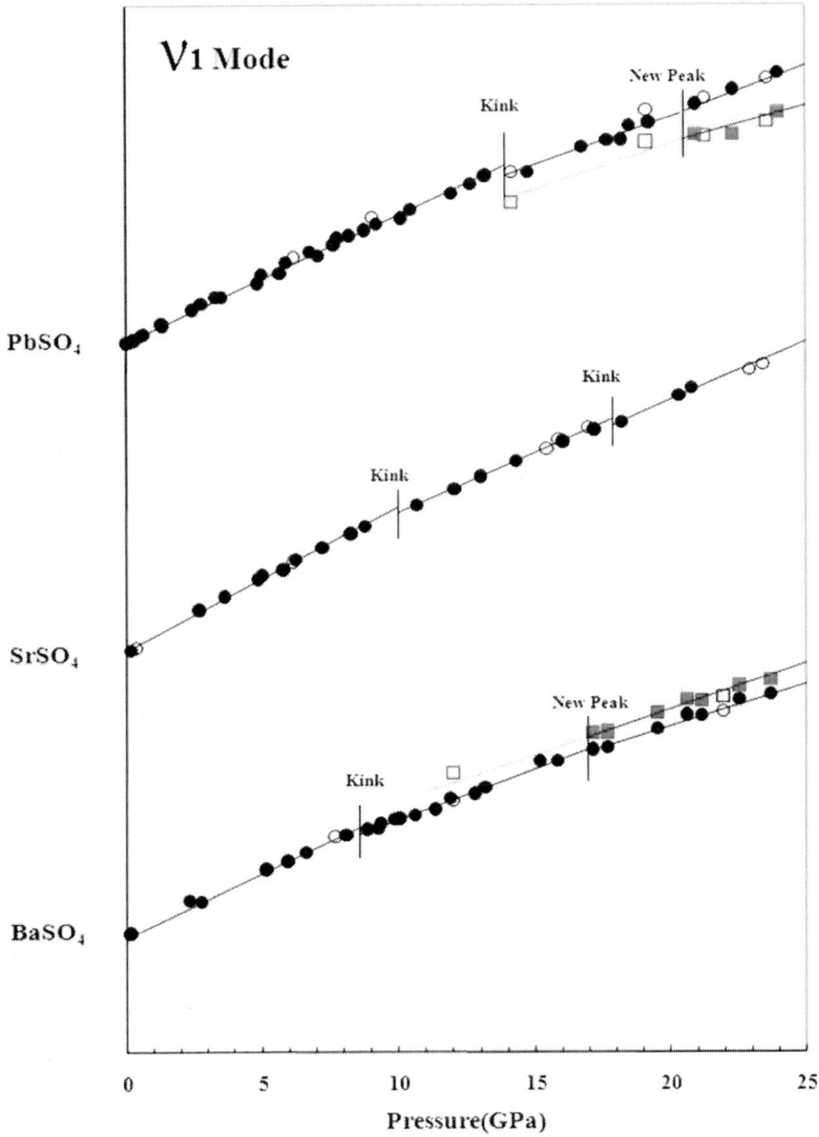

Figure 3. A plot shows the shifts of the wavenumber of Raman v_1 mode of $PbSO_4$, $SrSO_4$ and $BaSO_4$ as a function of pressure.

Miyake et al. [15] reported that the electronegativity of a metal and the Coulombic force between cation and oxygen may modify the symmetry of the SO_4 tetrahedron. For instance, the cation-oxygen bonding length in $PbSO_4$ is shorter than $BaSO_4$ due to their difference in electronegativity of M^{2+}. Though the size of Pb^{2+} is smaller than that of Sr^{2+}, the electronegativity of Pb^{2+} is much larger than Sr^{2+} (**Table 2**). Higher value in electronegativity fro a cation implies that it tends to form stronger bond with anions such as oxygen. Therefore, the bonding force of Pb-O in PbO_{12}polyhedra is higher than that of Sr-O in SrO_{12} polyhedra at the ambient conditions. The PbO_{12}polyhedra in $PbSO_4$ are more difficult to compress and the effect on the changes of force constants in S-O bonds would be expected to occur at pressure much higher than those in $SrSO_4$. That explains why the kinks and splitting in SO_4 stretching modes in $PbSO_4$ take place at pressure higher than those in $SrSO_4$.

5. CONCLUSION

Raman spectroscopic observation on $PbSO_4$ at high-pressure up to 35 GPa shows two stages of changes in the pattern of wavenumber-versus-pressure plot in most of the vibrational modes. The first stage of change is the kink in the slope of wavenumber-versus-pressure plot at about 14 GPa, indicating a change in the force constant of modes. Further compression up to 21 GPa causes the splitting in the S-O stretching modes, representing the distortion of SO_4 tetrahedron. Similar changes of Raman modes in $PbSO_4$ are present in members of the solid solution series of $BaSO_4$-$SrSO_4$. Transitions taking place at higher pressure for $PbSO_4$ is due to the relatively greater value of electronegativity for Pb^{2+}, as compared with M^{2+} in other complex compounds of similar structure.

REFERENCES

1. E. Huang, J. Xu, J. F. Lin and J. Hu, "Pressure-Induced Phase Transitions in Gypsum," High Pressure Research, Vol. 17, No. 1, 2000, pp. 57-75.
2. P. L. Lee, E. Huang and S. C. Yu, "Phase Diagram and Equations of State of Barite, $BaSO_4$," High Pressure Research, Vol. 21, No. 2, 2001, pp. 67-77.
3. P. L. Lee, E. Huang and S. C. Yu, "High-Pressure Raman and X-Ray Studies of Barite, $BaSO_4$," High Pressure Research, Vol. 23, No. 4, 2003, pp. 439-450.
4. E. Huang, R. J. Chen, A. Li, T. Yamanaka and J. Xu, "High Pressure Phase Transition in Al $(OH)_3$: Raman and X-Ray Observations," Geophysical Research Letters, Vol. 23, No. 22, 1996, pp. 3083-3086.
5. E. Huang, J. F. Lin, J. Xu, T. Huang, Y. C. Jean and H. S. Sheu, "Compression Studies of Gibbsite and Its High-Pressure Polymorph," Physics and Chemistry of Minerals, Vol. 26, No. 7, 1999, pp. 576-583.

6. R. W. James and W. A. Wood, "The Crystal Structure of Barite, Celestite and Anglesite," Proceedings of the Royal Society, Vol. 109A, No. 752, 1925, pp. 598-620.

7. J. S. Lee and S. C. Yu, "A Study of Hokutolite Synthesis and Its Mineralogical Implications to Natural Occurrences," Journal of the Geological Society of China, Vol. 39, No. 3, 1996, pp. 223-234.

8. B. Takano, M. Yanagisawa and K. Watanuki, "Structure Gap in $BaSO_4$-$PbSO_4$ Solid Solution Series," Mineralogical Journal, Vol. 6, No. 1, 1969, pp. 159-171.

9. Y. H. Chen, E. Huang, S. C. Yu and P. L. Lee, "Raman Spectroscopy and X-Ray Diffraction Studies on Celestite," Physica B: Condensed Matter, Vol. 405, No. 20, 2010, pp. 4386-4388.

10. Y. H. Chen, E. Huang and S. C. Yu, "High-Pressure Raman Study on the $BaSO_4$-$SrSO_4$Series," Solid State Communications, Vol. 149, No. 45-46, 2009, pp. 2050- 2052.

11. H. R. Wang, "The Study on the Flux Growth Method of Hokutolite Crystals," Master's Thesis, National Chen-Kung University, Tainan, 1998, p. 59.

12. W. P. Griffith, "Raman Spectroscopy of Minerals," Nature, Vol. 224, No. 5216, 1969, pp. 264-266.

13. H. K. Mao, J. Xu and P. M. Bell, "Calibration of the Ruby Pressure Gauge to 800 Kbar under Quasi-Hydrostatic Conditions," Journal of Geophysical Research, Vol. 91, No. 135, 1986, pp. 4673-4676.

14. P. Dawson, M. M. Hargreave and G. R. Wilkinson, "Polarized IR. Reflection, Absorption and Laser Raman Studies on a Single Crystal of $BaSO_4$," Spectrochimica Acta, Vol. 33A, No. 2, 1977, pp. 83-93.

15. M. Miyake, I. Minato, H. Morikawa and S. I. Iwai, "Crystal Structures and Sulphate Constants of Barite, Celestite, and Anglesite," American Mineralogist, Vol. 63, No. 5-6, 1978, pp. 506-510.

16. S. D. Jacobsen, J. R. Smyth, R. J. Swope and R. T. Downs, "Rigid-Body Character of the SO_4 Groups in Celestite, Anglesite and Barite," Canadian Mineral, Vol. 36, No. 4, 1998, pp. 1053-1060.

Chapter 5

Structural Deformation of $Sm@C_{88}$ under High Pressure

Jinxing Cui, Mingguang Yao, Hua Yang, Ziyang Liu, Fengxian Ma, Quanjun Li, Ran Liu, Bo Zou, Tian Cui, Zhenxian Liu, Bertil Sundqvist & Bingbing Liu

State Key Laboratory of Superhard Materials, Jilin University, No. 2699 Qianjin Street, Changchun 130012, P.R. China

ABSTRACT

We have studied the structural transformation of $Sm@C_{88}$ under pressure up to 18 GPa by infrared spectroscopy combined with theoretical simulations. The infrared-active vibrational modes of $Sm@C_{88}$ at ambient conditions have been assigned for the first time. Pressure-induced blue and red shifts of the corresponding vibrational modes indicate an anisotropic deformation of the carbon cage upon compression. We propose that the carbon cage changes from ellipsoidal to approximately spherical around 7 GPa. A smaller deformation of the carbon bonds in the area close to the Sm atom in the cage suggests that the trapped Sm atom plays a role in minimizing the compression of the adjacent bonds. Pressure induced a significant reduction of the band gap of the crystal. The HOMO-LUMO gap of the $Sm@C_{88}$ molecule decreases remarkably at 7 GPa as the carbon cage is deformed. Also, compression enhances intermolecular interactions and causes a widening of the energy bands. Both effects decrease the band gap of the sample. The carbon cage deforms significantly above 7 GPa, from spherical to a peanut-like shape and collapses at 18 GPa.

INTRODUCTION

Endohedral metallofullerenes (EMFs) have been attracting much attention because of their potential applications in the fields of medicine, materials science, physics and novel energy1. In EMFs, the metal atoms are trapped in fullerene cages to which they also transfer charge. The properties of EMFs depend strongly on the metal atoms trapped inside the cages2. The metal-cage bonding interaction in EMFs is different from the traditional metal ligand bond3'4. The metal atom is situated near the inner surface of the cage and interacts with its near hemisphere, leaving the other half almost unaffected. The coordination of the metal atom is thus uneven in space, which results in an anisotropic charge distribution around the atom, causing a series of special electronic transfer behaviors5. Heath *et al.* reported the formation of the first EMF, La@C_{60} 6, which is unfortunately sensitive to air7. The first EMF to be isolated is La@C_{82} which has a larger carbon cage7. Subsequently, numerous monometallic endohedral fullerenes containing lanthanide ions have been synthesized and isolated. However, the rather low production efficiency and the isomer mixtures of the high EMFs has hindered studies of the intrinsic structures and properties of EMFs. Recently, Yang *et al.* have developed an efficient method for the synthesis and isolation of high purity single isomers of various high EMFs containing a single samarium ion8'9'10. Very recently, we have isolated Sm@C_{88} from carbon soot obtained by electric arc vaporization of graphite rods containing Sm_2O_3. As yet, the fundamental structure and physical properties of Sm@C_{88} have not been investigated. This material offers an ideal model to study the internal structure and understand the metal-cage interaction.

Fullerenes have various potential applications, including use as photoconductive materials11'12, structural reinforcement materials13, and so on. In fact, fullerenes may often deform in the applications, and controllable deformations could result in novel structures and properties. High pressure is an effective way to modify the structure and thus tune the properties of materials14. The structural stability and evolution of fullerenes under high pressure has been widely studied15'16'17'18. For example, the C_{60} molecule deforms and forms links to nearest neighbors by [2 + 2] cycloaddition to form a 3D polymer at 15 GPa and 600 °C[18]. The hardness of this phase is comparable to that of cubic boron nitride (c-BN). In hydrogenated fullerene ($C_{60}H_{18}$), the hydrogen atoms can protect the fullerene cage and the structure of the material from distortion up to 32 GPa, much higher than for the pure C_{60} 19. When solvated fullerene (C_{60}*m-xylene) is compressed up to 60 GPa, the C_{60} cages collapse while the material still preserves a long-range ordered structure, containing amorphous carbon clusters as its basic units. This material is ultra-incompressible and hard enough to indent diamond20'21. These interesting phenomena not only contribute to our understanding on the crystalline structure of solids but also emphasize the importance of exploring the deformation of fullerene cages for better understanding the novel phase transitions. In particular, the deformation process of the fullerene cage and the real molecular morphology after the deformation are very important for understanding the resulting properties of the phases formed. Compared with C_{60}, high EMFs have a larger frame and lower symmetry, which may lead to a more obvious deformation of the carbon cage under high pressure and thus provide a good model to study the deformation of carbon cages. However, how EMFs deform under high pressure is still unknown and no information exists on how the confined metal atom affects the structural deformation of the carbon cage. High pressure may tune the metal-cage interaction and make it possible to study the effects of the metal atom on the molecular deformation to improve our understanding of the unique metal-cage interaction. This may bring a new insight into the structural stability and the deformation process of fullerenes at the atomic level, making the prediction of new carbon phases possible.

In this work, the infrared (IR) vibrational modes of Sm@C_{88} have been assigned with the help of theoretical simulations. The cage deformation, structural stability and the band gap change of Sm@C_{88} under high pressure have been further studied by IR spectroscopy. Classical molecular dynamics simulations on the Sm@C_{88} crystal and density functional theory (DFT) calculations on the molecule have been performed to understand the transformation of Sm@C_{88} upon compression. We present the structural evolution of Sm@C_{88} and the change of the band gap of the material under high pressure. It has also been found that the trapped metal atom supports the carbon cage against collapse under high pressure.

RESULTS

IR spectrum of Sm@C_{88}

The IR spectrum of the Sm@C_{88} crystals recorded at room temperature and atmospheric pressure is displayed in Fig. 1. No reports exist giving measured IR spectra or the assignment of the IR-active peaks of Sm@C_{88}. To understand the origin of the vibrational modes, we have calculated the IR spectrum for Sm@C_{88} with DMOL3. Based on this simulation, the IR spectrum can be roughly divided into three regions. (1) The bands from 200 to 450 cm^{-1} are attributed to the radial breathing vibrations of the carbon cage (for example, Fig. 2(a) shows the eigenvector of the vibrational mode at 419 cm^{-1}). (2) The bands between 450 and 810 cm^{-1} are from C-C/C=C radial bending vibrations. (3) The bands from 1000 to 1600 cm^{-1} are related to C-C/C=C tangential stretching vibrations on the spherical surface. For example, the mode at 1468 cm^{-1} arises from a stretching vibration of a five-membered ring, while the mode at 1453 cm^{-1} is correlated with the stretching vibration of a six-membered ring. The eigenvectors corresponding to these vibrational modes are shown in Fig. 2(c,d). Both spectra match well with each other in the number and the shape of the peaks except for slight differences in the frequencies in the far-infrared (FIR) range and the IR signals of toluene (marked with symbol "#") and bis(ethylenedithio)tetrthiafulvalene (ET) (marked with asterisks). This may be due to the interaction between $Sm@C_{88}$ and adjacent molecules. The above assignments of the IR-active vibrational modes are similar to the IR spectral analysis by Yang *et al.* for other endohedral fullerenes22'23.

Figure 3 shows IR spectra of the sample under high pressure up to 18 GPa. Most vibrational modes exhibit blue shifts except a few vibrational modes exhibit a red shift with increasing pressure. The blue shift of the vibrational modes is a common phenomenon in high-pressure studies and results from pressure-induced shrinkage of the bond length. The red shift of the vibrational modes may result from an extension of the corresponding bond lengths and changes of corresponding bond angles with increasing pressure. A pressure-induced red shift of some radial bending vibrations in C_{70} solvates has been observed in previous studies24.

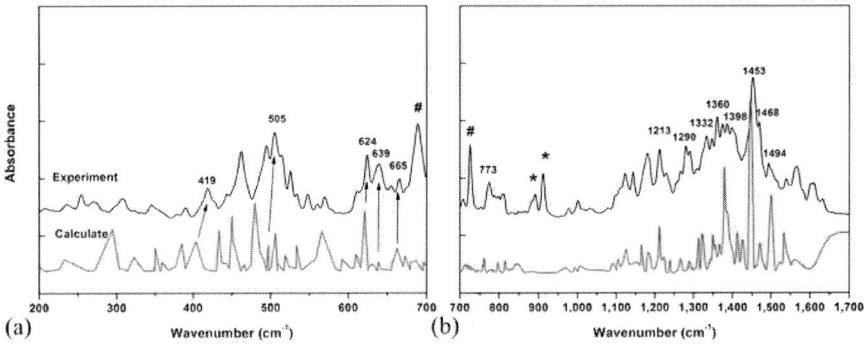

Figure 1: Experimental (black) and calculated (red) infrared spectra of $Sm@C_{88}$ at ambient pressure. (**a**) FIR spectra; (**b**) MIR spectra. The peaks from toluene are marked with symbol "#". The peaks from ET are marked with asterisks.

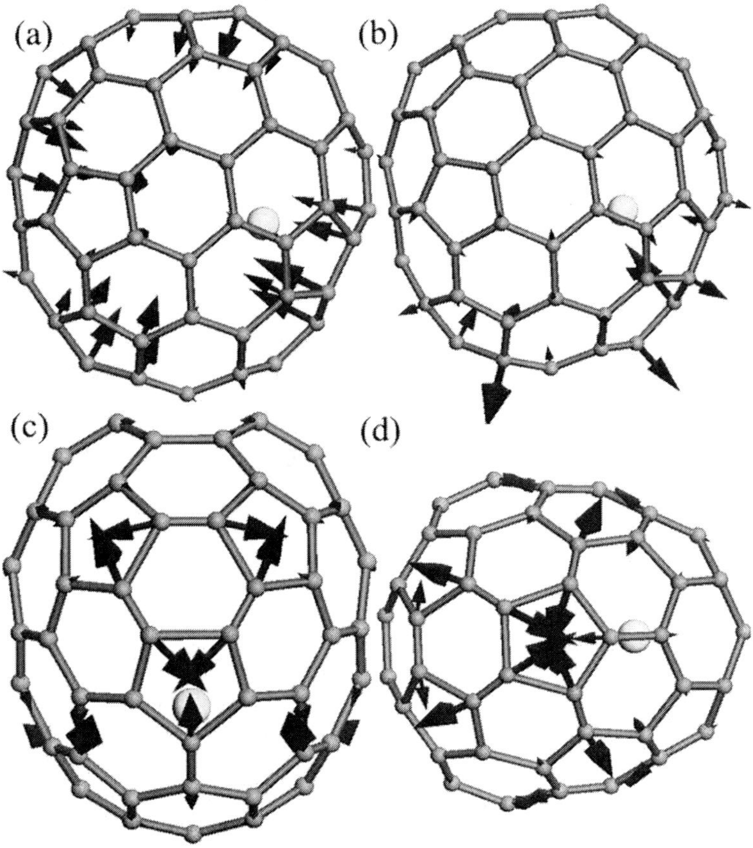

Figure 2: The eigenvectors of IR-active vibrational modes for the bands. (**a**) 419, (**b**) 773, (**c**) 1453 and (**d**) 1468 cm^{-1}.

With increasing pressure, the intensities of most modes decrease and the peaks broaden significantly. At 18 GPa all modes in the FIR region disappear and only a single broad band is detectable around 500 cm^{-1}. This indicates that the carbon cages start to lose the cage feature and collapse. Similar transitions have also been observed in other fullerenes. For example, Wasa *et al.* observed amorphization of C_{70} above 12 GPa25. We further used high resolution TEM (HRTEM) to study the detailed structure of the released sample. The HRTEM images of the samples at ambient pressure and decompressed from 18 GPa are shown in Fig. S2. The figures show that the sample at ambient pressure is an order phase, and the decompressed sample is amorphized. This indicates that the sample transforms to amorphization at 18 GPa and can be quenched to ambient pressure, in agreement with our IR results.

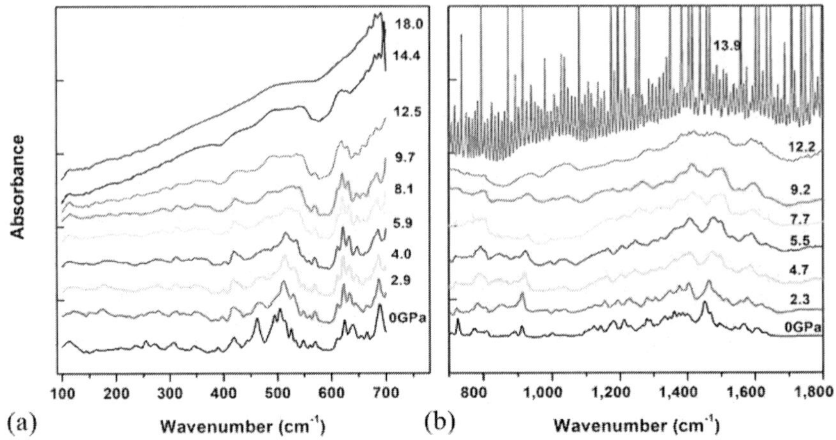

Figure 3: Absorbance spectra of Sm@C$_{88}$ under high pressure. (**a**) FIR region, (**b**) MIR to 1800 cm^{-1}. The IR absorbance spectra under high pressure were shifted for clarity.

It has been suggested that the pressure coefficient (dω/dp) reflects the pressure response of the corresponding vibrational modes, and changes in the pressure coefficients could reflect changes in molecular structure 19. We can thus analyze the pressure coefficients of the vibrational modes in different parts on the carbon cage to study the deformation of Sm@C$_{88}$ under high pressure. We used Lorentzian functions to fit the IR-active vibrational bands up to 12.5 GPa. The frequencies of four intense C-C/C=C radial bending vibrational bands at 505, 639, 665 and 773 cm^{-1} are plotted as a function of pressure inFig. 4(a). The modes at 505 and 773 cm^{-1}, which are related to the C-C/C=C radial bending vibrations of the top and bottom parts of the cage (the eigenvector of the vibrational mode at 773 cm^{-1} is shown inFig. 2(b)), shift to higher frequencies. However, the modes at 639 and 665 cm^{-1}, which are attributed to the C-C/C=C radial bending vibrations in the horizontal direction of the left and right parts of the cage (the carbon cage is seen from the view of Fig. 2(b)), shift to lower frequencies. Furthermore, the magnitudes of the pressure coefficients of the red-shift modes are smaller than those of blue-shift modes. The pressure coefficients of the vibrational modes start to change in the range 6–8 GPa. We also plotted the pressure dependences of five intense C-C/C=C tangential stretching vibrational modes at 1290, 1332, 1213, 1453 and 1398 cm^{-1} in Fig. 4(b). Comparing with the pressure coefficients of these vibrational modes, we find that the closer the relevant parts of the carbon cage are to the Sm atom, the smaller are the pressure coefficients of the corresponding C-C/C=C tangential vibrational modes. For example, the vibrational mode at 1290 cm^{-1}, which is from the part far away from the Sm atom, exhibits a larger pressure coefficient of 5.06 ± 0.31 cm^{-1} GPa^{-1}, while the vibrational mode at 1398 cm^{-1}, which is from the part close to the Sm atom, exhibits a much smaller pressure coefficient of 2.16 ± 0.16 cm^{-1} GPa^{-1}. We also plotted the modes from toluene and ET molecules as a function of pressure in Fig. 4(c). The modes at 688 and 726 cm^{-1} from toluene exhibit a slight red shift, which may result from pressure-induced enhancement of intermolecular van der Waals force between the carbon cage and toluene19. The red shift could also result from an extension of the corresponding bond lengths and changes of corresponding bond angles with increasing pressure. The modes at 891 and 911 cm^{-1} from ET exhibit a blue shift, which may be due to compression rather than charge-transfer interaction, because the pressure-induced charge transfer usually results in a red shift in the vibrational modes. Taking into account of the interaction between Sm@C_{88}, toluene and ET molecules, the change of the modes of Sm@C_{88}under high pressure may also result from an evolution of the interaction with the molecule environment. To determine the effect of the pressure medium on the transformations of our sample, we performed high-pressure mid-infrared (MIR) experiments with argon and KBr as pressure media, respectively. Both experiments gave very similar results on the pressure evolutions and coefficients of the corresponding vibration modes, which suggests that the observed structural deformations of the carbon cage are due to the effect of compression.

High-pressure absorption spectra shown in Fig. 5(a) display clear onset of a broad hump which corresponds to the absorption edge of the sample. This absorption edge gradually shifts to lower energy with increasing pressure and completely moves into FIR region at 13.9 GPa. The indirect band gap E_g of the sample can be estimated from the x-axis intercept by extrapolating the linear portion of the (αhv)2 versus hv plot to α = 0 (inset of Fig. 5(b)), where α and hv are the absorption coefficient and the incident photon energy. We plotted the band gap as a function of pressure in Fig. 5(b). The band gap decreases almost linearly from 0.73 eV at ambient pressure to 0.16 eV at 13.9 GPa with a pressure coefficient of −0.043 ± 0.001 $eVGPa^{-1}$. It is worth mentioning that the band gaps of solid C_{70}26 and C_{60}27 decrease as pressure increases, to a value of 1.4 eV before the molecular collapse, but never reach such low value as observed in our Sm@C_{88}. To determine whether the material experiences a metallization transformation, we measured the high-pressure infrared reflectivity spectra of the material as shown in Fig. 5(c). The reflectivity of the sample gradually increased with pressure, which is consistent with the result of the absorption spectra of the sample. However, neither an abrupt increase in reflectivity nor a Drude-Lorentz oscillator model feature in the reflectivity spectra was observed. That suggests that the band gap of the material only decreases to a small value, rather than closes.

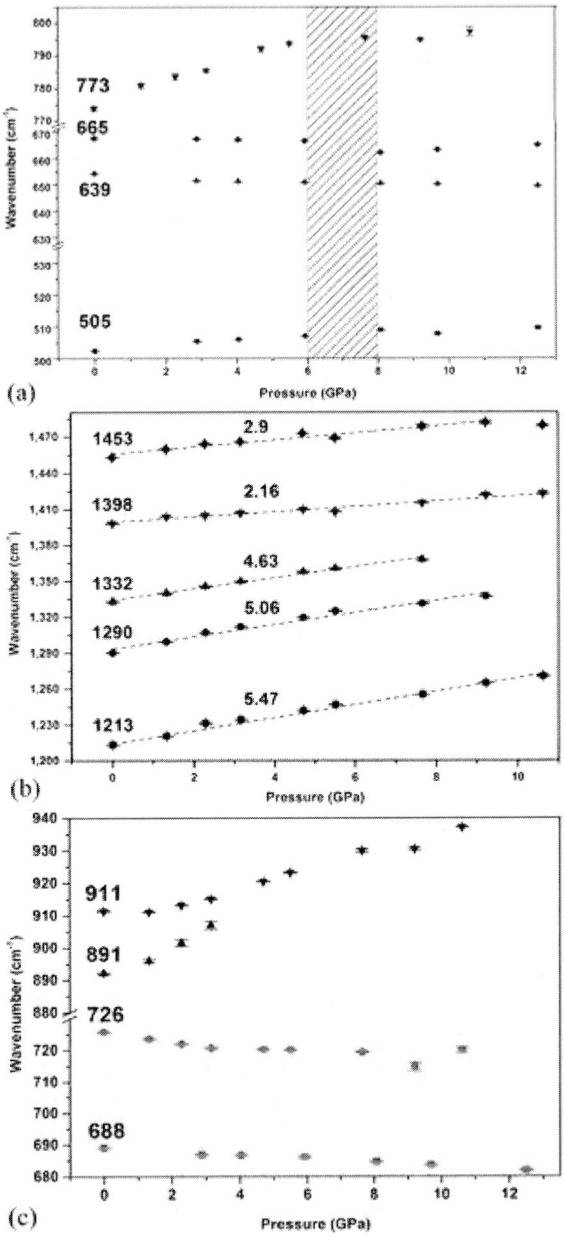

Figure 4: Frequencies of representative IR modes. (**a**) C-C/C=C radial bending vibrations and (**b**) C-C/C=C tangential stretching vibrations of the carbon cage as functions of pressure. The pressure coefficients are shown near the fitted lines. (**c**) The IR modes from ET (black) and toluene (red) as functions of pressure.

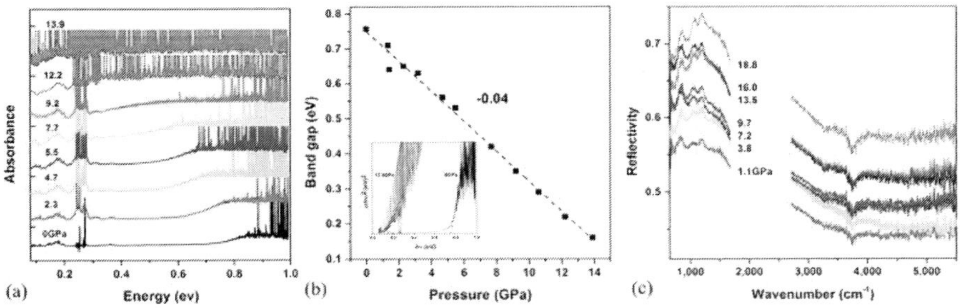

Figure 5 (**a**) The mid-IR region shows the absorption edge under high pressure. (**b**) The band gap as a function of pressure. Inset is the plots of (αhv)2 versus hv at ambient pressure and 13.9 GPa. (**c**) IR reflectivity spectra of the sample under high pressure.

Molecular Deformation of Sm@C_{88}

In order to explain the deformation process of our Sm@C_{88} under pressure, we performed classical molecular dynamics simulations on Sm@C_{88} crystal. Classical molecular dynamics simulations have been successfully used in studying the deformation of carbon nanotubes under pressure28'29. The results from the simulations agree well with the experiments. A sketch map of the Sm@C_{88} cage with carbon atoms numbered is shown in Fig. 6. Several characteristic distances that can be used to show the deformation of the carbon cage have been taken into account. The horizontal diameters (distance between C35-C45), the vertical diameters (distance between C1-C83) of the carbon cages and the distances of Sm-C45 (close to Sm) and Sm-C35 (far away from Sm) under high pressure were obtained from the simulations. We found that the pressure derivatives (dR/dp) of these distances start to change at around 7 GPa, as shown in Fig. 7(a). The horizontal diameter (C35-C45) of the carbon cage decreases slowly below 7 GPa, and then decreases at a faster rate above 7 GPa. The vertical diameter (C1-C83) of the carbon cage always shows a larger decrease than the horizontal one. Above 7 GPa, both distances of the carbon cage decrease at a faster rate with increasing pressure than before, demonstrating a more obvious deformation of the carbon cage. Based on the pressure evolutions of these diameters, we propose the following deformation process for the carbon cage. First, the deformation happens in the vertical direction of the cage. Then, the carbon cage changes from an ellipsoidal to an approximately spherical shape, further to a peanut-like shape at higher pressure, and finally collapses. The corresponding deformation process of Sm@C_{88} is shown in Fig. 6. It should point out that our simulations describe a less obvious deformation on the carbon cage than our experimental observations. This might be due to the fact that the classical potential is soft and the simulations usually underestimate the structural changes, as mentioned in the simulations of carbon nanotubes30. However, the classical method can give a qualitative description for the structural changes under pressure. Indeed, our simulations give results compatible with our experimental observations. The simulation pressure at which the pressure derivatives of the carbon cage change is consistent with that for an anomalous change in slope in the IR-active vibrational modes observed in our experiment. The anisotropic deformation of the carbon cage given by our simulation can explain the red and blue shifts of the experimental vibrational modes.

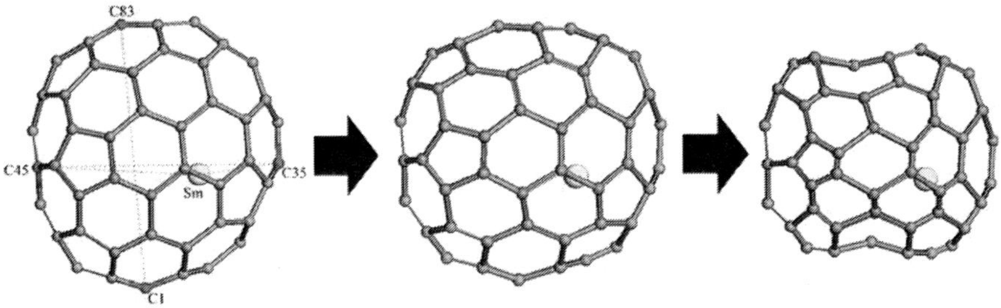

Figure 6: The deformation process of the Sm@C_{88} under high pressure. The major Sm (green) site along with the symmetry plane. C1, C35, C45 and C35 are marked as red.

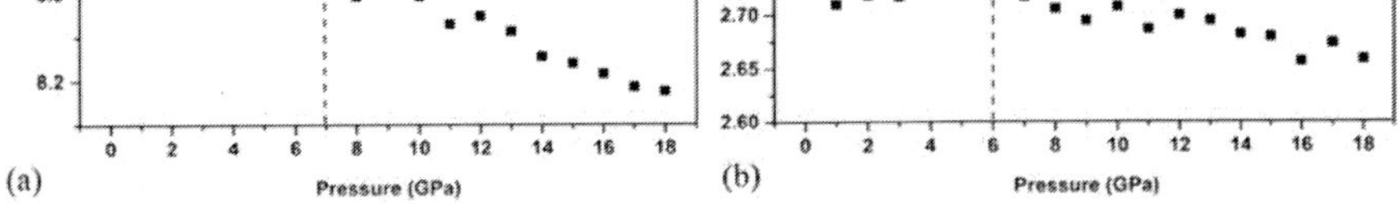

Figure 7: (**a**) Calculated pressure dependences of the C35-C45 and C1-C83 diameters of the carbon cage. (**b**) Calculated pressure dependences of the Sm-C35 and Sm-C45 distances.

From the pressure dependence of the simulated Sm-C distances, shown in Fig. 7(b), we observed that the Sm-C35 (close to Sm) distance decreases at a slower rate than the Sm-C45 (far away from Sm) distance, indicating that the area of the carbon cage closer to the Sm atom deforms less. This can explain the smaller pressure coefficients of the vibrational modes in the area close to the Sm atom. The unbalanced charge distribution on the carbon cage results in a stronger coulomb interaction between the Sm atom and the neighboring C atoms. Such 'support' from Sm to the near area of the carbon cage may restrain the change of the corresponding C-C/C=C vibrations with increasing pressure.

HOMO-LUMO gap of Sm@C_{88}

Let us now turn to understand the band gap reduction in our molecular crystals. In an EMF crystal, EMF molecules occupy the lattice sites. The electrons of the samarium atom are localized inside the cage and the electrons of the carbon cage are localized surrounding the carbon cage. The band structure of such a crystal can be estimated by a simple tight-binding approximation model, for which the EMF molecules are considered as a single cell unit. Taking the single atomic s-level as an example, the tight-binding approximation formula is

$$\varepsilon(\mathbf{k}) = E_S - \beta - \sum_{n} \gamma(\mathbf{R_n}) \cos \mathbf{k} \cdot \mathbf{R_n}, \qquad (1)$$

where $\varepsilon(\mathbf{k})$ is the Bloch electronic energy. E_S is the single atomic s-level electronic energy, which is the main component of $\varepsilon(\mathbf{k})$, and β is very small. The width of the energy band depends on $\gamma(\mathbf{R_n})$, which in turn depends on the interaction and the distance between the neighboring atoms31. A stronger interaction produces a wider energy band. Subsequently, in EMF crystals, the single EMF molecular levels broaden into a band due to the interaction between the neighboring molecules. The gap between the highest occupied molecular orbital (HOMO) and the lowest unoccupied molecular orbital (LUMO) determines the band gap between valence band and conduction band of the crystal formed. High pressure can deform the molecular structure and thus decrease the HOMO-LUMO gap of a molecule, to reduce the valence/conduction band gap of the formed crystal, which has been proved in the bisthiaselenazolyl radical dimer32. Therefore, the changes of the band gap of the Sm@C_{88} crystal should be related to the observed deformation of the Sm@C_{88} molecule.

To simulate the corresponding transformation, calculations using the DFT method by $DMOL^3$ were conducted on Sm@C_{88} molecules at different pressures. The relative eigenvalues and the gap between HOMO and LUMO of Sm@C_{88} at different pressures have been calculated. The plotted curve for the HOMO-LUMO gap as a function of pressure is shown in Fig. 8(a). It is clear that the HOMO-LUMO gap shows a significant drop (around 0.2 eV) in magnitude at around 7 GPa. Furthermore, we point out that the HOMO of

EMF is mostly localized on the cage, while the LUMO is centered on the endohedral metal atom. This has also been demonstrated in other EMF systems, such as $Sm_3@I_h$-C_{80}33. In our simulations, a significant deformation of the carbon cage and a decrease of the distance between the Sm atom and the carbon cage are observed at around 7 GPa (Fig. 7). Such changes can thus be reasonably related to the reduction of the HOMO-LUMO gap of $Sm@C_{88}$ under pressure. Therefore, the deformation of the carbon cage and the decrease of the Sm-cage distance results in a reduction of the HOMO-LUMO gap of the $Sm@C_{88}$ molecule.

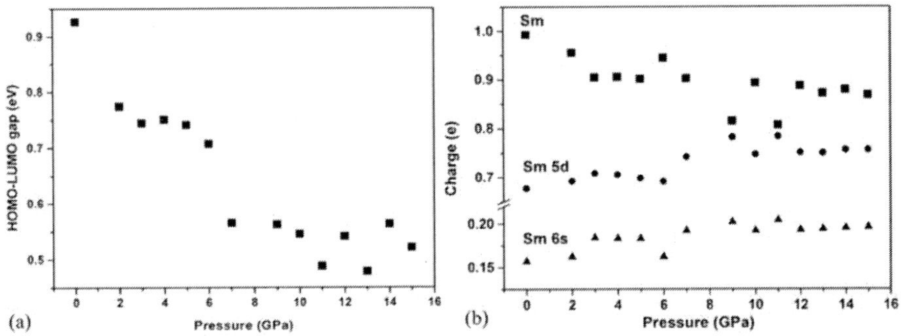

Figure 8: (**a**) Calculated HOMO-LUMO gap as a function of pressure. (**b**) Calculated samarium atomic charge and the electrons on 6s and 5d orbitals of a samarium atom as a function of pressure.

Above 7 GPa, the HOMO-LUMO gap of the $Sm@C_{88}$ molecule is varies slightly with increasing pressure. The reduction of the band gap could arise from the pressure-induced broadening of the valence and conduction bands of the crystal. This can be understood by the fact that the application of pressure compresses the crystal lattice and pushes $Sm@C_{88}$ molecules closer together, thus enhancing the interaction between neighboring molecules. A pressure-induced broadening of the energy bands takes place.

CHARGE TRANSFER OF SM@C_{88}

In addition to the cage deformation under high pressure, the degree of electron transfer between the entrapped metal and the carbon cage can also contribute to the change of the electronic properties10 of $Sm@C_{88}$. Thus we analyzed the Mulliken population of the $Sm@C_{88}$ molecule at different pressures by using *first-principles* with the local density approximation (LDA), shown in Fig. 8(b). The charge on the samarium atom decreases slowly with increasing pressure, suggesting that electrons transfer from the carbon cage to the samarium atom under pressure and that the valence of the carbon cage decreases. We also performed the simulations with the generalized gradient approximation (GGA). The results from the two simulations are similar to each other. To understand the change of the charge transfer, we analyze the composition of orbitals. The pressure dependences of the charge in the 6s and 5d orbitals are depicted in Fig. 8(b). At ambient condition, two 6s electrons of the samarium atom are transferred to the carbon cage, while some electrons from the occupied π orbitals of the cage are fed back to the 5d orbitals of samarium34. From Fig. 8(b), we can see that the electron density on 6s and 5d orbitals increases under high pressure. Thus each samarium atom donates fewer electrons, but receives more back-donation from the π orbitals of the carbon cage with increasing pressure. The change of the charge transfer between the Sm atom and the carbon cage may modify the electronic structure and the electron distribution in the material. In another similar system, the alkali-metal-doped fullerides (A_xC_{60}), the amount of charge transferred from alkali metal to carbon cage plays an important role for the electronic properties of the A_xC_{60} formed. A_4C_{60} is a small band-gap semiconductor35, while A_3C_{60} is metallic and even superconducting at low temperature36. It is also proposed that the metallicity of A_xC_{60} can be destroyed by a change in the valence of C_{60} and a reduction of the crystal symmetry37. The most obvious difference between EMFs and C_{60}/C_{70} fullerenes is that metal atoms are trapped inside the former. Thus the change of the charge transfer between the Sm atom and the carbon cage and the decrease of the Sm-cage distance are proposed to be important factors for the reduction of the band gap of Sm@C_{88}.

CONCLUSIONS

We have studied the structural stability and deformation of Sm@C_{88}under pressure up to 18 GPa by IR spectroscopy combined with theoretical simulations. We have assigned the experimentally measured IR vibrational modes of Sm@C_{88} at ambient conditions. The carbon cage exhibits anisotropic deformations upon compression and the shape of the carbon cage is proposed to change from ellipsoidal to approximately spherical. The presence of Sm inside the cage plays a role to support the carbon cage against the deformation. Moreover, we observed a significant reduction of the band gap of the material under high pressure. This should be related to the deformation of the carbon cage and the decrease of the Sm-cage distance, which cause a reduction of the HOMO-LUMO gap of the Sm@C_{88} molecule. The enhancement of the intermolecular interaction widens the energy bands, which gives additional contribution to the decrease of the band gap. In addition, the change of the charge transfer from the Sm atom to the carbon cage upon compression modifies the electronic structure and the electron distribution in the material. The carbon cage deforms significantly above 7 GPa, from a spherical to a peanut-like shape. Finally, the carbon cages collapses at 18 GPa.

Taking the EMF Sm@C_{88} as an example, we thus demonstrate the effect of the interaction between the trapped metal atom and the carbon cage on the structural evolution of the hosting carbon cage and the electrical properties under high pressure at the atomic level. This provides a very useful reference for future research on the structures and properties of EMFs and the possibility to create novel, EMF-based functional materials.

METHODS

The angle-dispersive synchrotron X-ray diffraction experiment at ambient condition was performed at 16BMD and 16IDB beamlines at the Advanced Photon Source (APS) of Argonne National Laboratory (ANL). The wavelength of the X-ray was 0.3065 Å. The scanning range was up to $2\theta_{max} = 38°$. Two-dimensional patterns were radially integrated using the software FIT2D38. The X-ray crystallographic study of Sm@C_{88} is shown in the Supplementary Information Part 1. The MIR and FIR high-pressure measurements were carried out at U2A beamline, National Synchrotron Light Source (NSLS), Brookhaven National Laboratory (BNL). The MIR spectra were collected in both transmission and reflection mode by a Bruker Vertex 80v FTIR spectrometer and a Hyperion 2000 IR microscope equipped with a nitrogen-cooled MCT detector. The MIR high-pressure measurements have been repeated with the same type of equipment at State Key Laboratory of Superhard Materials, Jilin University. The FIR absorption spectra were measured using a Bruker IFS 66v FTIR spectrometer and a custom IR microscope equipped with a liquid helium cooled bolometer. Spectral resolution of 4 cm^{-1} and 512 scans applied to all spectra. A symmetrical diamond anvil cell was used with liquid argon as the pressure medium. Samples were loaded into a 100 µm diameter hole drilled in the T301 stainless steel gasket. The pressure was measured with the ruby fluorescence technique39. HRTEM images were measured using electron microscopy (JEM-2200FS).

In this work, all *first-principles* calculations (including geometry optimizations, vibrational frequencies, and properties) were carried out by the DMOL3 method within the local density approximation (LDA). The classical molecular dynamics (MD) simulations were performed using the Forcite code in Materials Studio using the universal force field (UFF)40′41. The MD was performed in a NPT ensemble (constant pressure and temperature dynamics) during 0.2 ps (time step = 0.01 fs) in the pressure range from 0 to 18 GPa at room temperature.

ACKNOWLEDGEMENTS

We thank Prof. John S. Tse for helpful discussions. This work was supported financially by the National Basic Research Program of China (2011CB808200), the NSFC (51320105007, 11474121, 11179039, 11274283), and the Cheung Kong Scholars Programme of China. The use of the beamline U2A at NSLS is supported by NSF (DMR-0805056; EAR 06-49658, COMPRES) and DOE/NNSA (DE-FC03-03N00144, CDAC). NSLS is supported by the DOE/BES (DE-AC02-98CH10886). The use of beamlines 16BMD and 16IDB, APS is supported by the Carnegie Institute of Washington, Carnegie DOE Alliance Center, University of Nevada at Las Vegas, and Lawrence Livermore National Laboratory through funding from the DOE National Nuclear Security Administration, DOE Basic Energy Sciences, and National Science Foundation (NSF).

CONTRIBUTIONS

J.X.C., H.Y., Z.Y.L., Q.J.L., R.L. and Z.X.L. fabricated the devices, performed experiments. J.X.C. and F.X.M. performed theoretical simulations. J.X.C., M.G.Y., B.B.L., Z.X.L., B.S., B.Z. and T.C analyzed the results. M.G.Y. and B.B.L. contributed most of the ideas and design the whole research. J.X.C., M.G.Y., B.B.L. and B.S. wrote the paper. All authors discussed the results and commented on the manuscript.

COMPETING INTERESTS

The authors declare no competing financial interests.

REFERENCES

1. *Endofullerenes: A New Family of Carbon Clusters*. (eds Akasaka, T.,Nagase, S.) (Kluwer Academic Publishers, 2002).
2. Lu, X., Akasaka, T. & Nagase, S. Chemistry of endohedral metallofullerenes: the role of metals. *Chem. Commun.* **47**, 5942–5957 (2011).
3. Kessler, B. *et al.* Evidence for incomplete charge transfer and La-derived states in the valence bands of endohedrally doped La@C_{82}. *Phys. Rev. Lett.* **79**, 2289–2292 (1997).
4. Wang, K. D. *et al.* Unveiling metal-cage hybrid states in a single endohedral metallofullerene. *Phys. Rev. Lett.* **91**, 185504 (2003).

5. Shinohara, H. Endohedral metallofullerenes. *Rep. Prog. Phys.* **63**, 843–892 (2000).
6. Heath, J. R. *et al.* Lanthanum complexes of spheroidal carbon shells. *J. Am. Chem. Soc.* **107**, 7779–7780 (1985).
7. Chai, Y. *et al.* Fullerenes with metals inside. *J. Phys. Chem.* **95**, 7564–7568 (1991).
8. Yang, H. *et al.* Isolation and crystallographic identification of four isomers of Sm@C_{90}. *J. Am. Chem. Soc.* **133**, 6299–6306 (2011).
9. Yang, H. *et al.* Isolation of three isomers of Sm@C_{84} and X-ray crystallographic characterization of Sm@D_{3d}(19)-C_{84} and Sm@C_2(13)-C_{84}. *J. Am. Chem. Soc.* **134**, 5331–5338 (2012).
10. Yang, H. *et al.* Isolation and crystallographic characterization of Sm@C_{2v}(3)-C_{80} through cocrystal formation with Ni^{II}(octaethylporphyrin) or bis(ethylenedithio)tetrathiafulvalene. *Inorg. Chem.***52**, 1275–1284 (2013).
11. Wang, Y. Photoconductivity of fullerene-doped polymers. *Nature***356**, 585–587 (1992).
12. Wang, Y., West, R. & Yuan, C. H. Fullerene-doped polysilane photoconductor. *J. Am. Chem. Soc.* **115**, 3844–3845 (1993).
13. Sundqvist, B. Fullerenes under high pressures. *Adv. Phys.* **48**, 1–134 (1999).
14. Mujica, A., Rubio, A., Munoz, A. & Needs, R. J. High-pressure phases of group-IV, III-V, and II-VI compounds. *Rev. Mod. Phys.***75**, 863–912 (2003).
15. Wang, L. *et al.* Synthesis and high pressure induced amorphization of C_{60} nanosheets. *Appl. Phys. Lett.* **91**, 103112 (2007).
16. Wagberg, T., Jacobsson, P. & Sundqvist, B. Comparative Raman study of photopolymerized and pressure-polymerized C_{60} films.*Phys. Rev. B.* **60**, 4535–4538 (1999).
17. Soldatov, A. V. *et al.* Topochemical Polymerization of C_{70}Controlled by Monomer Crystal Packing. *Science* **293**, 680–683 (2001).
18. Yamanaka, S., Kini, N. S., Kubo, A., Jida, S. & Kuramoto, H.Topochemical 3D polymerization of C_{60} under high pressure at elevated temperatures. *J. Am. Chem. Soc.* **130**, 4303–4309 (2008).
19. Ma, H. L. *et al.* Synchrotron X-ray diffraction and infrared spectroscopy studies of $C_{60}H_{18}$ under high pressure. *J. Phys. Chem. Lett.* **1**, 714–719 (2010).
20. Wang, L. *et al.* Long-range ordered carbon clusters: a crystalline material with amorphous building blocks. *Science* **337**, 825–828 (2012).
21. Yao, M. G. *et al.* Pressure-induced transformation and superhard phase in fullerenes: the effect of solvent intercalation. *Appl. Phys. Lett.* **103**, 071913 (2013).

22. Yang, S. F., Troyanov, S. I., Popov, A. A., Krause, M. & Dunsch, L.Deviation from the Planarity—a Large Dy_3N Cluster Encapsulated in an I_h-C_{80} Cage: An X-ray Crystallographic and Vibrational Spectroscopic Study. *J. Am. Chem. Soc.* **128**, 16733–16739 (2006).
23. Yang, S. F., Liu, F. P., Chen, C. B., Jiao, M. Z. & Wei, T. Fullerenes encaging metal clusters—clusterfullerenes. *Chem. Commun.* **47**, 11822–11839 (2011).
24. Cui, W. *et al.* A new carbon phase constructed by long-range ordered carbon clusters from compressing C_{70} solvates. *Adv. Mater.* **26**, 7257–7263 (2014).
25. Wasa, S., Suito, K., Kobayashi, M. & Onodera, A. Pressure-induced irreversible amorphization of C_{70} fullerene. *Solid State Commun.***114**, 209–213 (2000).
26. Chandrabhas, N. *et al.* Reversible pressure-induced amorphization in solid C_{70}: Raman and photoluminescence study.*Phys. Rev. Lett.* **73**, 3411–3414 (1994).
27. Snoke, D. W., Syassen, K. & Mittelbach, A. Optical absorption spectrum of C_{60} at high pressure. *Phys. Rev. B.* **47**, 4146–4148 (1993).
28. Zang, J., Treibergs, A., Han, Y. & Liu, F. Geometric constant defining shape transitions of carbon nanotubes under pressure.*Phys. Rev. Lett.* **92**, 105501 (2004).
29. Elliott, J. A., Sandler, J. K., Windle, A. H., Young, R. J. & Shaffer, M. S. Collapse of single-wall carbon nanotubes is diameter dependent. *Phys. Rev. Lett.* **92**, 095501 (2004).
30. Sun, D. Y. *et al.* Pressure-induced hard-to-soft transition of a single carbon nanotube. *Phys. Rev. B.* **70**, 165417 (2004).
31. Ashcroft, N. W. & Mermin, N. D. *Solid State Physics*. (Harcourt College Publishers, 1976).
32. Tse, J. S. *et al.* Metallization of a hypervalent radical dimer: molecular and band perspectives. *J. Am. Chem. Soc.* **132**, 4876–4886 (2010).
33. Xu, W. *et al.* An experimentally observed trimetallofullerene $Sm_3@I_h$-C_{80}: encapsulation of three metal atoms in a cage without a nonmetallic mediator. *J. Am. Chem. Soc.* **135**, 4187–4190 (2013).
34. Jin, H. X. *et al.* Single samarium atoms in large fullerene cages. Characterization of two isomers of $Sm@C_{92}$ and four isomers of $Sm@C_{94}$ with the X-ray crystallographic identification of $Sm@C_1(42)$-C_{92}, $Sm@C_s(24)$-C_{92}, and $Sm@C_{3v}(134)$-C_{94}. *J. Am. Chem. Soc.* **134**, 10933–10941 (2012).
35. Sabouri-Dodaran, A. A. *et al. In situ* pressure study of Rb_4C_{60}insulator to metal transition by Compton scattering. *Phys. Rev. B.***72**, 085412 (2005).

36. Hebard, A. F. *et al.* Superconductivity at 18 K in potassium-doped C_{60}. *Nature* **350**, 600–601 (1991).
37. Iwasa, Y. & Takenobu, T. Superconductivity, Mott–Hubbard states, and molecular orbital order in intercalated fullerides. *J. Phys Condens. Matter* **15**, 495–519 (2003).
38. Hammersley, A. P., Svensson, S. O., Hanfland, M., Fitch, A. N. &Häusermann, D. Two-dimensional detector software: from real detector to idealized image or two-theta scan. *High Press. Res.* **14**, 235–248 (1996).
39. Mao, H. K., Xu, J. & Bell, P. M. Calibration of the ruby pressure gauge to 800 kbar under quasi-hydrostatic conditions. *J. Geophys. Res.* **91**, 4673–4676 (1986).
40. Rappe, A. K., Casewit, C. J., Colwell, K. S., Goddard, W. A. & Skiff, W. M. UFF, a full periodic table force field for molecular mechanics and molecular dynamics simulations. *J. Am. Chem. Soc.***114**, 10024–10035 (1992).
41. Root, D. M., Landis, C. R. & Cleveland, T. Valence bond concepts applied to the molecular mechanics description of molecular shapes. 1. Application to nonhypervalent molecules of the P-block. *J. Am. Chem. Soc.* **115**, 4201–9 (1993).

Chapter 6

Study of Nanomaterials under High Pressure

Madan Singh, Moruti Kao

Department of Physics and Electronics, National University of Lesotho, Roma, Lesotho

ABSTRACT

A simple theory is proposed to predict the effect of pressure for study of volume expansion of nanomaterials. Different possible forms of equation of state are discussed with their correlations. Only two input parameters, namely, the bulk modulus and its first pressure derivative, are required for calculations. We have considered a wide variety of nanomaterials, such as, CdSe (4.2 nm), Fe-Cu (14 nm), γ-Al_2O_3 (67 nm), γ-Al_2O_3 (37 nm), Ni (20 nm), Fe (10 nm), CeO_2 (cubic Fluorite phase) (15 nm), CeO_2 (Orthorhombic Phase) (15 nm), CuO (24 nm) and TiO_2 (rutile phase) (10 nm) to analyze the effect of pressure on them. The theoretical predictions for the given nanomaterials agree with the experimental results and the other theoretical models.

Keywords: Equation of State; Nanomaterials; Murnaghan EOS

1. INTRODUCTION

The study of nanocrystalline materials with dimension less than 100 nm is the active area of research in physics, chemistry and engineering [1]. Nanomaterial is a field that takes a material science based on the approach to nanotechnology. Nanomaterials differ significantly from the bulk materials by virtue of their small size. Often, the behavior of nanomaterials may depend more on its surface area than particle's composition itself. Electrical, optical and chemical properties of nanomaterials are very different at large scale. More than half of the material atoms are on the surface, increasing relative surface area and quantum effects. These factors can change or enhance the properties such as reactivity, strength, electrical and optical characteristics. Various physical properties such as hardness, melting temperature, sintering ability and electronic structure depend upon particle size [2- 7]. Nanomaterials, nanoparticles, nanowires and nanotubes have been reported to show physical, chemical and mechanical properties that are noticeably different from their corresponding bulk counterparts.

For example, phonon frequency blue shift in nanosemiconductors and nanometals, increase of elastic modulus in thin films and nanoparticles, decrease of melting temperature and thermal conductivity in nanocrystals are associated with the particle size of the material under consideration [8,9]. There are four classes of materials that are used to make nano-sized products including metal, carbon, composites and dentrimers. These products have a variety of application and are used in the automotive industry, biomedical industry, in films and much more. Metal-based nanomaterials are those such as nanogold, quantum dots and nanosilver. Quantum dots come in several sizes with a maximum size of several hundred nanometers. The dot contains semiconductor crystals that are packed together so thousands of atoms are all in a very small area. Other metal-based chemicals used at this small scale include types of metal oxides. Carbon-based nanomaterials are those that are mostly made of carbon. The carbon is shaped into hollow tubes, ellipsoids or spheres. Ellipsoidal and spherical carbon nanomaterials are called fullerenes. Nanomaterials are expected to be the turning point of the next technological revolution in solid-state electronics, to emerge as new structural materials, to serve as systems for controlling drug delivery and to have a considerable impact in practically all domains of science. These materials, remarkable for their extremely small size, have the potential for wide ranging industrial, biomedical, and electronic applications. Miguel [10] has critically reviewed the expected effects and benefits of the effect of pressure on nanomaterials. Pressure and temperature dependence of volume V/V_0, bulk modulus B, and coefficient of volume thermal expansion was investigated by varying the temperature from room temperature to 2000 K, and pressure from room pressure to 200 kbar. Due to high pressure on nanomaterials, many effects happen such as pressure ionization, modification in electronic properties etc. Therefore, pressure volume relation of condensed matter known as equation of state (EOS) is an important idea. As many EOS exist in literature, there is still a need to explain the effect of pressure at constant temperature for nanomaterials. Lithium-alumino-silicate ceramics have gained considerable commercial attention because of very low thermal expansion, transparency, high chemical durability and strength.

They are very sensitive to the external parameters like pressure and temperature. The physical properties of these materials depend strongly on structure and interatomic distances. Pressure can vary these distances, which implies that we can study relationships between structure and properties of the materials [11]. Therefore, it is not just a matter of great interest, but also a necessity, to study the thermodynamic properties of nanocrystalline (NC) materials.

Investigation of the behavior of NC materials under high pressure can provide valuable information about their intrinsic microstructural characteristics. The study of nanomaterials with varying pressure can help us to study a wide range of solid state materials. High temperature-pressure applications have the potential to constitute a unique way for the elaboration of new materials in a controlled manner. Pressure application, as for bulk materials, allows the continuous modification of the interatomic interactions of the nano-object and constitutes an invaluable tool to explore physico-chemical interactions at the nanoscale and their link with physical properties of interest.

The physical properties of materials depend strongly on the structure and interatomic distances. High pressure can vary these distances, which implies that we can study relationships between structure and properties of the materials. Due to the high pressure on nanomaterial, many effects occur, such as 1) transformation of nano-constitutive elements, 2) transformations of the interaction between nano-objects, and 3) modification of interactions between the nano-object and pressure transmitting medium. Due to such applications, the effects of pressure on nanomaterials have attracted the attention of the researchers. In many applications, the understanding of material properties that are influenced by pressure becomes crucial. Therefore, investigation of the behavior of nano-sized materials under high pressure is indispensable, especially due to the immense potential of applications associated with these materials. Several experiments have been performed to study and explain the behavior of nonmaterial [7-10,12-18]. It is then not surprising that high-pressure investigations of nanomaterials develop in parallel to the growth of nanoscience either for better understanding of the properties of nanomaterials or providing alternative methods for nanostructuration. Experimental data [7-10,12-18] used by different techniques such as X-Ray diffraction under hydrostatic pressure in a diamond anvil cell (DAC) method are obtained, using high energy Synchrotron radiation and Raman spectroscopic technique, High pressure X-Ray diffraction and X-Ray absorption, Neutron diffraction studies of compressibility method, Synchrotron X-Ray diffraction method and Laser induced vapour phase reaction.

An equation of state (EOS) recently proposed for nanomaterials is discussed critically. Different possible forms of the EOS are discussed with their correlations. We have considered 20 nanomaterials for this purpose, viz. CdSe (4.2 nm), Fe-Cu (14 nm), γ-Fe_2O_3 (10 nm), γ-Al_2O_3 (67 nm), γ-Al_2O_3 (37 nm), Ni

(20 nm), Fe (10), CeO_2 (cubic Fluorite phase) (15 nm), CeO_2(Orthorhombic Phase) (15 nm), CuO (24 nm), TiO_2 (rutile phase) (10 nm), TiO_2 (anatase) (40 nm), 3C-SiC (30 nm), AlN (Hexagonal wurtzite Phase) (10 nm), γ-Si_3N_4 (10 nm), AlN (Cubic rocksalt Phase) (10 nm), ZnO (wurtzite Phase) (10), ZnO (Rocksalt Phase) (10), SnO_2 (14 nm) and SnO_2 (8 nm). The results have been found to present a good agreement with the available experimental data.

2. METHOD OF ANALYSIS

The dependence of volume on pressure can be written as,

$$P = m(1 - V/V_0) + n(1 - V/V_0)^2 + q(1 - V/V_0)^3 + \cdots, \quad (1)$$

where, V/V_0 is the relative change in volume, and m, n and q are size dependent parameters, which may be determined from the definition of bulk modulus, and its first and second order pressure derivatives, respectively. In Equation (1), the higher order terms may be ignored because of the smaller contributions. This is advantageous also because higher order pressure derivatives of bulk modulus, which are not available for nanomaterials, are involved in these higher terms.

The Bulk modulus is defined as,

$$B = -V\left(\frac{\partial P}{\partial V}\right)^{T}$$

Using this definition of bulk modulus, Equation (1) may be written as,

$$B = (V/V_0)\left[m + 2n(1 - V/V_0)\right] \quad (2)$$

First order pressure derivative of bulk modulus is defined as,

$$B' = \left(\frac{\partial B}{\partial P}\right)^{T} = -\left(\frac{V}{B}\right)\left(\frac{\partial B}{\partial V}\right)^{T} \text{ or, } B' = -\frac{1}{B}(V/V_0)\left[m + 2n\{1 - 2(V/V_0)\}\right]$$

Using all the above equations and applying the boundary conditions, B = B_0, when V = V_0, we obtain, m = B_0 and $n = \frac{B_0(B_0' + 1)}{2}$.

Substituting these values of m and n in Equation (1), we get the EoS as,

$$P = B_0\{1 - (V/V_0)\} + \frac{B_0(B_0' + 1)}{2}\{1(V/V_0)\}2 \quad (3)$$

Interestingly, Equation (3) can also be derived using the theory of Mie-Gruneisen EoS, which defines pressure as,

$$P = -\frac{\partial U}{\partial V} + P_{Th}$$

where, P_{Th} is the thermal pressure and U, the lattice potential energy. This Equation gives the relation for thermal expansion as:

$$(V/V_0) - 1 = \frac{1 - \left\{1\left[-2\left(B_0' + 1\right) P_{Th} / B_0\right]^{1/2}\right\}}{B_0' + 1} \tag{4}$$

It has been claimed that Equation (4) may be written as when $P \neq 0$,

$$(V/V_0) - 1 = \frac{\left\{1 - \left[1 - 2\left(B_0' + 1\right)\left(P_{Th} - P\right) / B_0\right]^{1/2}\right\}}{B_0' + 1} \tag{5}$$

When thermal pressure is zero (P_{Th} = 0), Equation (5) may be written as,

$$(V/V_0) - 1 = \frac{\left\{1 - \left[1 + 2\left(B_0' + 1\right) P / B_0\right]^{1/2}\right\}}{B_0' + 1} \tag{6}$$

On rearranging we get,

$$P = B_0\left(1 - (V/V_0)\right) + \left\{\frac{B_0\left(B_0' + 1\right)}{2}\right\}\left(1 - (V/V_0)\right)^2 \tag{7}$$

It is the same as Equation (3), which supports the legitimacy of Equation (3). We are including some other equation of state (EOS) for comparison purpose such as Murnaghan EOS and Vinet EOS.

On assuming Anderson parameter, δ_T, independent of V/V_0, Murnaghan EOS [19] looks as

$$P = \left(B_0 / B_0'\right)\left\{\exp\left[-B_0' \ln \frac{V}{V_0}\right] - 1\right\} \tag{8}$$

Also one can recited Vinet EOS [20] as

$$P = 3\left(1-(V/V_0)^{1/3}\right)$$
$$B_0\left\{\exp\left[\frac{3(B_0'-1)}{2}\right]\left(1-(V/V_0)^{1/3}\right)\right\}(V/V_0)^{1/2} \quad (9)$$

3. RESULT AND DISCUSSION

In Equation (1), the next higher order terms have not been considered, we have considered only up to the second term of the series. Because inclusion of higher terms requires second and higher order pressure derivatives of the bulk modulus, which are still not available for most of the nanomaterials. Moreover, it is legitimate and may be useful to discuss the effect of higher order terms in Equation (1). The values of volume expansion (V/V_0) have been calculated as a function of pressure, using Equations (7), (8) and (9), respectively for CdSe (4.2 nm), Fe-Cu (14 nm), γ-Fe_2O_3(10 nm), γ-Al_2O_3 (67 nm), γ-Al_2O_3 (37 nm), Ni (20 nm), Fe (10), CeO_2 (cubic Fluorite phase) (15 nm), CeO_2 (Orthorhombic Phase) (15 nm), CuO (24 nm), TiO_2 (rutile phase) (10 nm), TiO_2(anatase) (40 nm), 3C-SiC (30 nm), AIN (Hexagonal wurtzite Phase) (10 nm)γSi_3N_4 (10 nm), AIN (Cubic rocksalt Phase) (10 nm), ZnO (wurtzite Phase) (10), ZnO (Rocksalt Phase) (10), SnO_2 (14 nm), SnO_2 (8 nm) nanomaterials. We have used Equation (1) to predict the compression behavior of these nanomaterials. To make the analysis more rigorous, we also included some more EOSs, i.e. Murnagham EOS, Vinet EOS. The results obtained are shown in figures and compared with experimental data. The deviation occur from experimental values at high pressure, it may be because of that we are ignoring higher order terms in Equation (1).

We have selected these materials because of the fact that the experimental data are available. The input data required are given in **Table 1**. The results obtained are reported in Figures 1-20 along with the available experimental data [14-18] for the sake of comparison. It is found that the results obtained from Equation (7) are in good agreement with the experimental data. It may be concluded that Equation (1) is quite satisfactory for these materials in the pressure range considered in the present paper. This clearly demonstrates that the approximation used in Equation (1) is justified. It should also be understood that all equations of state discussed in this work (Murnaghan and Vinet) have matched with the experimental data. There was a good agreement between the theoretical values of V/V_0 as a function of P, obtained from Equation (7) as has been found with the corresponding experimental data for CdSe (4.2 nm), Fe-Cu (14 nm), Ni (20 nm), γ-Fe_2O_3 (10 nm), Fe (10 nm), and γ-Al_2O_3 (37 nm) nanomaterials [14-18]. It may be concluded from the above discussion that the EoS obtained in this work [Equation (7)], in general, can produce much better pressure dependent volume expansion data, as compared to other equations of state, for a wide range of nanomaterials.

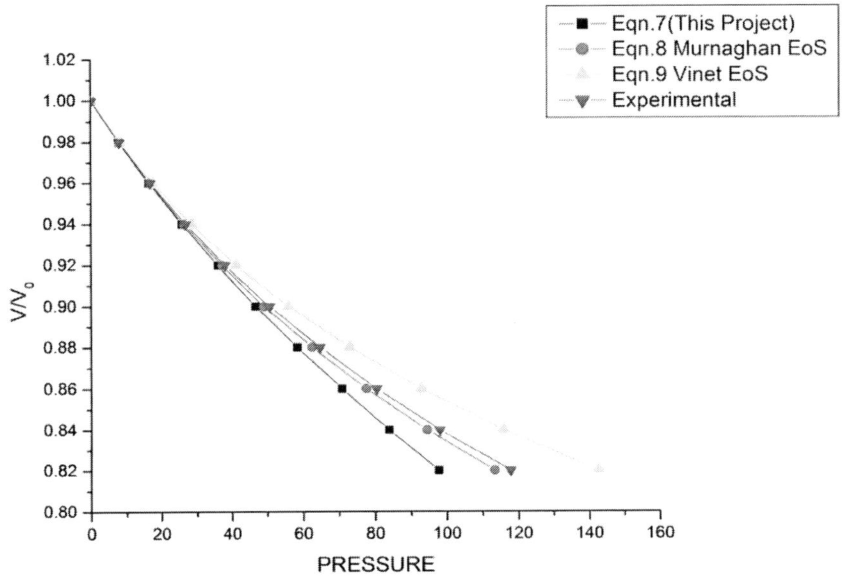

Figure 1. Volume expansion (V/V_0) as a function of pressure (GPa) for Fe (10 nm).

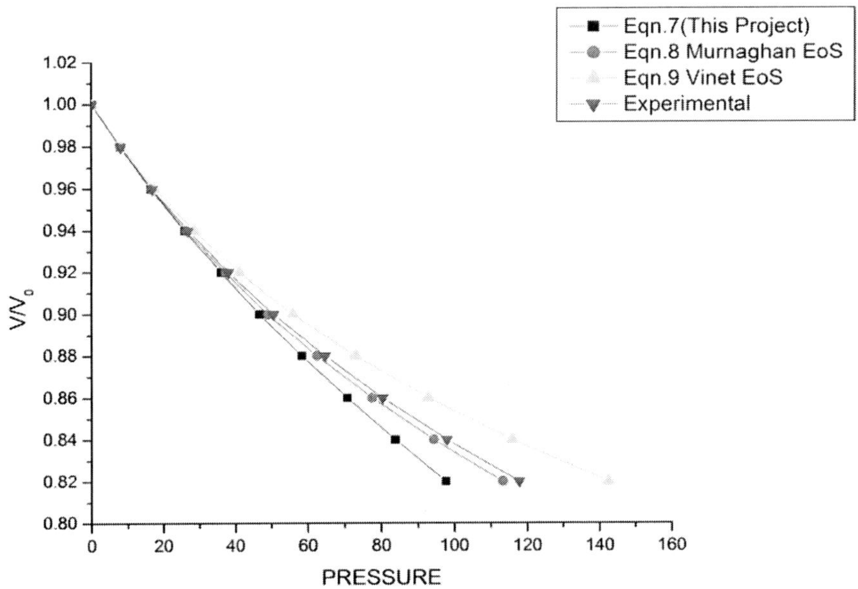

Figure 2. Volume expansion (V/V_0) as a function of pressure (GPa) for y-Al_2O_3 (37 nm).

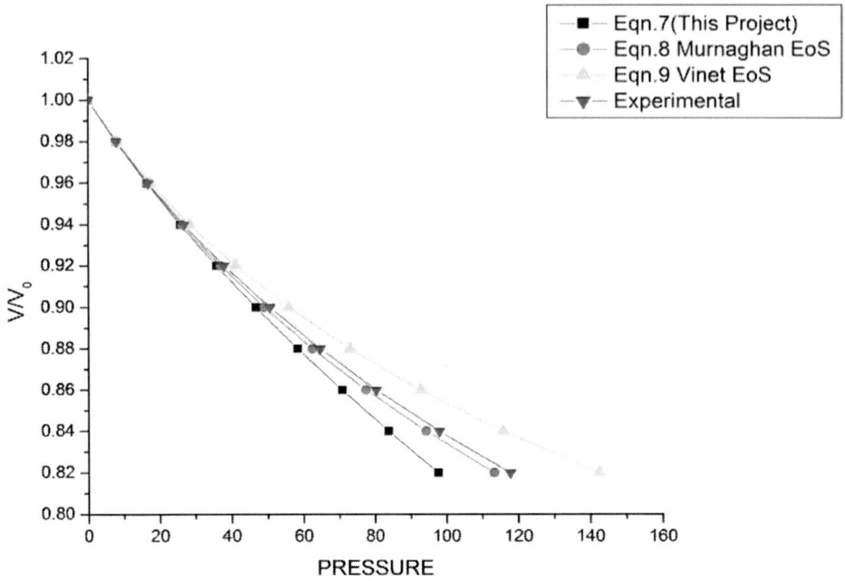

Figure 3. Volume expansion (V/V_0) as a function of pressure (GPa) for γ-Fe_2O_3 (10 nm).

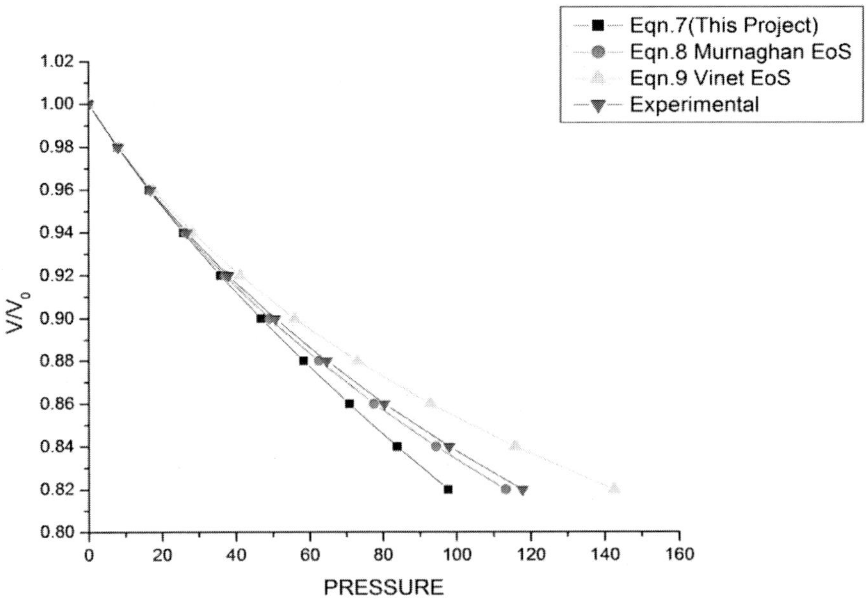

Figure 4. Volume expansion (V/V_0) as a function of pressure (GPa) for CdSe (4.2 nm).

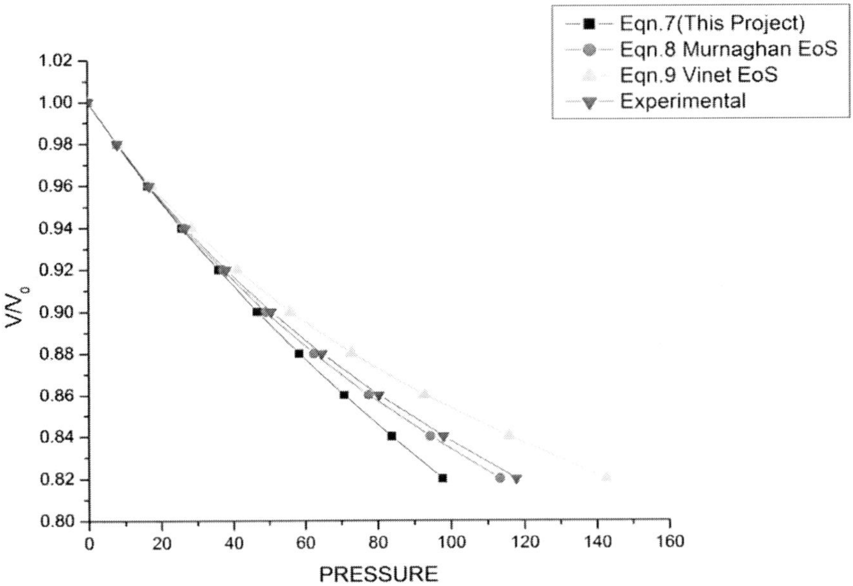

Figure 5. Volume expansion (V/V_0) as a function of pressure (GPa) for Fe-Cu (14 nm).

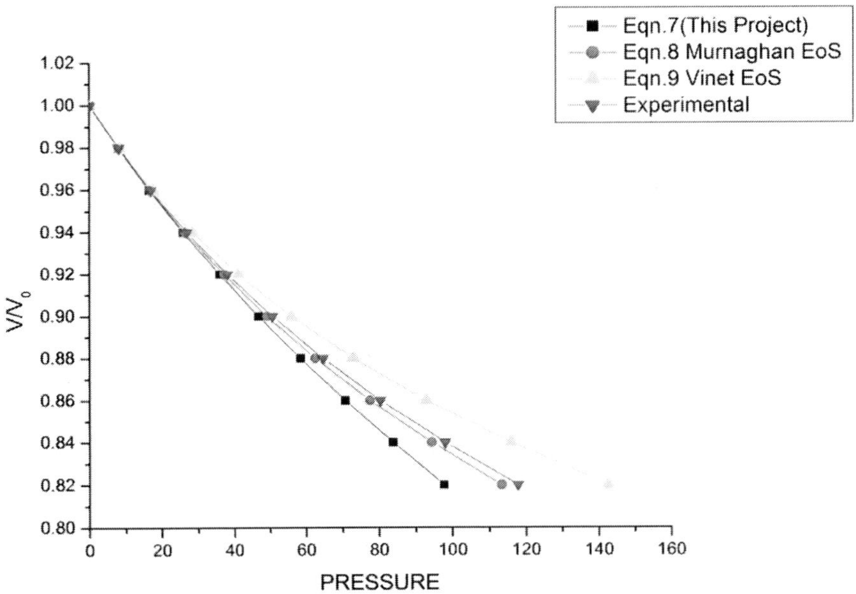

Figure 6. Volume expansion (V/V_0) as a function of pressure (GPa) for Ni (20 nm).

Table 1. Input parameters [7-9,12-18] used in theoretical calculations.

S.No.	Nanomaterials	B_0 (GPa)	B_0'
1	CdSe (4.2 nm)	74	4
2	Fe-Cu (14 nm)	129	4
3	Ni (20 nm)	185	4
4	γ-Fe_2O_3 (10 nm)	374	4
5	Fe (10 nm)	179	3.6
6	CeO_2 (cubic Fluorite Phase) (15 nm)	328	4
7	CeO_2 (Orthorhombic Phase) (15 nm)	326	4
8	γ-Al_2O_3 (67 nm)	238	4
9	γ-Al_2O_3 (37 nm)	172	4
10	CuO (24 nm)	81	4
11	TiO_2 (rutile phase) (10 nm)	211	4
12	TiO_2 (anatase) (20 nm)	190.4	2.9
13	3C-SiC (30 nm)	245	4
14	AlN (Hexagonal wurtzite Phase) (10 nm)	321	4
15	AlN (cubic rocksalt Phase) (10 nm)	359	4
16	γ-Si_2N_4 (10 nm)	339	4
17	ZnO (wurtzite Phase) (10 nm)	294	4
18	ZnO (Rocksalt Phase) (10 nm)	264	4
19	SnO_2 (14 nm)	248	4
20	SnO_2 (8 nm)	246	4

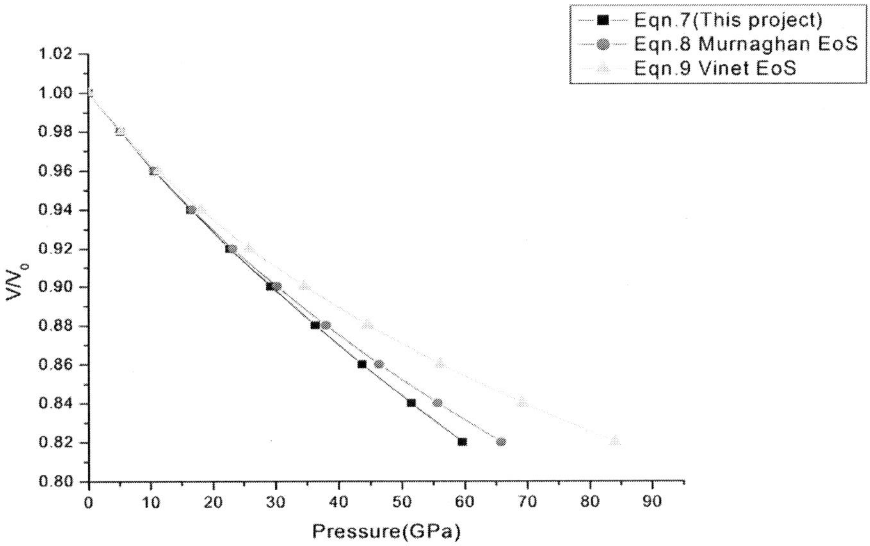

Figure 7. Volume expansion (V/V_0) as a function of pressure (GPa) for 3C-SiC (3O nm).

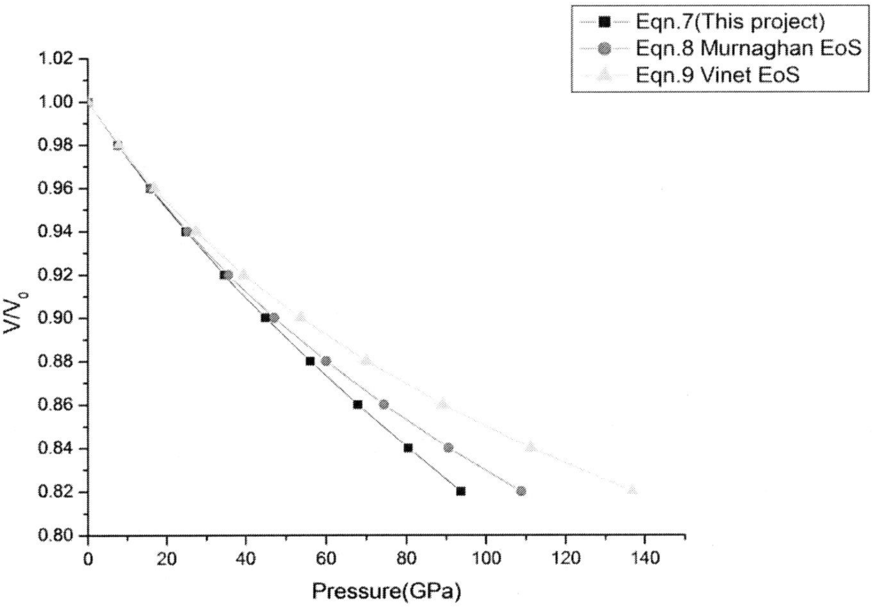

Figure 8. Volume expansion (V/V_0) as a function of pressure (GPa) for AIN (Cubic rocksalt Phase) (10 nm).

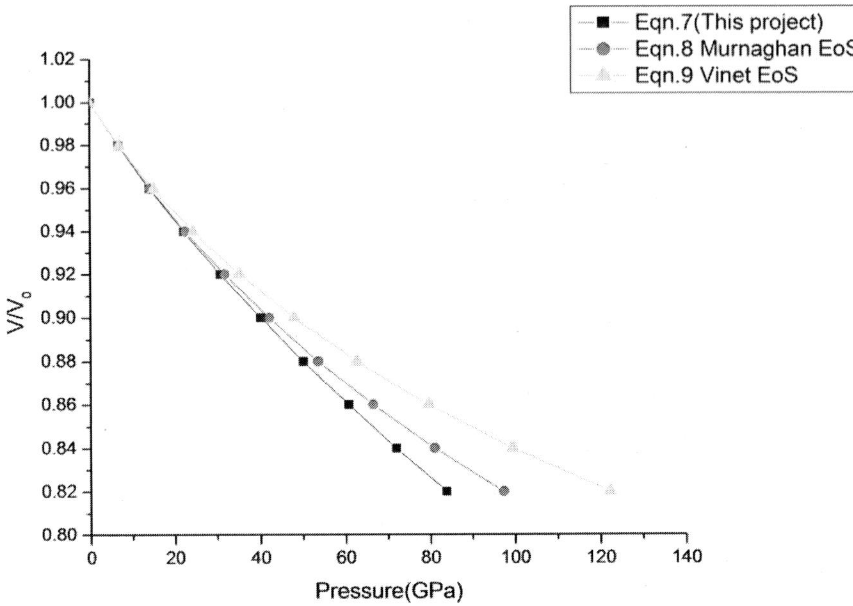

Figure 9. Volume expansion (V/V_0) as a function of pressure (GPa) for AIN (Hexagonal wurtzite Phase) (10 nm).

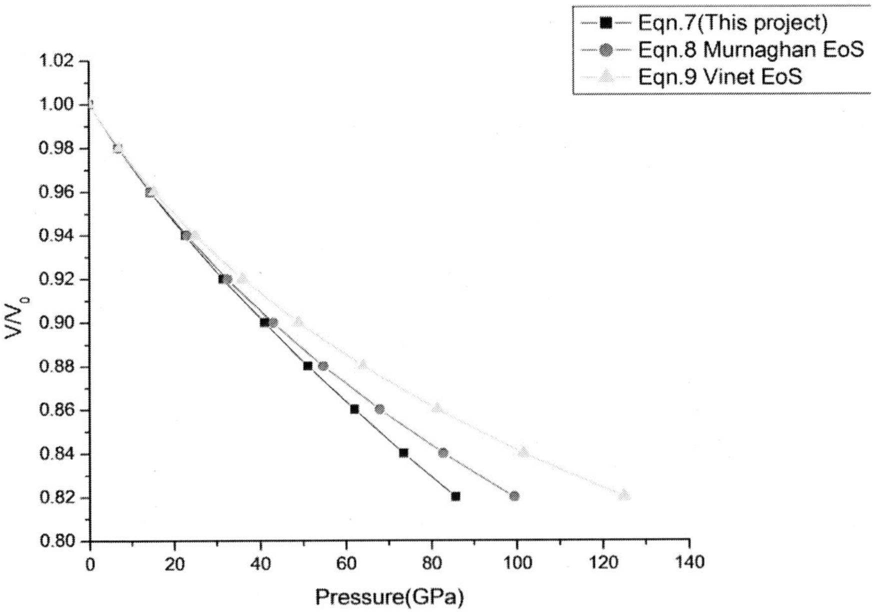

Figure 10. Volume expansion (V/V_0) as a function of pressure (GPa) for CeO_2 (cubic Fluorite) (15 nm).

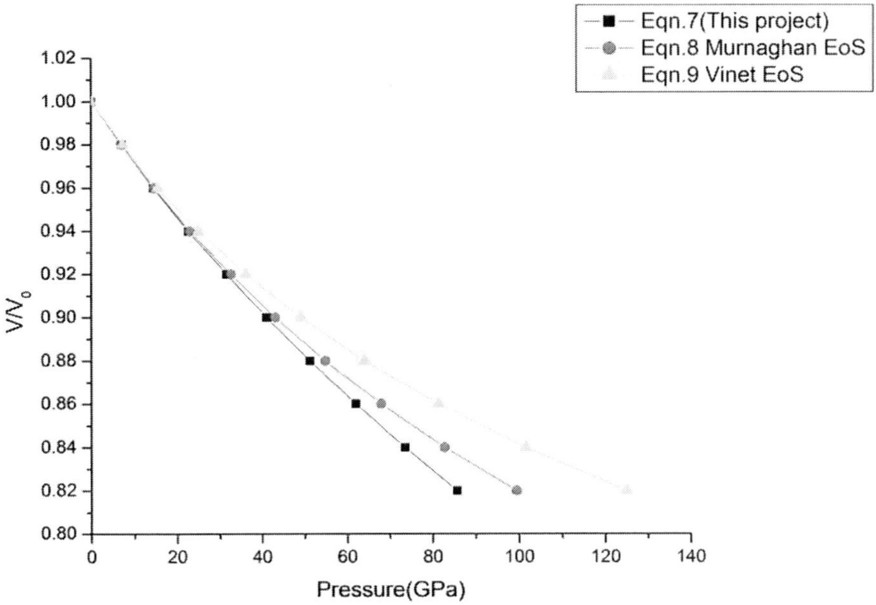

Figure 11. Volume expansion (V/V_0) as a function of pressure (GPa) for CeO_2 (Orthorhombic Phase) (15 nm).

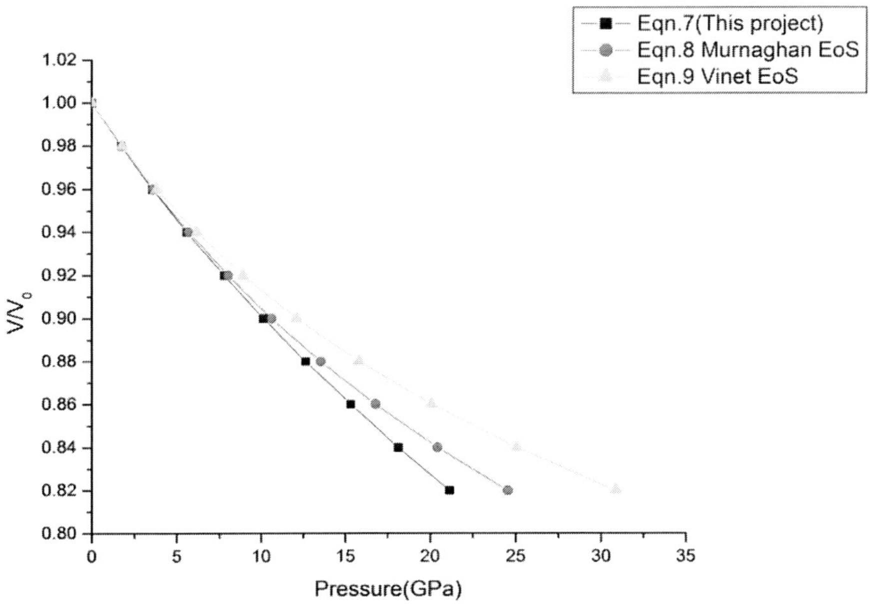

Figure 12. Volume expansion (V/V_0) as a function of pressure (GPa) for CuO (24 nm).

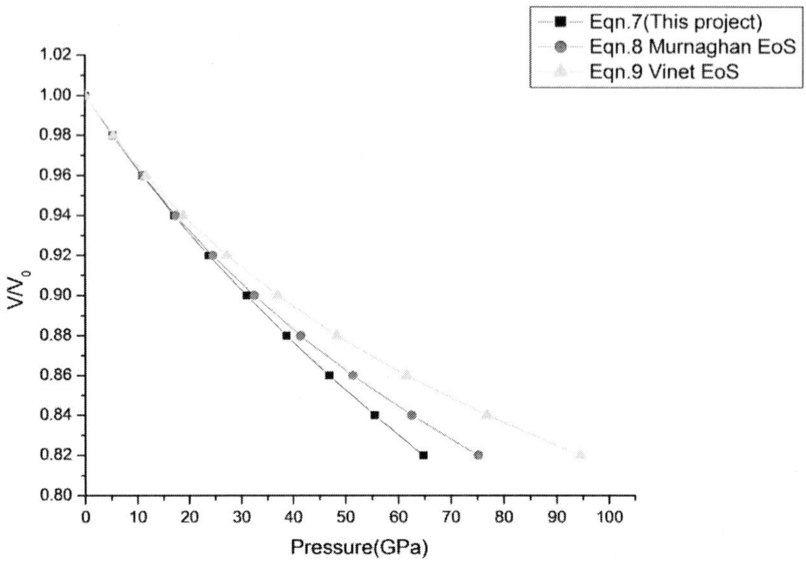

Figure 13. Volume expansion (V/V_0) as a function of pressure (GPa) for SnO_2 (14 nm).

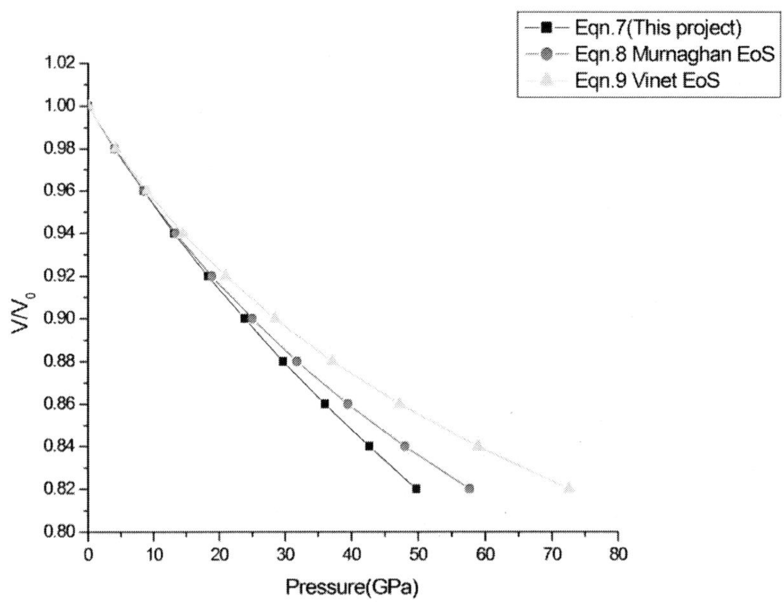

Figure 14. Volume expansion (V/V_0) as a function of pressure(GPa) for TiO_2 (anatase) (40 nm).

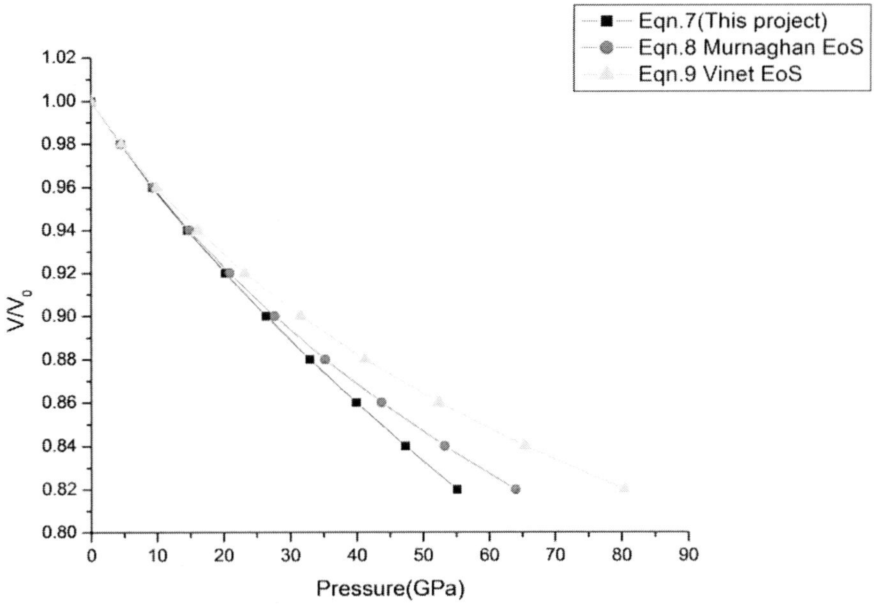

Figure 15. Volume expansion (V/V_0) as a function of pressure (GPa) for TiO_2 (rutile phase) (10 nm).

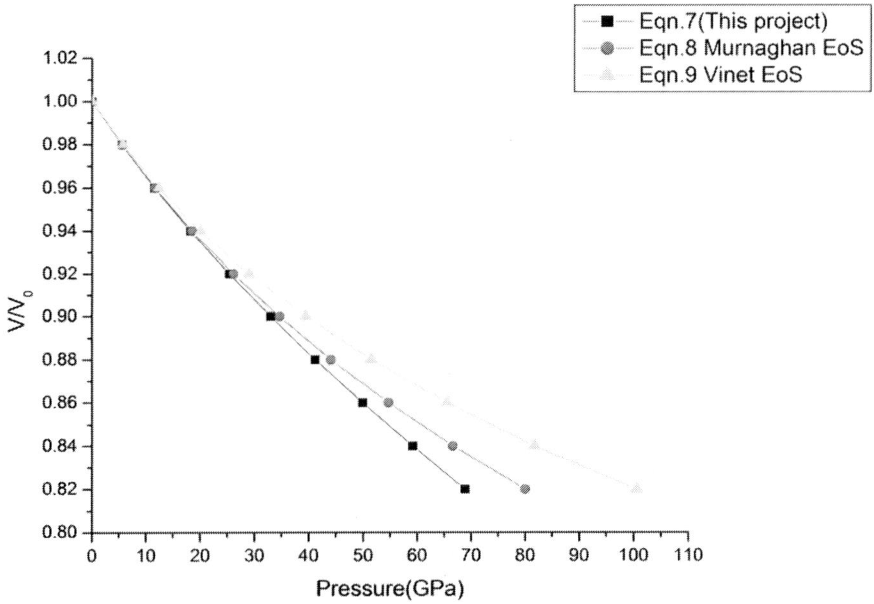

Figure 16. Volume expansion (V/V_0) as a function of pressure (GPa) for ZnO (Rocksalt Phase) (10 nm).

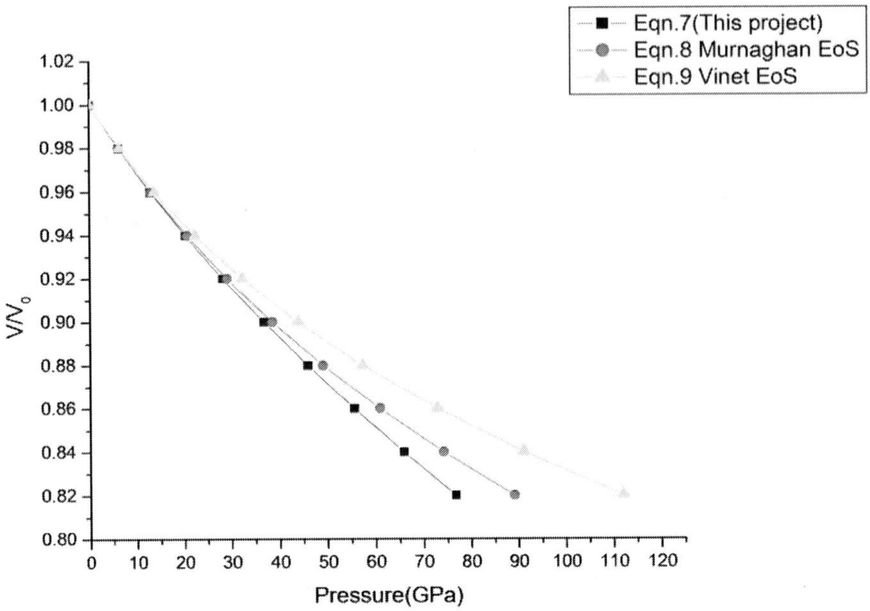

Figure 17. Volume expansion (V/V_0) as a function of pressure (GPa) for ZnO (Wurtzite Phase) (10 nm).

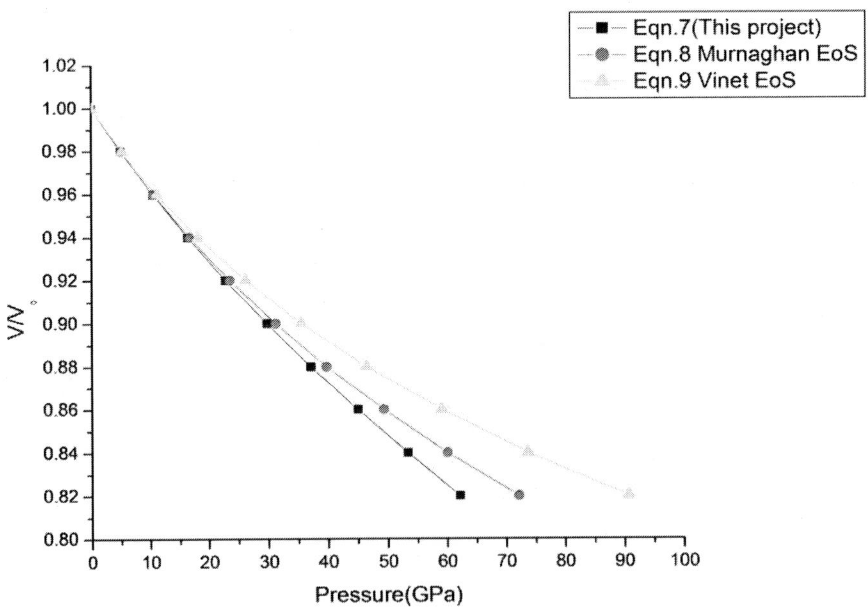

Figure 18. Volume expansion (V/V_0) as a function of pressure (GPa) for y-Al_2O_3 (67 nm).

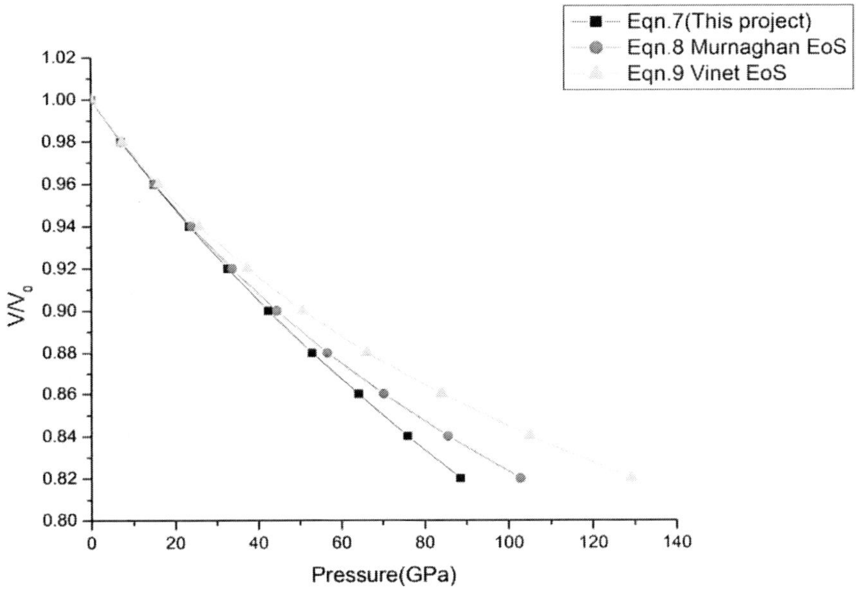

Figure 19. Volume expansion (V/V_0) as a function of pressure (GPa) for γ-Si_3N_4 (10 nm).

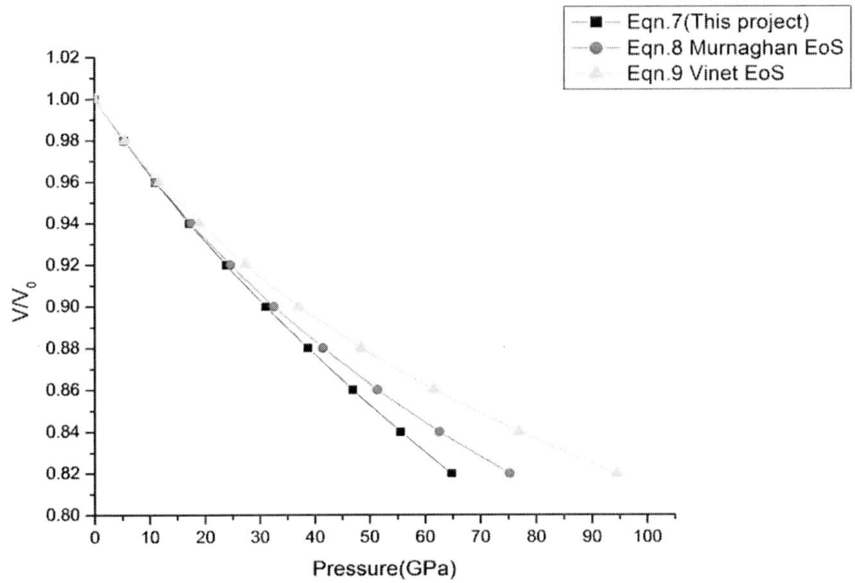

Figure 20. Volume expansion (V/V_0) as a function of pressure (GPa) for SnO_2 (8 nm).

4. CONCLUSION

The behavior of volume expansion as a function of pressure for some nanomaterials is studied. The results obtained are in good agreement with the available experimental data [14-18] as compared with the other EOSs such as Vinet's [20]. Very significant agreement has been observed in the low pressure values. Our theory agrees with the experiments throughout the pressure range investigated for a given nanomaterial, whereas there is some deviation, in general, towards low V/V_0 values, especially at high pressure. We have extended the formulation for a number of nanomaterials and found good agreement with the experimental data, which demonstrate the validity of the formulation developed. Thus, the present work is a simple method to study and analyze the effect of pressure on nanomaterials.

REFERENCES

1. H. Gleiter, "Nanostructured Materials, Basic Concept and Microstructure," Acta Materialia, Vol. 48, No. 1, 2000, pp. 6243-6248.
2. Q. Jiang, L. H. Liang and D. S. Zhao, "Lattice Contraction and Surface Stress of Fcc Nanocrystals," Journal of Physical Chemistry B, Vol. 105, No. 27, 2001, pp. 6275- 6277.
3. F. H. Streitz, K. Sieradzki and R. C. Cammarata, "Elastic Properties of Thin Fcc Films," Physical Review B, Vol. 41, No. 17, 1990, pp. 12285-12287.
4. L. H. Liang, J. C. Li and Q. Jiang, "Size-Dependent Elastic Modulus of Cu and Au Thin Films," Solid State Communications, Vol. 121, No. 8, 2002, pp. 453-455.
5. J. H. Zhang, Q. A. Huang, H. Yu and J. Wang, "The Influence of Surface Effects on Size-Dependent Mechanical Properties of Silicon Nanobeams at Finite Temperature," Journal of Physics D, Vol. 42, No. 4, 2009, Article ID: 045409.
6. M. Fujii, T. Nagareda, S. Hayashi and K. Yamamoto, "Low Frequency Raman Scattering from Small Silver Particles Embedded in SiO_2 Film," Physical Review B, Vol. 44, No. 12, 1991, pp. 6243-6248.
7. W. F. Zhang, Y. He, M. S. Zhang, Z. Yin and Q. Chen, "Raman Scattering Study on Anatase TiO_2 Nanocrystals," Journal of Physics D, Vol. 33, No. 8, 2000, pp. 912.
8. H. Liang and B. Li, "Size-Dependent Thermal Conductivity of Nanoscale Semiconducting Systems," Physical Review B, Vol. 73, No. 15, 2006, Article ID: 153303.

9. H. W. Kroto, J. R. Heath, S. C. O. Brien, R. F. Curl and R. E. Smalley, "C_{60}: Buckminsterfullerene," Nature, Vol. 318, No. 6042, 1985, pp. 162-163.

10. A. S. Miguel, "Nanomaterials under High Pressure," Chemical Society Reviews, Vol. 35, No. 10, 2006, pp. 876-889.

11. M. Singh, P. Singh, B. Gupta and M. Kumar, "Temperature and Pressure Dependence of Elastic Constants," High Temperature-High Pressures, Vol. 33, No. 2, 2001, pp. 199-206.

12. S. Karmakar, S. M. Sharma, P. V. Teredesai and A. K. Sood, "Pressure-Induced Phase Transitions in Iron-Filled Carbon Nanotubes: X-Ray Diffraction Studies," Physical Review B, Vol. 69, No. 16, 2004, Article ID: 165414.

13. B. Kiefer, S. R. Shieh and T. S. Duffy, "Strength, Elasticity, and Equation of State of the Nanocrystalline Cubic Silicon Nitride γ-Si_3N_4 to 68 GPa," Physical Review B, Vol. 72, No. 1, 2005, Article ID: 014102.

14. S. H. Tolbert and A. P. Alivisatos, "The Wurtzite to Rock Salt Structural Transformation in CdSe Nanocrystals under High Pressure," The Journal of Chemical Physics, Vol. 102, No. 11, 1995, pp. 4642-4656.

15. J. Z. Jiang, J. S. Olsen, L. Gerward and S. Mørup, "Compressibility of Nanostructured Fe-Cu Materials Prepared by Mechanical Milling," Nanostructured Materials, Vol. 12, No. 5-8, 1999, pp. 847-850.

16. B. Chen, D. Penwell and M. B. Kruger, "The Compressibility of Nanocrystalline Nickel," Solid State Communications, Vol. 115, No. 4, 2000, pp. 191-194.

17. J. Zhao, L. Guo, J. Liu, Y. Yang, et al., "High Bulk Modulus of Nanocrystal γ-Fe_2O_3 with Chemical Dodecyl Benzene Sulfonic Decoration under High Pressure," Chinese Physics Letters, Vol. 17, No. 2, 2000, p. 126.

18. B. Chen, D. Penwell, M. B. Kruger, A. F. Yue and B. Fultz, "Nanocrystalline Iron at High Pressure," Journal of Applied Physics, Vol. 89, No. 9, 2001, p. 4794.

19. O. L. Anderson, "Equation of State for Geophysics and Ceramic Sciences," Oxford University Press, Oxford, 1995.

20. P. Vinet, J. Ferrente, J. R. Smith and J. H. Rose, "A Universal Equation of State for Solids," Journal of Physics C: Solid State Physics, Vol. 19, No. 20, 1986, p. L467.

Chapter 7

Fabrication of Diamond Based Sensors for Use in Extreme Environments

Gopi K. Samudrala *, Samuel L. Moore and Yogesh K. Vohra

Department of Physics, University of Alabama at Birmingham, Birmingham, AL 35294, USA

ABSTRACT

Electrical and magnetic sensors can be lithographically fabricated on top of diamond substrates and encapsulated in a protective layer of chemical vapor deposited single crystalline diamond. This process when carried out on single crystal diamond anvils employed in high pressure research is termed as designer diamond anvil fabrication. These designer diamond anvils allow researchers to study electrical and magnetic properties of materials under extreme conditions without any possibility of damaging the sensing elements. We describe a novel method for the fabrication of designer diamond anvils with the use of maskless lithography and chemical vapor deposition in this paper. This method can be utilized to produce diamond based sensors which can function in extreme environments of high pressures, high and low temperatures, corrosive and high radiation conditions. We demonstrate applicability of these diamonds under extreme environments by performing electrical resistance measurements during superconducting transition in rare earth doped iron-based compounds under high pressures to 12 GPa and low temperatures to 10 K.

Keywords: diamond anvils; chemical vapor deposition; superconductor; diamond based sensors

1. INTRODUCTION

The static high pressure study on materials generally requires the use of single crystal diamonds in an opposed anvil configuration in diamond anvil cell devices. The high shear strength of diamond allows for the generation of ultra high pressure conditions at the tip of these anvils and transparency of diamond to a variety of electromagnetic radiations allows for a number of spectroscopic and diffraction measurements to be performed on materials under extreme conditions. In high pressure research, studying the electrical and magnetic properties of materials is accomplished by placing electrical probes in the sample region. These probes are in the region where shearing stress exceeds 10 GPa [1], and temperatures are very high or very low. Therefore, protecting the probes from these extreme conditions is very important to gain reliable and meaningful data. Designer diamond anvils facilitate this by encapsulating electrical probes under chemical vapor deposition grown diamond. It has been shown that designer diamond anvils can be used to study materials at extreme conditions such as megabar pressures, very low temperatures [2,3,4,5]. It has been shown previously that using a diamond anvil as a substrate, designer anvils can be produced by the use of lithography, laser pantogrpahy and CVD growth techniques [6]. These methods require the use of highly customized and expensive tools. Mask aligners need to be used for this method and physical masks tailored to each pattern to be drawn on diamond anvils need to be prepared and replaced after few uses. Such difficulties can be overcome by the use of maskless lithography. The use of maskless lithography allows us to eliminate the use of multiple systems to draw electrical circuits on diamond anvils. The maskless lithography not only allows us to fabricate designer diamond anvils but also makes it possible to fabricate other diamond based sensors such as thermocouples that can function in extreme environments.

As mentioned earlier, several important studies have been carried out in high pressure research utilizing designer diamond anvils. The use of designer diamond anvils in diamond anvil cell allows us to use a metallic gasket for sample containment and for precise four-probe electrical resistance measurements. This is particularly helpful in observing how superconductivity changes as a function of pressure in compounds such as 1-2-2 iron (Fe)-based materials AFe_2As_2 (122) [A = Ba, Sr, Ca, Eu]. High pressure superconductivity in a rare-earth-doped $Ca_{0.86}Pr_{0.14}Fe_2As_2$ single-crystalline sample has been studied up to 12 GPa and temperatures down to 11 K using the designer diamond anvil previously [7]. These superconducting compounds are of particular interest because under pressure, superconducting transition temperature (T_c) as high as ~51 K at 1.9 GPa has been observed, presenting the highest $T_c$reported in the intermetallic class of 1-2-2 iron-based superconductors. In this paper we report results from our study on a rare-earth doped iron-based superconductor $Ca_{0.9}Pr_{0.1}Fe_2As_2$ using a newly fabricated designer diamond anvil.

2. EXPERIMENTAL SECTION

Type-Ia and Type-IIa gem quality diamonds are utilized to fabricate the designer diamond anvils. High resolution and highly customizable circuit patterns have been imprinted onto single crystal diamond (SCD) substrate anvil surfaces prior to being entirely encapsulated in SCD utilizing the following instrumentation: DC-sputter deposition, maskless lithography and microwave plasma chemical vapor deposition (MPCVD). The DC-sputter deposition system is AJA International Inc.'s (Scituate, MA, USA) Orion sputtering system with a sputter down configuration. The maskless lithography system is SF-100 Xcel system from Intelligent Micro Patterning LLC (St. Petersburg, FL, USA). The maskless system allows us to draw patterns on samples with extreme topography such as diamond anvils which have angles between surfaces ranging from 7° to 45°. This is very important because we need to extend electrical probes down onto facets of diamond anvils as shown in Figure 1. The MPCVD system is a custom built system at University of Alabama at Birmingham. The anvil geometries utilized consist of original culet sizes ranging from 10 to 600 microns, with bevel angles (angle between the culet and facet) ranging from 7° to 50°. Two lithographic process techniques have been used to achieve the desired results and are outlined below:

Method 1: The top row (TR) of graphics in Figure 1 outline the steps of obtaining encapsulated metallic circuits on diamond anvils with the use of positive (or negative) tone resist and wet etching. In this process, the bare diamond anvil substrate (TR-A) has been etched in a gentle RF-Plasma in the sputter deposition chamber prior to sputter coating. During the sputtering process, an RF bias has been maintained to ensure we deposit good quality metal film on our substrates. The substrate then has a uniform tungsten metal layer sputter deposited onto its surface so that the entire substrate is coated with a metallic layer (TR-B). The thickness of the metallic coating is in the range of 0.1–2 microns. After the metal has been sputter coated, photoresist (both positive tone and negative tone resists have been utilized to achieve the final result) is applied to the diamond using a spin coater with angular velocity ranging from 1000 to 12000 rotations per minute (rpm). The angular velocity is varied according to the anvil geometry in order to apply a uniform resist coating (TR-C). The resist has been processed according to manufacturer's recommendation by baking it at suitable temperature.

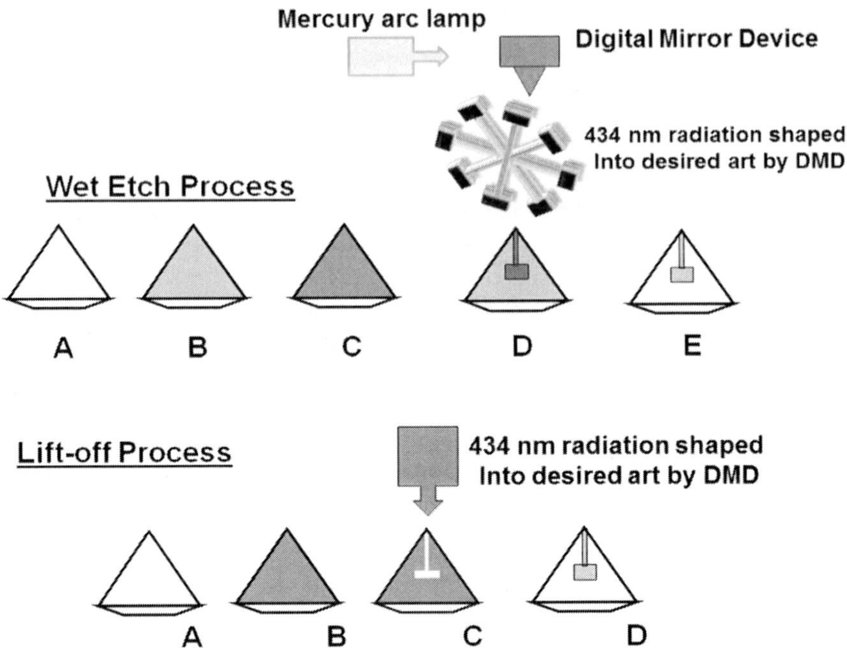

Figure 1. The top row shows the steps involved in fabricating a designer diamond through the wet etch process. The bottom row shows the steps involved in fabricating designer diamond anvils through the lift-off process.

The maskless photolithography instrument is then implemented to expose the substrate in very specific regions with visible light in the 360–450 nm range in a very high resolution pattern (micron scale resolution is achieved) via the digital micro mirror device (DMD) contained as an internal component of the maskless lithographic instrument (TR-D). Depending on the tone of the resist (positive resist becomes soluble when exposed to radiation, whereas negative resist becomes insoluble), the artwork loaded into the lithographic system software is designed specifically to meet the final circuit dimension specifications. The exposed resist is then placed in a developer solution and a pattern is rendered in the resist layer. The resist that remains is essentially a protective layer for the underlying tungsten film during the next phase of wet etching the substrate in a weakly acidic tungsten etchant. Once the etchant has completely dissolved the excess tungsten in the base layer, the substrate is removed from the solution and the photoresist layer is dissolved in a solvent. The result is a diamond anvil with a metallic pattern drawn on it (TR-E). The diamond anvil is then transferred into a 1.2kW MPCVD chamber and SCD is grown at temperatures ranging from 800 to 1300 °C with a CH_4/H_2 ratio of 1%–10%. This results in the growth of CVD diamond with a thickness of 10–70 microns. The anvil is then polished until electrical contacts on the culet surface are exposed and the diagnostic contact pads on the facets are all exposed, facilitating the connection of external laboratory equipment.

Method 2: The bottom row (BR) of Figure 1 outlines the liftoff process. Lift off process is essentially the reverse process of the wet etch procedure previously described. In the lift off process the same instrumentation, materials and process parameters outlined in the wet-etch method are used to perform the fabrication process. However, in this method resist coating and all lithographic process steps occur on the bare SCD substrate prior to sputter deposition (BR-A,B). As a result, once the lithographic process is completed the diamond surface regions in which metal will be deposited to render the final circuit pattern are exposed to incident tungsten atoms (BR-C). During sputter deposition tungsten particles coat these regions and adhere to the diamond substrate surface. After sputter deposition, the resist layer is stripped from the diamond anvil, and the excess tungsten deposited on top of the resist layer will be removed and only the final circuit pattern remains (BR-D). Resolution enhancement has been achieved by depositing an interstitial under layer of resist prior to applying an imaging resist layer. In this bi-layer method the interstitial layer (base layer of resist deposited directly onto the diamond) has a slightly higher dissolution rate than the imaging resist that is deposited on top of it, this results in an "overhang" effect during the development phase which leads to improved resolution as sputter deposited material is less likely to delaminate from the diamond surface due to attachment of tungsten to the resist coating during the resist stripping phase. Once sputter deposition and resist stripping is complete, the anvil undergoes the same steps as in the encapsulating MPCVD and polishing phase described in method 1.

3. RESULTS AND DISCUSSION

A diamond anvil with a central flat size of 70 microns in diameter, beveled at 7.5 degrees to a culet size of 350 microns in diameter has been chosen as base substrate for the fabrication of a designer diamond anvil. A tungsten film of 0.5 microns thick has been sputter deposited onto this diamond anvil. This diamond was then coated with Shipley 1827 positive photoresist. Utilizing an eight electrical probe design graphic file as an input for the maskless lithography system, we have removed photoresist from unwanted areas. The first step in the fabrication of this designer diamond was completed after developing the photoresist, and wet etching step to remove tungsten from unwanted areas. Figure 2a shows the resulting 8 probe pattern made of tungsten metal on the diamond anvil substrate. The width of metallic probes is 10 microns and their thickness is 0.5 microns. Figure 2b shows the CVD diamond grown on top the eight probe pattern to encapsulate it. The fabrication of the designer diamond anvil was completed by polishing the CVD diamond layer and exposing the probes in the sample region.Figure 2c shows a fully finished designer diamond anvil with eight electrical probes. The final culet size of this designer diamond anvil after polishing is 370 microns. The diameter of the circle of probes that have been exposed is 85 microns. These dimensions allow for researchers to include high volume of material in the high pressure research experiments. Designer diamond anvils with this geometry have been utilized in studying the electrical and magnetic properties of materials such as rare earth elements gadolinium, dysprosium [4,5].

Figure 2. (**a**) Eight electrical probes lithographed onto diamond anvil; (**b**) CVD diamond grown on top of eight probe pattern; (**c**) A fully fabricated designer diamond anvil with a culet size of 370 microns with a probe circle diameter of 85 microns.

The 1-2-2 Fe-based materials AFe_2As_2 (122) [A = Ba, Sr, Ca, Eu] have $ThCr_2Si_2$ type tetragonal crystal structure at ambient conditions and are among the widely studied class of Fe-based superconductors. These compounds exhibit their superconductivity upon application of high pressure or chemical doping once the magnetic ordering is suppressed [8].

The superconducting transition temperature (T_c) in the 1-2-2 family has been limited to low temperatures with the maximum being 38 K observed in hole-doped $Ba_{1-x}K_xFe_2As_2$ [9]. However, unusually high superconducting transition temperature characterized by a mysterious small volume fraction which superconducts up to 45–49 K, has been recently realized in single crystalline rare-earth (RE) doped $Ca_{1-x}Pr_xFe_2As_2$ [10,11,12]. It is even more surprising to note that the superconducting state for $Ca_{1-x}Pr_xFe_2As_2$ with an onset T_C of 49 K is non-bulk or filamentary, highly anisotropic, and displays a Josephson coupling characteristic and double resistive transition as evidenced from the resistive and magnetic results of single crystalline sample at ambient pressure [10,11]. Presently, limited high pressures studies have been reported [12] on these compounds and it is not well known whether high pressure and electron doping play a similar role in optimizing the superconducting properties of the RE doped $CaFe_2As_2$ material.

Figure 3a shows the iron-based superconducting sample in a diamond anvil cell sample chamber assembly. The sample chamber in a stainless steel gasket is 120 microns in diameter and has an initial thickness of 70 microns and contains $Ca_{0.9}Pr_{0.1}Fe_2As_2$ sample along with a ruby pressure marker surrounded by an electrically insulating pressure medium steatite. The pressure was continuously monitored utilizing the spectral location of Ruby R_1 fluorescence emission at high pressures and low temperatures [7]. Figure 3b shows the four probe electrical resistance as a function of temperature at various pressures. The onset temperature of superconductivity (T_c) is marked by a sharp downturn in electrical resistance and the insert in Figure 3b illustrates our methodology for the determination of T_c.Figure 3c shows a plot of T_c as a function of pressure and illustrates a gradual decrease of T_c with increasing pressure and can be fitted to the following quadratic equation.

$$T_c \text{ (in Kelvin)} = 45.2 - 0.272\, P - 0.202\, P^2 \text{ (P is in GPa units)} \quad (1)$$

The present study shows that the superconducting transition temperature (T_c) in Pr-doped $CaFe_2As_2$is a monotonically decreasing function of pressure for a 10% doping level and the highest value of T_c of 45 K is observed at ambient pressure. The rare-earth doped $CaFe_2As_2$ materials continue to show the highest superconducting transition temperatures amongst the family of 1-2-2 iron-based superconductors [7,11,12,13]. In addition, a two-step superconductivity behavior with two different transition temperatures observed for 14% doping level in Pr-doped $CaFe_2As_2$ [7] is not reproduced in the 10% sample. This points to the need for *simultaneous* measurements of

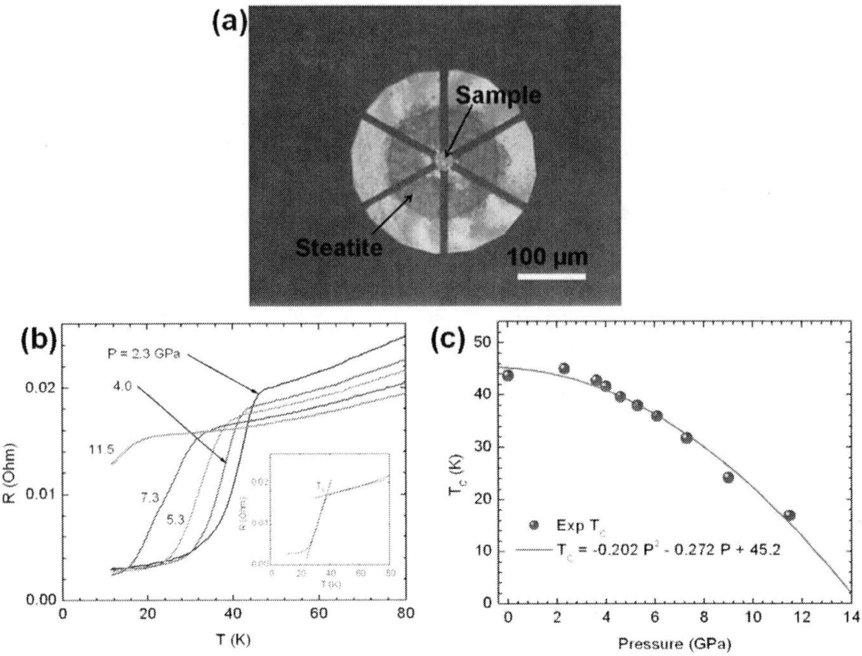

Figure 3. (**a**) $Ca_{0.9}Pr_{0.1}Fe_2As_2$ sample loaded in a diamond anvil cell along with steatite pressure medium, and ruby for pressure calibration; (**b**) Electrical resistance data collected with a designer diamond anvil. The insert shows our method of determining the onset of superconductivity (T_c); (**c**) T_c as a function of pressure for $Ca_{0.9}Pr_{0.1}Fe_2As_2$.

electrical properties (or T_c) and crystal structures on the same sample to correlate observed superconducting behavior and the crystalline phases at a given temperature and pressure. Such *simultaneous* measurements of superconductivity and crystal structures at high-pressure and low-temperatures are possible with the designer diamonds fabricated in this study.

We have described a novel method for fabrication of designer diamond anvils utilizing a combination of maskless lithography, sputtering and MPCVD techniques. However, the versatility of the maskless system allows us to fabricate a variety of other diamond based sensors. One sensor that is currently under development is a thermocouple that can function in any extreme environment. Thin-film thermocouples have been fabricated in earlier studies on substrates such as ceramics and superalloys [14]. These materials have their limitations as thermocouple metals exposed to extreme environments will undergo oxidations and will suffer mechanical damage compromising their integrity. Such hazards can be avoided by sputter deposition of thermocouples on diamond substrates and encapsulating them within a single crystalline CVD diamond layer. The high thermal conductivity; chemical, radiation inertness of diamond make it an ideal candidate as a base material for building thermocouples. By utilizing sputtering, maskless lithography; thermocouple alloys can be sputter deposited on diamond substrates. Patterns made of thermocouple alloys with features down to 5 microns in width and 0.5 microns thickness can be fabricated on diamond substrates utilizing the methods described in this paper. These alloys can be encapsulated under CVD grown diamond and can be exposed in strategic locations to make contact with laboratory equipment.

4. CONCLUSIONS

In summary, a versatile and easy to use method for fabricating designer diamond anvils using maskless lithography is developed. The methodology demonstrated in this paper not only allows for the fabrication of sensors for high pressure research, but has the potential to develop other diamond based sensors such as thermocouples which can function in extreme environments. As an application of this technology, superconducting behavior of a rare-earth doped iron-based compound has been studied to high pressures of 12 GPa and temperatures down to 10 K. The materials fabrication processes described in this paper can extend the use of diamond-based sensors in a wide range of extreme conditions of pressures, temperatures, radiation and chemical corrosive environments.

ACKNOWLEDGMENTS

This material is based upon the work supported by the National Science Foundation Partnership for Innovation (NSF: PFI) program under Grant number IIP-1317210, and by the Department of Energy-National Nuclear Security Administration under Award Number DE-NA0002014. We would like to thank Georgiy M. Tsoi for his assistance in high pressure experiments, and Athena Sefat, Oak Ridge National Laboratory for providing the sample of iron-based superconductor.

AUTHOR CONTRIBUTIONS

Yogesh K. Vohra and Gopi K. Samudrala discussed and developed the overall concept of combining maskless lithography with chemical vapor deposition technology for fabrication of diamond-based sensors. Samuel L. Moore and

Gopi K. Samudrala worked on optimization of process parameters for the maskless lithography process.

CONFLICTS OF INTEREST

Authors have intellectual property/patents in the area of research described in this paper.

REFERENCES

1. Hemley, R.J.; Mao, H-K.; Shen, G.; Badro, J.; Gillet, P.; Hanfland, M.; Hauserman, D. X-ray Imaging of Stress and Strain of Diamond, Iron, and Tungsten at Megabar Pressures. *Science* **1997**, *276*, 1242–1245.
2. Velisavljevic, N.; MacMinn, K.M.; Vohra, Y.K.; Weir, S.T. Electrical measurements on praseodymium metal to 179 GPa using designer diamond anvils. *Appl. Phys. Lett.* **2004**, *84*, 927–929.
3. Jackson, D.D.; Malba, V.; Weir, S.T.; Baker, P.A.; Vohra, Y.K. High-pressure magnetic susceptibility experiments on the heavy lanthanides Gd, Tb, Dy, Ho, Er, and Tm. *Phys. Review B* **2005**, *71*.
4. Samudrala, G.K.; Tsoi, G.K.; Weir, S.T.; Vohra, Y.K. Structural and magnetic phase transitions in gadolinium under high pressures and low temperatures. *High Press. Res.* **2014**, *34*, 385–391.
5. Samudrala, G.K.; Tsoi, G.K.; Weir, S.T.; Vohra, Y.K. Magnetic ordering temperatures in rare earth metal dysprosium under ultrahigh pressures. *High Press. Res.* **2014**, *34*, 266–272.
6. Weir, S.T.; Akella, J.; Ruddle, C.A.; Vohra, Y.K.; Catledge, S.A. Epitaxial diamond encapsulation of metal microprobes for high pressure experiments. *Appl. Phys. Lett.* **2000**, *77*, 3400–3402.
7. Uhoya, W.O.; Cargill, D.; Gofryk, K.; Tsoi, G.K.; Vohra, Y.K.; Sefat, A.S.; Weir, S.T. High pressure effects on the superconductivity in rare-earth-doped $CaFe_2As_2$. *High Press. Res.* **2014**, *34*, 49–58.
8. Johnston, D.C. The Puzzle of High Temperature Superconductivity in Layered Iron Pnictides and Chalcogenides. *Adv. Phys.* **2010**, *59*, 803–1061.
9. Rotter, M.; Tegel, M.; and Johrendt, D. Superconductivity at 38 K in the Iron Arsenide $(Ba_{1-x}K_x)Fe_2As_2$. *Phys. Rev. Lett.* **2008**, *101*.
10. Chu, C.W. A Possible Approach from BCS through HTS to RTS with Three Examples. *Phys. C***2012**, *482*, 33–44.

11. Lv, B.; Deng, L.; Gooch, M.; Wei, F.; Sun, Y.; Meen, J.K.; Xue, Y-Y.; Lorenz, B.; Chu, C.W. Unusual Superconducting State at 49 K in Electron-Doped $CaFe_2As_2$ at Ambient Pressure. *Proc. Nat. Acad. Sci.* **2011**, *108*, 15705–15709.

12. Saha, S.R.; Butch, N.P.; Drye, T.; Magill, J.; Ziemak, S.; Kirshenbaum, K.; Zavalij, P.Y.; Lynn, J.W.; Paglione, J. Structural Collapse and Superconductivity in Rare-Earth-Doped $CaFe_2As_2$. *Phys. Rev. B* **2012**, *85*.

13. Qi, Y.; Gao, Z.; Wang, L.; Wang, D.; Zhang, X.; Yao, C.; Wang, C.; Wang, C.; Ma, Y. Transport properties and anisotropy in rare-earth doped $CaFe_2As_2$ single crystals with Tc above 40 K.*Supercond. Sci. Technol.* **2012**, *25*.

14. Martin, L.C.; Holanda, R. Applications of Thin Film Thermocouples for Surface Temperature Measurement. Available online: http://www.grc.nasa.gov/WWW/sensors/PhySen/docs/TM-106714.pdf (accessed on 10 January 2015).

Chapter 8

Nonmetallization and band inversion in beryllium dicarbide at high pressure

Henan Du, Wanxiang Feng, Fei Li, Dashuai Wang, Dan Zhou & Yanhui Liu

Department of Physics, College of Science, Yanbian University, Yanji 133002, China

ABSTRACT

Carbides have attracted much attention owing to their interesting physical and chemical properties. Here, we systematically investigated global energetically stable structures of BeC_2 in the pressure range of 0–100 GPa using a first-principles structural search. A transition from the ambient-pressure α-phase to the high-pressure β-phase was theoretically predicted. Chemical bonding analysis revealed that the predicted phase transition is associated with the transformation from sp^2 to sp^3 C-C hybridization. The electrical conductivity of the high-pressure phase changed from a metal (α-phase) to a narrow bandgap semiconductor (β-phase), and the β-phase had an inverted band structure with positive pressure dependence. Interestingly, the β-phase was a topological insulator with the metallic surface states protected by the time-reversal symmetry of the crystal. The results indicate that pressure modulates the electronic band structure of BeC_2, which is an important finding for fundamental physics and for a wide range of potential applications in electronic devices.

INTRODUCTION

Carbides draw much attention because of their unusual physical and chemical properties1'2'3. The transition between various carbides has long been a topic of interest because these phases have very different electronic properties, owing to the change from sp^2 to sp^3 carbon hybridization during the transition from graphite to diamond under sufficient pressure. Alkaline earth metal carbides have been the focus of research for more than half a century, and have been the subject of extensive theoretical and experimental studies of advanced functional materials4′5′6. Many alkaline earth metal dicarbides possess a linear crystal structure at normal condition. Previous experimental results show that magnesium dicarbide (MgC_2) crystallizes in a tetragonal lattice with the space group $P4_2/mnm$7′8. Experimental and theoretical studies reported that the structure of CaC_2 belongs to the *I4/mmm*space group. Recently, the structural evolution of CaC_2 was determined using first principles calculations9. Especially, the high-pressure phase of CaC_2 (*Immm*) of the intercalated graphite structure has a high superconducting critical temperature (Tc) of 7.9–9.8 K, which is comparable to Tc CaC_6 (11.5 K)10′11′12′13′14′15. However, there are few studies exploring the crystal structure of BeC_2, and its high-pressure structures are poorly understood, which will impede further exploration of this important class of material. A pressing task is to understand the crystal and electronic structures of this special matter under external pressure condition that may alter the underlying fundamental physics, which has motivated us to carry out the work.

In this study, we therefore explored the hitherto unknown high-pressure phase diagram of beryllium dicarbide via a first-principles swarm-intelligence structure search and established a comprehensive understanding of the high-pressure evolution of BeC_2. Our work shows that BeC_2 crystallizes in the monoclinic *C2/m* (α-phase) structure at ambient pressure and undergoes isostructural transitions to the β-phase at high pressure. We identified two thermodynamically stable phases associated with the transition from sp^2 to sp^3 C-C hybridization and predicted that the β-phase is a topological insulator. Our result represents a significant step toward understanding the structural and electrical behavior of alkaline earth metal carbides under extreme condition.

RESULT AND DISCUSSION

Crystal structures of BeC_2

We performed structure prediction for BeC_2 with variable-cell simulation cell sizes of 1–4 formula units at pressure of 0–100 GPa. At ambient pressure, our simulation reveals that the *C2/m* phase denotes as the *α*-structure/phase has lower enthalpies than all the other candidates, indicating that the *α*-structure is in the thermodynamic ground state. Over the whole pressure range, a new low-enthalpy structure denotes as the β-structure/phase is explored which has the same space group as the *α*-structure. The enthalpy-pressure relationship for the candidate structures is shown in Fig. 1. It has been seen that, for compressed BeC_2, the ambient-pressure α-phase is the most stable structure below 56.7 GPa, but the β-phase becomes more favorable from 56.7 to 100 GPa. The pressure evolution of the unite cell volume of BeC_2 in the α- and β-structure is depicted in inset of Fig. 1. The volume collapses about 9.4% around 56.7 GPa, indicating the first-order nature of the phase transition in BeC_2. No imaginary frequency is observed throughout the whole Brillouin zone, which indicates the two novel phases are dynamically stable in the pressure region from this study.

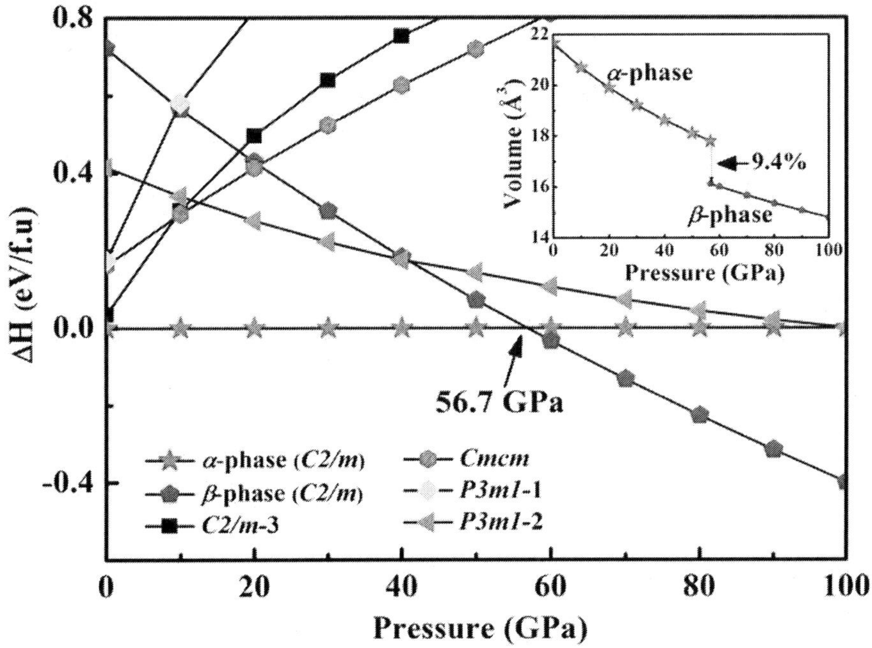

Figure 1: Enthalpies per formula unit of various structures as a function of pressure with respect to the ambient pressure of the α-phase. Inset: Calculated equations of states for the predicted structures.

The atomic arrangements of the competing structures are shown in Fig. 2. The calculated lattice constants of the ambient-pressure are calculated. The α-phase has a monoclinic symmetry, with a = 6.223 Å, b = 2.535 Å and c = 5.904 Å, β = 111.5° at ambient pressure. The Be and C atoms occupy Wyckoff 4i position in the unit cell with Be at (0.672, 0.5, 0.11), C at (0.431, 0.5, 0.227) and (0.467, 1.000, 0.374). The α-phase is similar to that the recently predicted for MgC_2 of *C2/m* phase with carbon atoms polymerized into ribbon with a six-membered ring. Planes of Be atoms separate the hexagonal honeycomb layers of carbon atoms, which have the nearest distance with the outer C atom of the ribbon.

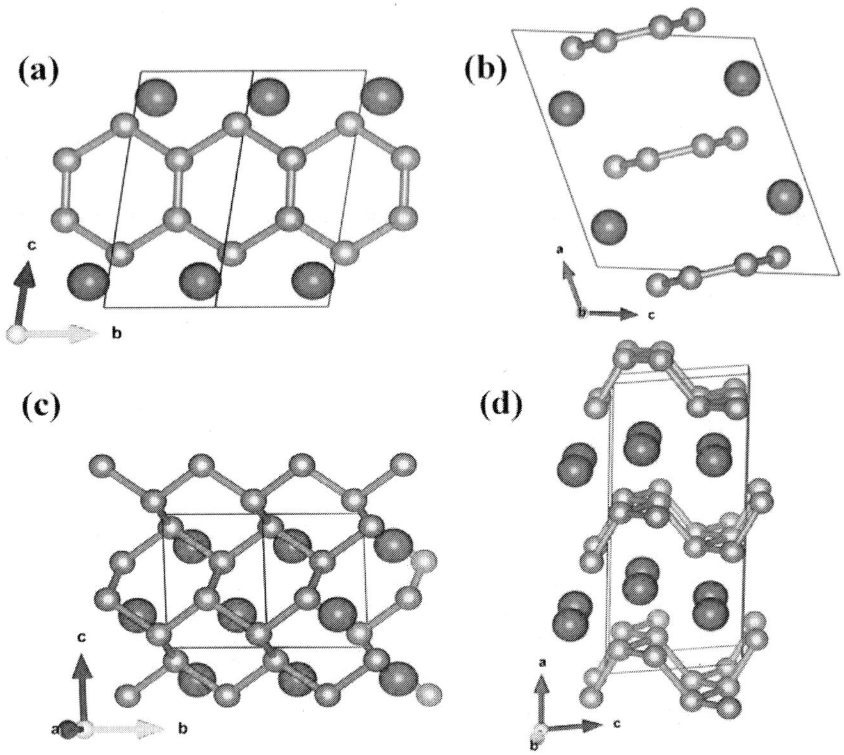

Figure 2: Structures of predicted stable BeC_2 crystal. (**a**,**b**) α-phase and (**c**,**d**) β-phase. The large (red) and small (blue) spheres are Be and C atoms, respectively.

The β-phase has a monoclinic symmetry, with lattice parameters of $a = 7.879$ Å , $b = 2.404$ Å, $c = 3.415$ Å, and $\beta = 84.12°$ at 56.7 GPa. The Be and C atoms occupy Wyckoff 4*i* positions in the unit cell with Be at (0.212, 0.5, 0.758), C at (0.420, 0.5, 0.9) and (0.418, 1.000, 0.637). It is obvious that the β-phase consists of bulky honeycomb-layer of carbon atoms separated by planes of Be atoms. Figure 3 shows the calculated lattice parameter versus pressure curves for the α- and β-phase. In α-phase, the slope of *a*is much more steeper than that of *b* and *c*. The results show that the inter-atomic forces among the interlayer molecules are weaker than the forces among in the intralayer molecules. For the β-phase, the slopes in different axes are nearly identical. In the inset of Fig. 3, the change of β-phase induced by the pressure is attributable to the elongating of the interlayer (*a* axis) and shortening of the intralayer (*c* axis) in α-phase, which also caused the well-organized six-ring carbon rings collapse into the puckered quasi 2-dimension graphite sheet. Due to the occurrence of graphite sheet between neighboring Be atomic layers, the β-phase can be regarded as one of the puckered graphite intercalation compounds.

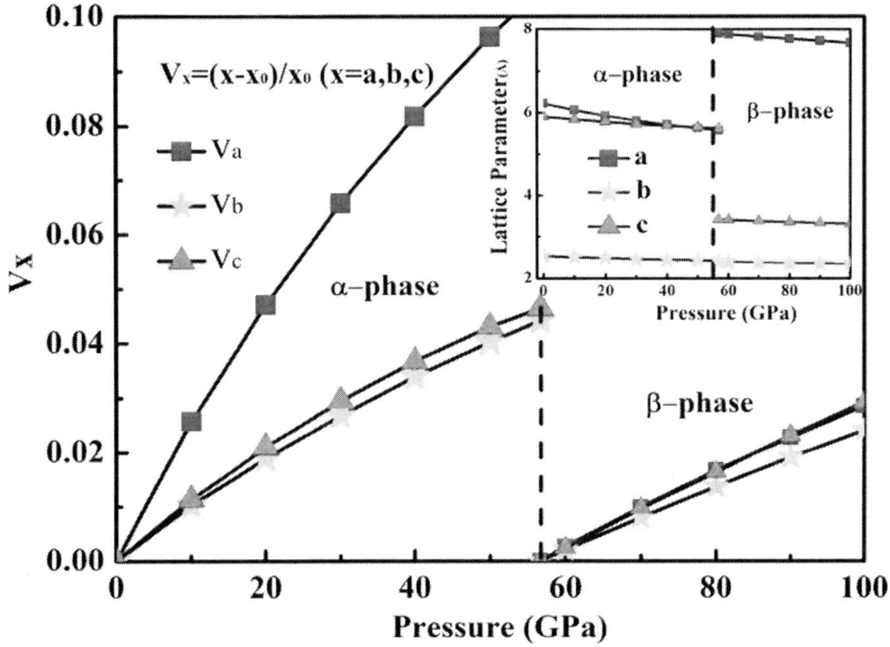

Figure 3: Calculated lattice parameter change on the pressure in the α- and β- phase. (X_0 indicate the lattice parameters for the α- and β- phase at 0 GPa and 56.7 GPa, respectively). Inset: The Lattice parameter in the α- and β-phase as a function of pressure.

Nonmetallization of BeC_2

The calculated band structure and projected density of states (DOS) are shown in Fig. 4. The predicted overlap between the conduction and the valence bands shows the α-phase with sp^2 hybridization is metallic, along with Be- and *C*- *p* electrons near the Fermi level. At 56.7 GPa, the β-phase characterized by sp^3 hybridization becomes an insulator with a direct band gap of 0.03 eV (without spin-orbit coupling). Generally, metals under pressure exhibit decreasing interatomic distance and increasing metallic behavior, whereas BeC_2 undergoes an unexpected and often counterintuitive pressure-induced metal-to-insulator transition. This surprising behavior can also be seen in Sodium which shows an optically transparent phase at 200 GPa16.

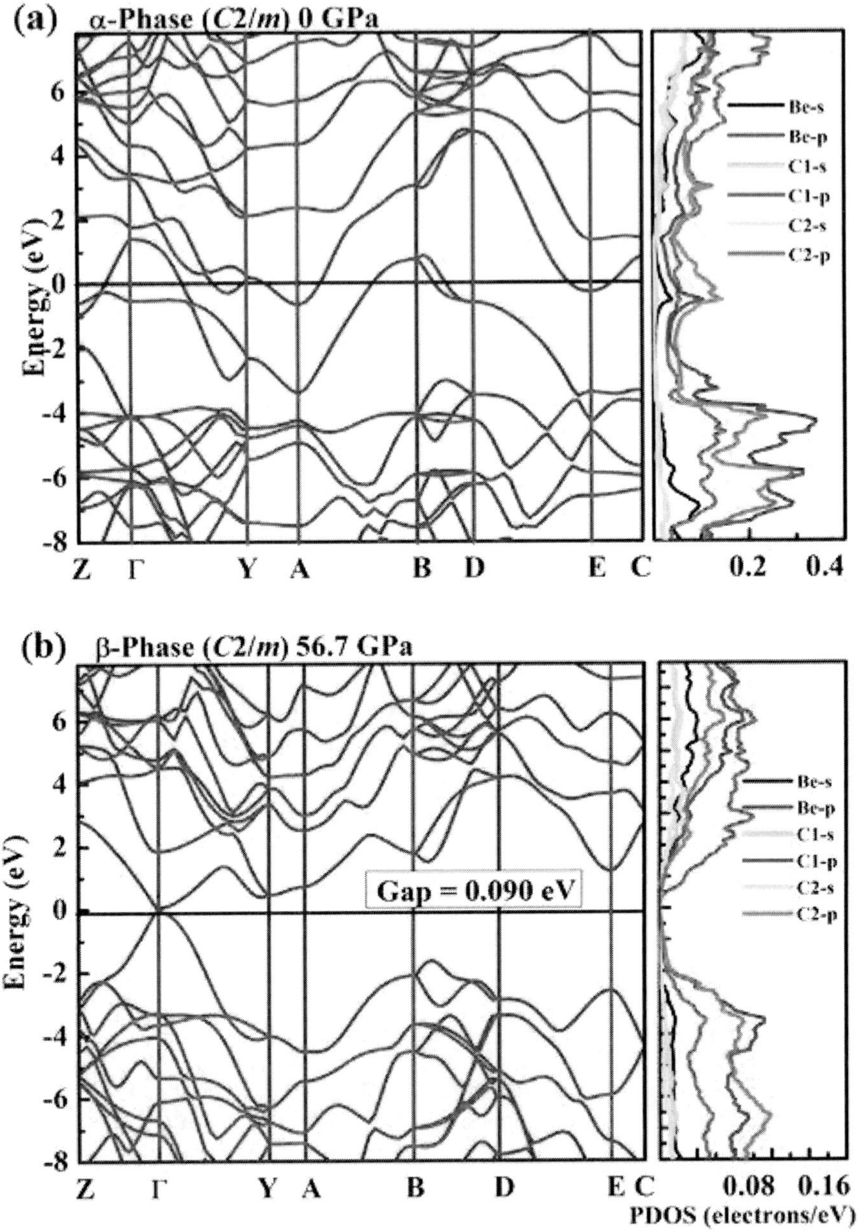

Figure 4: Electronic band structure and projected DOS of BeC_2. (**a**) α-phase and (**b**) β-phase. Zero of energy is at the Fermi level.

To examine the nature of these bonding states, we plotted the electron localization function (ELF) isosurfaces of BeC_2 (Fig. 5)17. The α-phase contains a ribbon of six-member carbon rings, which have sp^2-like properties. Each inner C atom forms three C-C covalent bonds and each outer C atom has one electron lone pair along with two C-C covalent bonds. The remaining $2p_z$ electrons of the C_6 ring form a delocalized πsystem. In the β-phase, the puckered honeycomb-layered structure lacks a system of delocalized π bonds with mobile electrons and has more localized electrons resulting in sp^3 hybridization. Each C atom has one electron lone pair and three covalent C-C bonds, resembling diamond structure, accompanied by the appearance of an insulator state. The significant evolution of electronic structures from the sp^2 tosp^3 hybridization of the carbon atoms between the α- and β-phases is similar to the transformation of graphite to diamond under sufficient compression.

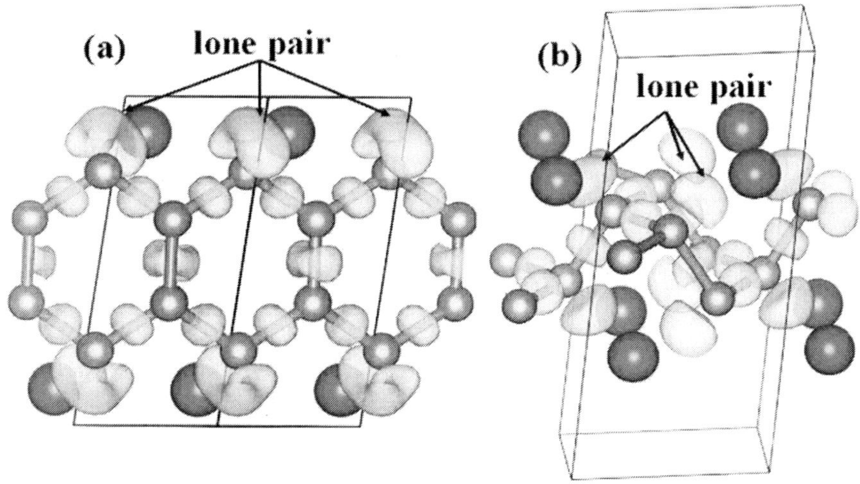

Figure 5: ELF isosurface (value = 0.8) illustrated on a plane containing C_6 rings in BeC_2. (**a**) α-phase and (**b**) β-phase.

Band inversion and topological insulator state in β-phase

Remarkably, the size of the band gap of the β-phase depends positively on pressure in Fig. 6. This peculiar behavior stems from the requirement for symmetry thus forbidding the crossing inverted conduction minimum and valence maximum at Γ point; consequently, the conduction minimum and valence maximum will repel each other more strongly under higher pressure. It is important to check whether the β-phase is a topological insulator at high pressure. Due to the presence of inversion symmetry, the parity criterion can be used to justify the topological properties of the β-phase18′19′20. Following the method addressed in ref. 18, we calculated the parities δ_1 at eight time-reversal invariant momenta (TRIMS) (i = 1–8) in three-dimensional Brillouin zone in Table 1. At 40 GPa, only the Γ point has negative parity, whereas the other TRIMS have positive parity. At 50 and 60 GPa, all TRIMS have positive parity. This can also be observed from the band structures of the β-phase around the critical pressure of the phase transition, as shown in Fig. 6. One can clearly observe that the parities of band edges are exchanged when above 40 GPa. The Z_2 topological invariants can then be evaluated from the parities using the following equations,

$$(-1)^{\nu_0} = \prod_{i=1}^{8} \delta_i \qquad (1)$$

$$(-1)^{\nu_k} = \prod_{n_k=1, n_{j\neq k}=0,1} \delta_{i=(n_1 n_2 n_3)} \qquad (2)$$

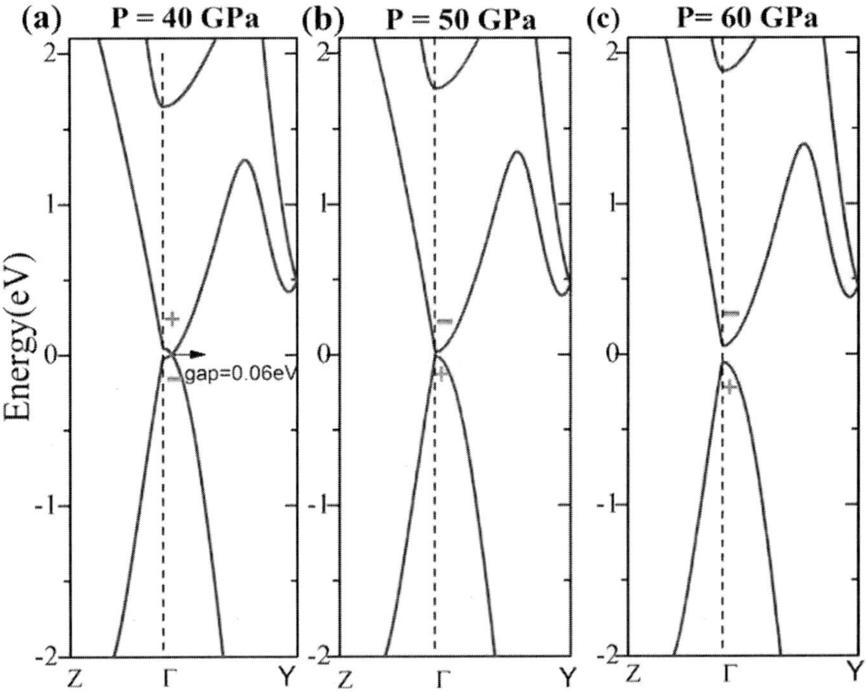

Figure 6: Electronic band structures for the β-phase under different pressures. The topological phase in (**a**) is a topological insulator, those in (**b**,**c**) are normal insulators. Zero of energy is at the Fermi level.

Table 1: Parities at eight time-reversal invariant momenta and Z_2 topological invariants for tne β-phase around the critical pressure of the phase transition.

Pressure (GPa)	Parities								Z_2 index
	(0, 0, 0)	(0, π, 0)	(π, 0, 0)	(π, π, 0)	(0, 0, π)	(0, π, π)	(π, 0, π)	(π, π, π)	
40	−	+	+	+	+	+	+	+	1; (111)
50	+	+	+	+	+	+	+	+	0; (000)
60	+	+	+	+	+	+	+	+	0; (000)

The Z_2 numbers are 1; (111) for 40 GPa and 0; (000) for 50 and 60 GPa. Together with the evolution of band energy as a function of pressure (Fig. 6), we can confirm that the β-phase is a strong topological insulator when below 40 GPa, whereas it is a normal insulator when above 50 GPa.

CONCLUSION

In summary, we have determined the high-pressure structural evolution of BeC_2 by using the particle swarm optimization technique with first-principles electronic structure calculation. We predicted a pressure-induced phase transition from the α- to β-phase at 56.7 GPa. The structural, dynamical, and electronic properties of BeC_2 under pressure were systemically investigated. The electrical calculation shows that the β-phase has an inverted band structure with a positive pressure dependence. Interestingly, the β-phase is a topological insulator with metallic surface states protected by the time-reversal symmetry of the crystal. These results show that pressure has a strong effect on the fundamental crystal and electronic structure of BeC_2, and that pressure tuning of the electronic properties offers an effective tool to modulate a wide range of physical properties for its potential applications.

METHOD

We performed structure predictions through a global minimization of free energy surfaces merging *ab initio* total-energy calculations implemented in CALYPSO (crystal structure analysis by particle swarm optimization) code21'22'23'24'25. The method has successfully predicted the high-pressure structures of various systems, ranging from elements to binary and ternary compounds26'27. The first-principles energetic calculations were carried out using density functional theory (DFT) with the Perdew-Burke-Ernzerhof of exchange-correlation as implemented in the Vienna *ab initio* simulation package (VASP)28. The projector augmented wave method has been adopted, with $2s^2$ and $2s^2\ 2p^2$ treated as valence electrons for Be and C atoms, respectively29'30. For Brillouin zone integration, we used the Monkhorst-Pack scheme and checked convergence of ground state calculations with uniformly increasing *k*-point meshes for each structure31. We used cutoff energy of 800 eV for the expansion of the wave function into the plane-wave basis-set. Monkhorst-Pack *k*-point meshes with a grid of 0.03 $Å^{-1}$ for Brillouin zone sampling were chosen to achieve the total energy convergence of less than 1 meV/atom. The phonon calculations were carried out by using a supercell approach as implemented in the PHONOPY code32. This method uses the Hellman-Feynman forces calculated from the optimized supercell through VASP. Ultrasoft pseudopotentials for C and Be were used, and convergence tests concluded that suitable calculation parameters are 80 Ry for kinetic energy cutoff, 20 × 20 × 12*k*-point sampling mesh, and 5 × 5 × 3 *q*-mesh in Brillouin zone.

ADDITIONAL INFORMATION

How to cite this article: Du, H. *et al*. Nonmetallization and band inversion in beryllium dicarbide at high pressure. *Sci. Rep*. **6**, 26398; doi: 10.1038/srep26398 (2016).

ACKNOWLEDGEMENTS

This work was supported by the Natural Science Foundation of China under No. 11474125, 11274136, 51202084, 11104104, 11164031, 11404035,11374033 and the 2012 Changjiang Scholars Program of China, Changjiang Scholarand Innovative Research Team in University (IRT1132). The Specialized Research Fund for the Doctoral Program of Higher Education of China (Grant No. 20131101120052), the Basic Research Fund of Beijing Institute of Technology (Grant No. 20141842004).

CONTRIBUTIONS

The idea was conceived by Y.H.L. The simulation was performed by Y.H.L. and H.N.D. and the data analyses were performed by Y.H.L., W.X.F., F.L., D.S.W. and D.Z. This manuscript was written by Y.H.L., H.N.D. and D.Z. All authors discussed the results and contributed to the paper.

COMPETING INTERESTS

The authors declare no competing financial interests.

REFERENCES

1. Hoffmann, R., Janiak, C. & Kollmar, C. A chemical approach to the orbitals of organic polymers. *Macromolecules* **24**, 3725–3746 (1991).
2. Zhong, X. *et al.* Pressure stabilization of long-missing bare C_6hexagonal rings in binary sesquicarbides. *Chem. Sci.* **5**, 3936–3940 (2014).
3. Zhang, M. *et al.* Superhard BC_3 in Cubic Diamond Structure. *Phys. Rev. Lett.* **114**, 015502–015507 (2015).
4. Liu, H. Y., Gao, G. Y., Li, Y. W., Hao, J. & S. Tse, John. Crystal Structures and Chemical Bonding of Magnesium Carbide at High Pressure. *J. Phys. Chem. C* **119**, 23168–23174 (2015).
5. Wang, D. S. *et al.* Evolution of crystal and electronic structures of manesium dicarbide at high pressure. *Sci. Rep.* **5**, 17815–17823 (2015).
6. Wang, D. H. *et al.* BaC: a thermodynamically stable layered superconductor. *Phys.Chem. Chem.Phys.* **16**, 20780–20784 (2014).
7. Fjellvaag, H. & Karen, P. Crystal structure of magnesium sesquicarbide. *Inor. Chem.* **31**, 3260–3263 (1992).
8. Karen, P., Kjekshus, A., Huang, Q. & Karen, V. L. The crystal structure of magnesium dicarbide. *J. Alloy. Comp.* **282**, 72–75 (1999).
9. Xiao, B., Feng, J., Chen, J. & Yu, L. Crystal structures and electronic properties of MC_2 (M = Mg, Ca, Sr, Ba) by comparative studies based on *ab-initio* calculations. *Chem. Phys. Lett.* **448**, 35–40 (2007).
10. Rahnejat, K. C. *et al.* Charge density waves in the graphene sheets of the superconductor CaC_6. *Nat. Commun.* **2**, 558–564 (2011).
11. Weller, T. E., Ellerby, M., Saxena, S. S., Smith, R. P. & Skipper, N. T.Superconductivity in the intercalated graphite compounds C_6Yb and C_6Ca. *Nat. Phys.* **1**, 39–41 (2005).

12. Emery, N. *et al.* Superconductivity of Bulk CaC_6. *Phys. Rev. Lett.***95**, 087003–087007 (2005).
13. Tütüncü, H. M. &Srivastava, G. P. A comparative *ab initio* study of superconductivity in the body centered tetragonal YC_2 and LaC_2.*J. Appl. Phys.* **117**, 153902–153910 (2015).
14. Li, Y. L. *et al.* Pressure-induced Superconductivity in CaC_2. *Proc. Natl. Acad. Sci. USA* **110**, 9289–9294 (2013).
15. Li, Y. L. *et al.* Formation of nanofoam carbon andre-emergence of superconductivity incompressed CaC_6. *Sci. Rep.* **3**, 3331–3338 (2013).
16. Savin, P. D. A., Nesper, P. D. R., Wengert, D. S. & Fässler, D. T. F. ELF: The Electron Localization Function. *Angew. Chem. Int. Ed. Engl.* **6**, 1808–1832 (1997).
17. Ma, Y. M. *et al.* Transparent dense sodium. *Nature* **458**, 182 (2009).
18. Feng. W. X. *et al.* First-principles calculation of Z_2 topological invariants within the FP-LAPW formalism. *Comput. Phys. Commun.* **183**, 1849–1859 (2012).
19. MaHsieh, T. H. *et al.* Topological crystalline insulators in the SnTe materials class. Transparent dense sodium. *Nature commun*. **3**, 982–987 (2012).
20. Zhou, D. *et al.* Pressure-Driven Enhancement of Topological Insulating State in Tin Telluride. *J. Phys. Chem. C* **117**, 8437 (2013).
21. Wang, Y. C., Lv, J., Zhu, L. & Ma, Y. M. Crystal structure prediction viaparticle-swarm optimization. *Phys. Rev. B* **82**, 094116–094124 (2010).
22. Wang, Y. C., Lv, J., Zhu, L. & Ma, Y. M. CALYPSO: A method for crystal structure prediction. *Comput. Phys. Commun.* **183**, 2063–2070 (2012).
23. Lv, J., Wang, Y. C., Zhu, L. & Ma, Y. M. Predicted novel high-pressure phases of lithium. *Phys. Rev. Lett.* **106**, 015503–015507 (2011).
24. Zhu, L. *et al.* Substitutional alloy of Bi and Te at high pressure.*Phys. Rev. Lett*, **106**, 145501–145505 (2011).
25. Wang, Y. C. *et al.* High pressure partially ionic phase of water ice.*Nat. Commun.* **2**, 563–568 (2011).
26. Li, Q. *et al.* Global structural optimization of tungsten borides.*Phys. Rev. Lett.* **110**, 136403–136408 (2013).
27. Peng, F. *et al.* Crystalline LiN_5 Predicted from First-Principles as a Possible High-Energy Material. *J. Phys. Chem. Lett.* **6**, 2363–2367 (2015).
28. Blöchl, P. E. Projector augmented-wave method. *Phy. Rev. B* **50**, 17953–17979 (1994).
29. Kresse, G. & Joubert, D. From ultrasoft pseudopotentials to the projector augmented-wave method. *Phys. Rev. B* **59**, 1758–1775 (1999).

30. Kresse, G. & Furthmüller, J. Efficient iterative schemes for *ab initio* total-energy calculations using a plane-wave basis set.*Phys. Rev. B* **54**, 11169–11186 (1996).
31. Monkhorst, H. J. & Pack, J. D. Special points for Brillouin-zone integrations. *Phys. Rev. B* **13**, 5188–5192 (1976).
32. Baroni, S., Giannozzi, P. & Testa, A. Green's-function approach to linear response in solid. *Phys. Rev. Lett.* **58**, 1861–1864 (1987).

Chapter 9

Densification of Lithium Disilicate under High Pressure Investigated by XPS

Silvio Buchner[1], Cláudio Radtke[2], Naira Maria Balzaretti[1]

[1]*Institute of Physics, Universidade Federal do Rio Grande do Sul, Porto Alegre, Brazil*

[2]*Institute of Chemistry, Universidade Federal do Rio Grande do Sul, Porto Alegre, Brazil*

ABSTRACT

The aim of this work was to investigate by X-ray photoelectron spectroscopy the effect of high pressure on the chemical environments of Si 2p, O 1s and Li 1s in lithium disilicate glass ceramic with stoichiometric composition $Li_2O{\cdot}2SiO_2$ (LS_2). A group of samples was processed at 2.5 GPa, 4 GPa and 7.7 GPa at room temperature and a second group was crystallized under high pressure and high temperature. Large shifts of the binding energy toward higher energies were observed in the X-ray photoelectron spectroscopy spectra for samples of the first group after densification at 2.5 and 4 GPa. For samples processed at 7.7 Gpa, the major component of the binding energy for the Si 2p environment remained practically unchanged compared to the pristine sample but new components, with smaller intensities, appeared in the spectra, indicating the existence of distinct Q-species induced by high pressure. This behavior may be related to changes in the number of bridged and non-bridged oxygen atoms in the glass structure. The results for the second group of samples, crystallized under high pressure, showed evidences of three binding energies for the O atoms, one of them related to non-bridged and two of them to bridged O atoms.

Keywords: XPS; Glass Ceramic; Lithium Disilicate; High Pressure; Binding Energy

1. INTRODUCTION

Structural analysis of glasses under high pressure is important to understand the effect of densification mechanisms on the glass properties [1]. Bridgman et al. [2] pointed out that the permanent densification after applying high pressure and high temperature (HPHT) is a phenomenon observed only for glassy materials due to their structural freedom.

Kitamura et al. [1] investigated the densification of lithium disilicate $Li_2O \cdot 2SiO_2$ (LS_2) glasses at 6 GPa and 400°C. Based on Raman spectroscopy and radial distribution function, they attributed the densification to an increase of the packing density of SiO_4 tetrahedra due to a decrease of the Si-O-Si bond angle between the tetrahedra. They showed the Si-O-Si bond angle decreased from 109.8° to 108.8° and the Si-O length increased by 0.001 nm.

Schmelzer et al. [3] discussed the pressure dependence of viscosity and free volume of liquids in frozen systems in non-equilibrium states. According to them, the viscosity of simple liquids typically increased with pressure.

The crystallization of LS_2 under HPHT has been studied in the past few years [1,4-11]. In previous works, the effect of densification on the crystallization process and thermal properties of LS_2 glass was investigated [4], as well as the effect of pressure and temperature on the mechanical [5] and optical [6] properties. It was observed that the hardness and elastic modulus of the sample processed at HPHT increased noticeably compared to the samples crystallized at atmospheric pressure. The refractive index of LS_2 was also significantly affected by HPHT.

Nuclear Magnetic Resonance (NMR) results obtained by Fuss et al. [9] indicated the formation of lithium metasilicate solid solution (LS-ss) for samples submitted to 6 GPa and thermal treatment below 750°C. According to the authors, the difference between lithium metasilicate (MS) and LS-ss was not detectable by X-ray diffraction since there were no changes in the crystal symmetry of both structures. The difference would be related to the number of bridged (BO) and non-bridged (NBO) oxygen atoms. Buchner et al. [6], on the other hand, showed the formation of crystalline MS after processing LS_2 at 7.7 GPa and 455/500°C for nucleation followed by 610°C for crystal growing.

Adams & De Jong [12] investigated the early stages of crystallization of a slightly hyper stoichimetric LS_2 sample (34.5 mol% Li_2O) at 450°C using X-ray photoelectron spectroscopy (XPS) and they ascribed the presence of an extra Li 1s peak in the XPS spectrum (~51.7 eV) to the formation of a MS phase. Soares et al. [13], however, repeated the XPS analysis at the same conditions and concluded that this extra Li 1s peak was related to contamination. Moreover, they concluded that XPS was not a suitable technique to detect structural changes in LS_2 when the volume fraction of the crystals was very low. On the other hand, the formation of the MS phase in the early stages of nucleation was confirmed by them using TEM analysis [14].

Nesbitt et al. [15] investigated the chemical environment around Na 1s, O 1s and Si 2p atoms in Na_2O-SiO_2 glass as a function of the glass composition. Six compositions of Na_2O-SiO_2 glass ranging from 100 to 45 mol% SiO_2 were investigated by XPS and NMR. The Si $2p_{3/2}$ peak maximum shifted to lower binding energy (BE) with increasing Na_2O content. According to the authors, the BE shift indicated increased valence electron density on the Si nuclei with increased Na_2O concentration. Each Q-species should yield a peak at slightly different BE since the electron density over each Si atom should be different, depending on the number of BO and NBO bonded to it.

According to Sawyer et al. [16], the BE of the Q-species are strongly affected by the number of NBO bonded to the Si atom, with BE decreasing systematically from Q^4 to Q^0 species. They observed the shift of the peak toward lower BE with increasing K_2O content in K_2O-SiO_2 glass. Two doublets in the Si 2p spectra of different glass compositions were identified. According to the authors, the strongly asymmetric nature of the Si 2p spectra of the potasic glasses may result from two or more Q-species contributions to the Si 2p spectra and the intensity of the doublet would depend on the amount of the particular Q-specie.

In this context, the aim of this work was to investigate the effect of high pressure on the chemical environment of Si 2p, O 1s and Li 1s atoms of the LS_2 glass structure using XPS to help elucidate the mechanisms responsible for the early stages of the formation of the metasilicate structure induced by densification. For comparison, the chemical environment of these atoms was also investigated after complete crystallization of the glass structure under high pressure and high temperature.

2. EXPERIMENTAL PROCEDURE

2.1. Sample Preparation

Lithium disilicate glass of stoichiometric composition $Li_2O \cdot 2SiO_2$ (LS_2) was prepared using standard reagent grade Li_2CO_3 (Aldrich Chem. Co., 99+%) and ground quartz (<99.9% SiO_2). The 200 g batch was melted in a Pt crucible at 1450°C during 2 hours in an electric furnace. The melt was poured on a steel plate, annealed below the glass transition temperature, at 430°C, during 1 h and cooled down slowly to room temperature.

2.2. High Pressure Experiments

The HP experiments were performed in a toroidal high pressure apparatus. The specific configuration and type of high-pressure chamber used in the experiments were optimized for providing quasi-hydrostatic pressure [17]. The pressure calibration was carried out by the "fixedpoints" technique using Bi and Yb, which permitted the calibration of the pressure at the following points: Bi with phase transitions at 2.5 GPa and 7.7 GPa, and Yb with a phase transition at 4.0 GPa. A detailed description of the high-pressure calibration method has been provided elsewhere [17].

During the HPHT experiments, a monolithic LS_2 sample was placed inside a hexagonal boron nitride (hBN) capsule which acted as a soft pressure transmitting medium. The hBN capsule was placed inside a graphite cylinder which acted as a heater element. For experiments performed at HP and room temperature (HPRT), the capsule of hBN was replaced by a lead capsule, which is a softer pressure transmitting compared to hBN.

In the first group, the samples were submitted to 2.5 GPa, 4 GPa and 7.7 GPa at room temperature during 5 min inside the lead capsule. In the second group, the samples were submitted to HP (2.5, 4.0 and 7.7 GPa) inside the hBN capsule and, simultaneously, to the following thermal treatment: 500°C during 2 h for nucleation followed by 610°C during 0.5 h for crystal growing. For comparison, a pristine glass sample and a sample submitted to the same heat treatment at 1 atm were also investigated.

According to a previous work [6], the processing at 2.5 GPa and high temperature induced the crystallization of the monoclinic phase of lithium disilicate ($Li_2Si_2O_5$). At 4.4 GPa it was observed the formation of the orthorhombic phase with the same stoichiometry and, at 7.7 GPa the glass structure crystallized as MS (Li_2SiO_3).

After HPRT and HPHT processing, the surface of the samples was grounded on SiC abrasive paper up to #1200 and polished with CeO_2 slurry for XPS analyses.

2.3. Analytical Techniques—XPS

XPS analyses were performed in an Omicron-SPHERA station using Al Kα radiation (1486.6 eV). The anode was operated at 225 W (15 kV, 15 mA). Survey spectra were recorded with a 50 eV pass energy. Si 2p, O 1s, and Li 1s regions were recorded with high resolution (pass energy of 20 eV). The detection angle of the photoelectrons (Θ) with respect to the sample surface (take-off angle) was fixed at 53° for all measurements. The C 1s signal from adventitious carbon at 284.6 eV was used as an internal energy reference. All spectra were fitted assuming a Shirley background. Lines were fitted by 70%

Gaussian + 30% Lorentzian functions with set values of full width at half maximum for each line.

3. RESULTS AND DISCUSSION

3.1. Densification at Room Temperature (HPRT)

Figures 1-3 show high resolution XPS spectra for Si 2p, O 1s and Li 1s, respectively, after processing at high pressure compared to the pristine sample. It is clearly seen that the BE of the three elements changed irreversible after densification, revealing the strong influence of high pressure in the chemical environment due to the structural freedom of the glassy LS_2.

According to **Figure 1**, Si 2p presents two doublets from 2.5 to 7.7 GPa, corresponding to Si $2p_{3/2}$and Si $2p_{1/2}$. However, the intensity and BE of each doublet changed with pressure. **Figure 2**shows that there is practically only one BE component for O 1s up to 2.5 GPa, while after processing at 4 and 7.7 GPa there are two components. In these cases, the component with higher intensity corresponds to BO while the lower intensity component corresponds to NBO. For Li atoms (**Figure 3**), only at 7.7 GPa it is possible to identify a second component of BE.

Table 1 shows the values of the BE obtained in this work by fitting the experimental data compared to the results found in the literature for LS_2.

3.2. Crystallization under High Pressure (HPHT)

Figures 4-6 show the XPS results for the samples crystallized under high pressure. In this case there was practically no shift in the BE as a function of pressure.

Figure 4 shows that the formation of doublets for Si $2p_{3/2}$ and Si $2p_{1/2}$ now happens only after processing at 7.7 GPa. **Figure 5** shows that there were three components of BE for O 1s up to 7.7 GPa, one associated to NBO and two associated to BO. The intensity of the component with higher BE decreases with increasing pressure. For Li atoms (**Figure 6**), only at 7.7 GPa it was possible to identify a component of the BE. Soares et al.

[18] already reported that the detection of the Li 1s XPS signal from crystalline samples of LS_2 was difficult.

The BE measured for the pristine glass sample and for the crystalline sample obtained at atmospheric pressure are similar to the results found in the literature [12,13,18], as shown in**Table 1**. The permanent shift of the BE toward higher energy and the change in the intensity of the peaks for Li 1s, O 1s and Si 2p observed in Figures 1-3 indicated that the densification affects irreversibly the glass structure, even without crystallization.

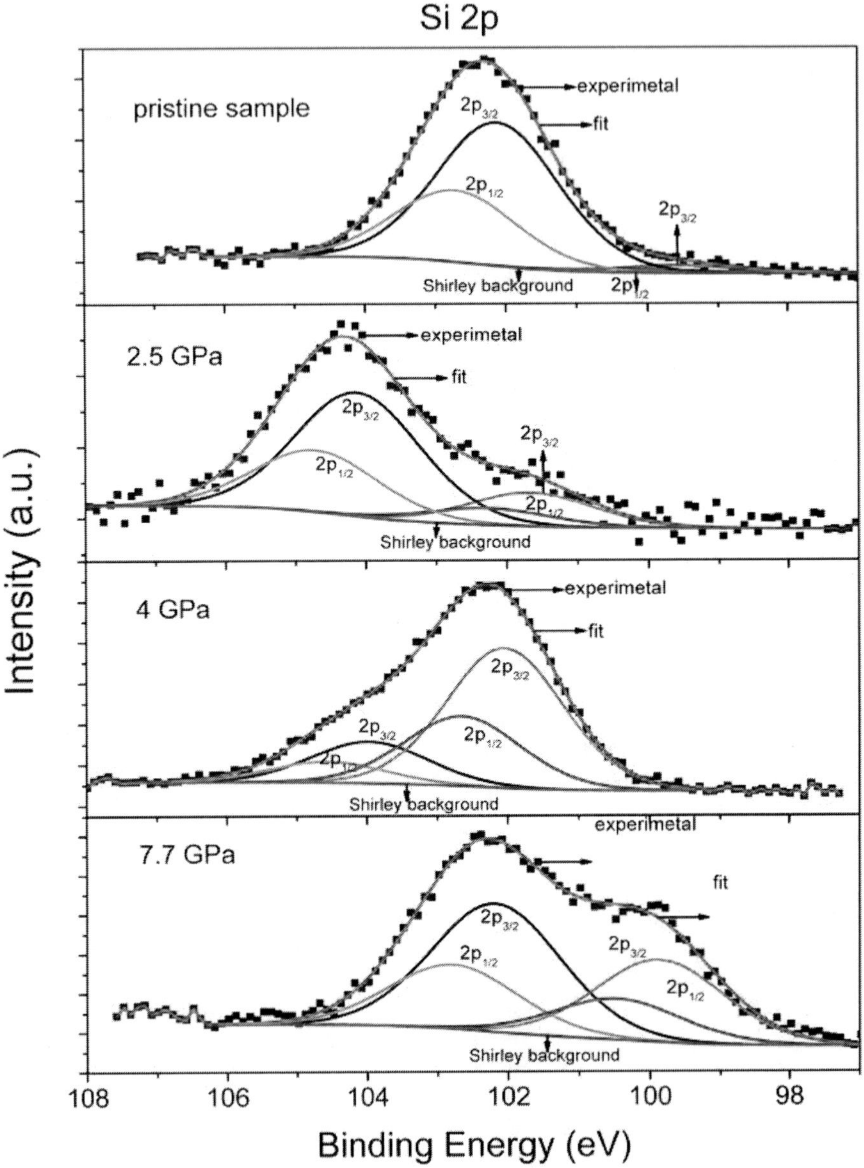

Figure 1. Si 2p region of XPS spectrum of the LS_2 pristine sample and after processing at 2.5 GPa, 4 GPa, and 7.7 GPa at room temperature.

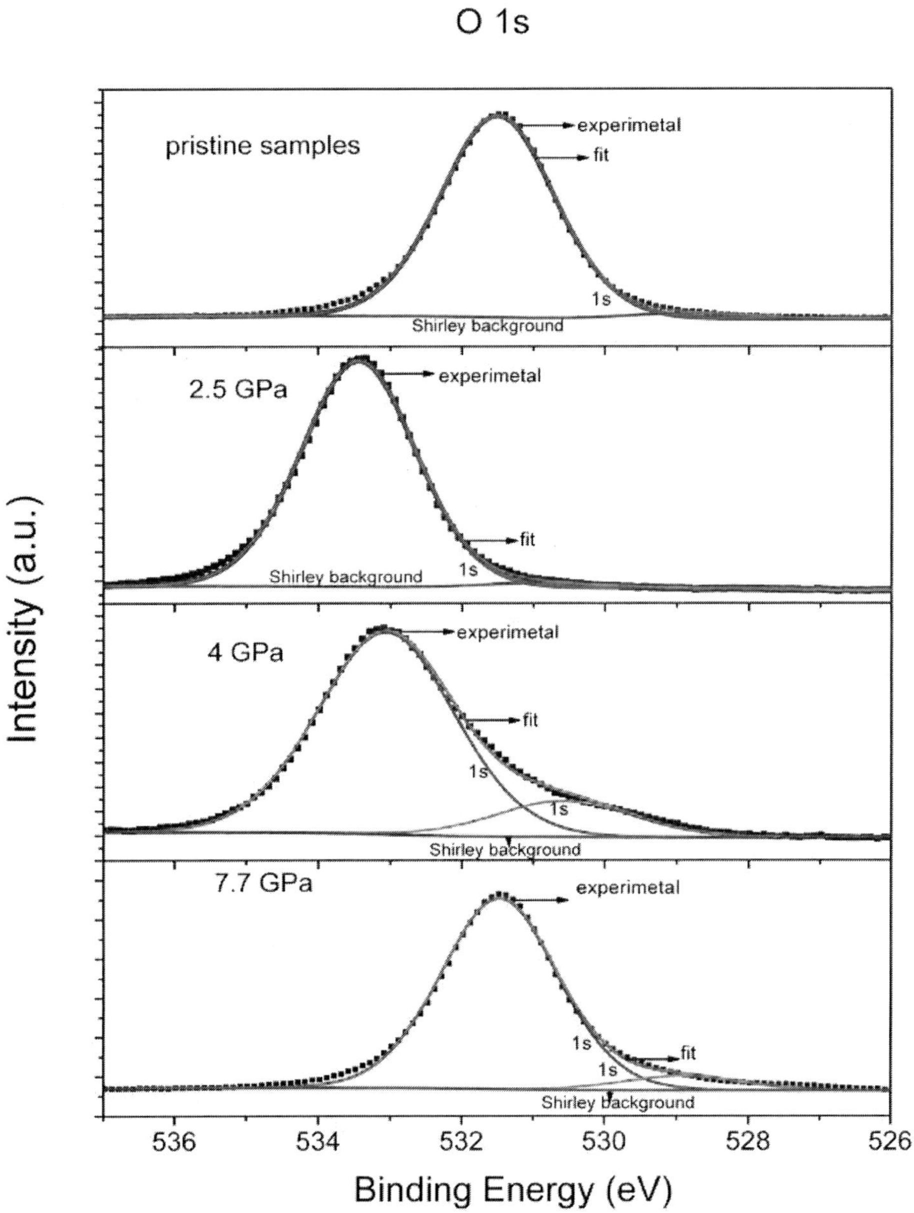

Figure 2. O 1s region of XPS spectrum of the LS_2 pristine sample and after processing at 2.5 GPa, 4 GPa, and 7.7 GPa at room temperature.

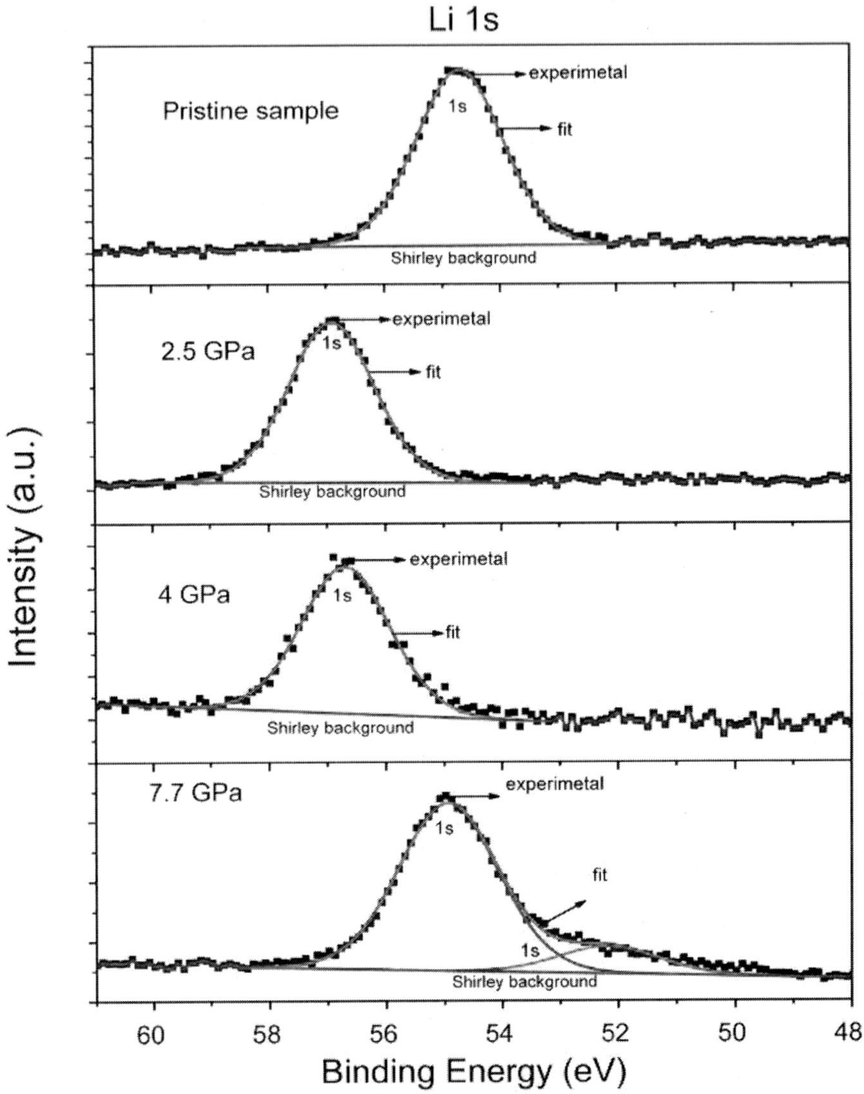

Figure 3. Li 1s region of XPS spectrum of the LS_2 pristine sample and after processing at 2.5 GPa, 4 GPa, and 7.7 GPa at room temperature.

The large shift of the BE toward higher energies observed for Si 2p after processing at 2.5 and 4 GPa (Figure 1) suggests a stronger interaction with the surrounding oxygen atoms compared to the pristine glass. According to Sawyer et al. [16], the BE decreases systematically from Q^4 to Q^0species for Si 2p. According to **Figure 1**, two Q-species were observed in all samples for Si 2p, but for pristine samples the intensity of the Q-species of lower energy was very small. Probably in pristine glass samples the Q^3 specie was dominant and the small peak of lower energy corresponded to residual Q^2 species.

Table 1. Binding energy values obtained from the fitting of the XPS experimental data for the samples processed under high pressure at room temperature (glass) and processed under high pressure and high temperature (glass ceramic) compared to the results found in the literature at 1 atm.

Glass										
Pressure	Heat treatment	Binding Energy (eV) Si 2p		O 1s			Li 1s		Ref.	Obs
1 atm		102.1	99.5	531.4	528.8		54.7	-	This work	
2.5 GPa		104.1	101.6	533.4	530.9		56.9	-	This work	
4.0 GPa		104.0	102.1	533.0	530.5		56.7	-	This work	
7.7 GPa		102.2	99.8	531.5	528.9		54.9	52.1	This work	
1 atm	Room temperature	102.7	-	530.7	532.4		51.6	55.7	[13]	$Li_2O \cdot 2SiO_2$
1 atm		103.2		530.1	531.6		51.0	54.9	[18]	$Li_2O \cdot 2SiO_2$ 32.0-32.5 %mol Li_2O
1 atm		102.1		530	532.01		-	-	[15]	$Na_2O \cdot SiO_2$
1 atm		102.93	101.73	531.86	529.71		-	-	[16]	$K_2O \cdot SiO_2$
Glass ceramic										
Pressure	Heat treatment	Binding Energy (eV) Si 2p		O 1s			Li 1s		Ref	Obs
1 atm	455°C/2 h + 610°C/2 h	101.95		533.01	531.70	529.51	-	-	This work	$Li_2Si_2O_5$
2.5 GPa	500°C/2 h + 610°C/0.5 h	102.56		533.43	531.87	529.67	-	-	This work	$Li_2Si_2O_5$
4.0 GPa	500°C/2 h + 610°C/0.5 h	102.55		533.13	531.70	529.44	-	-	This work	$Li_2Si_2O_5$
7.7 GPa	500°C/2 h + 610°C/0.5 h	102.36	99.56	532.94	531.60	529.75	55.2	-	This work	Li_2SiO_3 + amorphous SiO_2
1 atm	450°C/5 h	102.7		532.2	530.4		55.6	51.7	[13]	$Li_2Si_2O_5$

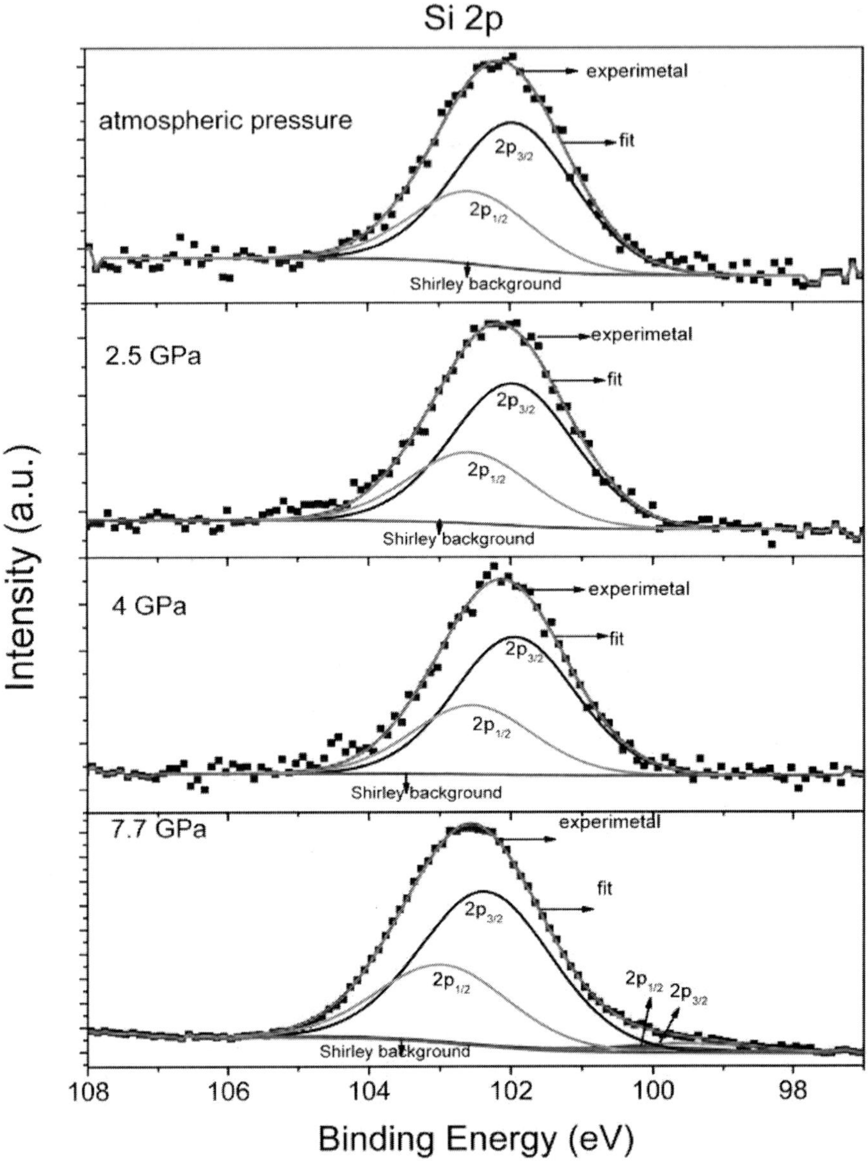

Figure 4. Si 2p XPS spectra obtained from crystalline samples produced at atmospheric pressure, 2.5 GPa, 4 GPa, and 7.7 GPa. The thermal treatment was: 500˚C during 2 h for nucleation followed by 610˚C during 0.5 h for crystal growing.

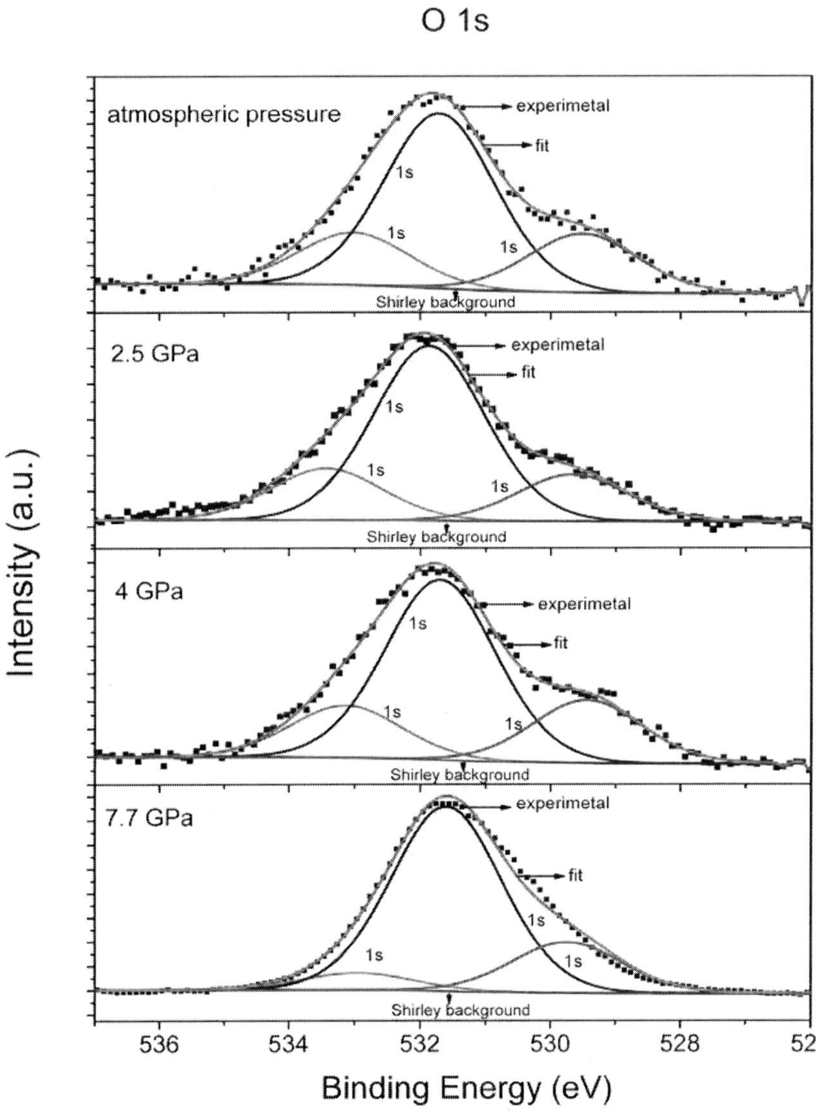

Figure 5. O 1s XPS spectra obtained from crystalline samples produced at atmospheric pressure, 2.5 GPa, 4 GPa, and 7.7 GPa. The thermal treatment was: 500˚C during 2 h for nucleation followed by 610˚C during 0.5 h for crystal growing.

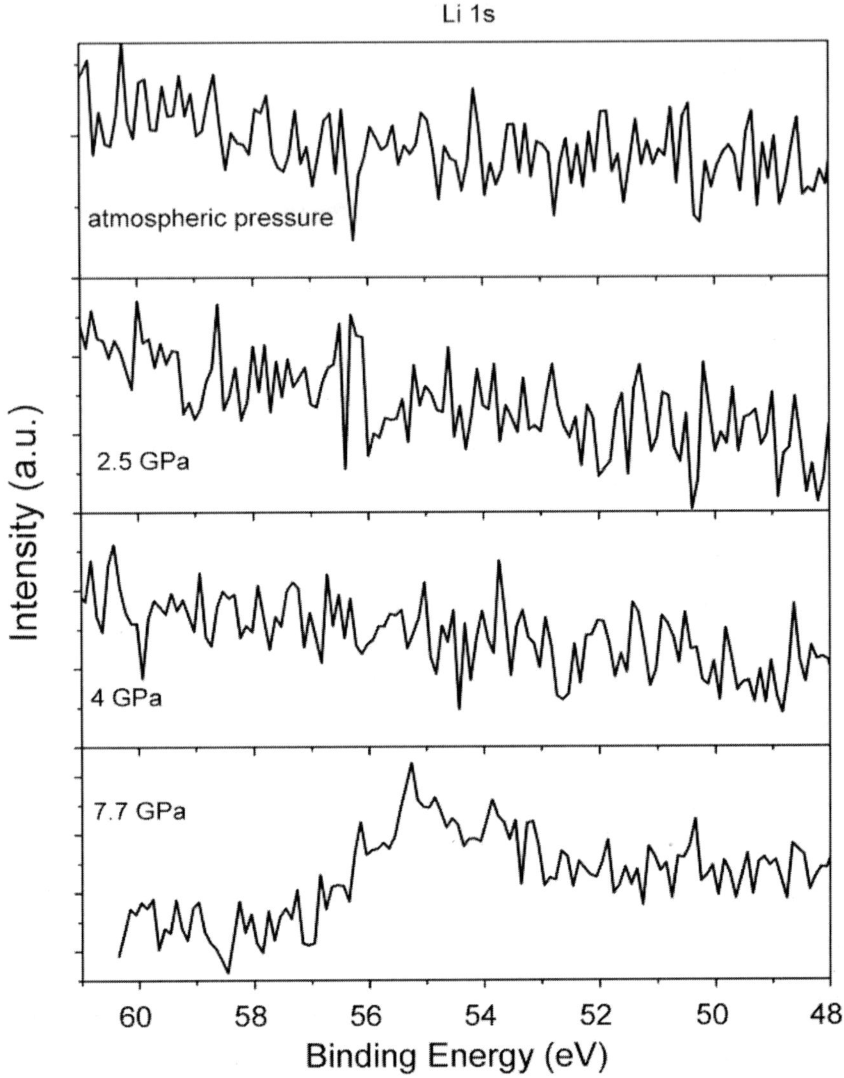

Figure 6. Li 1s XPS spectra obtained from crystalline samples produced at atmospheric pressure, 2.5 GPa, 4 GPa, and 7.7 GPa. The thermal treatment was: 500°C during 2 h for nucleation followed by 610°C during 0.5 h for crystal growing.

In the case of O 1s atoms (**Figure 2**), the peak at higher energy (~531 eV) for the pristine sample corresponds to BO atoms (Si-O-Si) while the other component (~528.8 eV) corresponds to the NBO atoms (Si-O-Li) [13,19]. After processing at 2.5 and 4 GPa the BE of the BO atoms shifted irreversibly to higher energies, indicating a larger interaction with the Si atoms, in agreement with the behavior of the Si 2p XPS spectrum. The intensity of the peak related to the NBO atoms was very low for the pristine sample and after processing at 2.5 GPa the intensity of this peak slightly increased after processing at 4 and 7.7 GPa, probably indicating a decomposition of the Q^3 into the Q^2 species at higher pressure.

Figure 3 shows that the BE for Li 1s also shifted after processing at 2.5 GPa and 4 GPa, and a second component at 52.1 eV was detected after processing at 7.7 GPa. According to Adams & De Jong [12,20], this component may be related to a local arrangement that would be favorable to the formation of the metastable phase during heat treatment.

Considering the XPS spectra for crystalline samples produced under high pressure, **Figure 4** shows that the Si 2p presents only one chemical environment related to Q^3 species formed during crystallization of lithium disilicate ($Li_2{\cdot}Si_2O_5$). The Si $2p_{3/2}$ spectrum for the sample crystallized at 7.7 GPa, however, presents a doublet indicating the formation of two structures, one of them related to lithium metasilicate $Li_2{\cdot}SiO_3$ and the other, to residual SiO_2 to satisfy stoichiometric restrictions [6].

For O 1s (**Figure 5**), the smaller BE corresponds to the NBO atoms (Si-O-Li) [13,19] and it was observed for all pressure range investigated. For pressures up to 4 GPa, the intensity and energy of this peak remained practically unchanged, compatible with the formation of crystalline lithium disilicate. At 7.7 GPa, it was observed a small shift of this component toward higher energy and a slight decrease of its intensity. This result is in agreement with the changes observed in the spectrum for Si 2p at the same pressure, indicating a distinct configuration of Q-species at higher pressures.

Two peaks related to BO atoms, of higher BE, were observed for all pressure range in the crystalline samples. The most intense component was probably related to Si-O-Si and the other one may be related to BO-Li. Ching et al. [21] also observed two types of BO atoms in crystalline Na disilicate. According to them, one type was assigned to Si-O-Si as in vitreous silica whereas the other type of BO atom was also bonded to a Na atom with the BO-Na bond length being similar to the NBONa bond length [22]. Du and Cormack [23] and Mountjoy [24] investigated by molecular dynamics simulations the formation of two BO in sodic glasses. The experimental results obtained by Nesbitt et al. [15] for sodic glass also provided evidences for two types of BO, one bonded to two Si atoms and one bonded to a Na atom as well as to two Si atoms. Sawyer et al. [16] reported that the width of the O 1s XPS peak was highly variable and wider than that of the NBO peak for potasic glass. The variable width would be a consequence of the presence of two types of BO atoms with somewhat different BE; one type bridged to two Si atoms while the other type bridged to three atoms, two Si and one K atom. According to the authors, the abundance of the two types of BO atoms was similar to the abundance of Q^4 and Q^3 species of previous NMR studies.

Fuss et al. [9] observed that at 4.5 GPa and high temperature, LS_2 crystallized as the lithium disilicate phase, consisting only of Q^3 tetrahedra, but the NMR results indicated the presence of a small amount of Q^2 SiO_4 tetrahedra, which could be a glassy or crystalline phase acting as imperfections. In the present work, the crystallization at 4 GPa induced the formation of only the Q^3species in the Si 2p spectrum, without any evidence of residual Q^2. At 6 GPa, the results obtained by Fuss et al. suggested that the LS_2 glass densified under pressure and crystallized as a single new phase with a higher density, resembling a slightly deformed MS phase. For this sample, they observed one sharp Q^2 peak in the NMR spectrum associated with crystalline MS, a broad Q^4 band related to glassy SiO_2 and a broad Q^3 band associated with the residual glass. Additionally, they found four more peaks related to Q^3 and Q^4 tetrahedra that could not be related to residual glass. Based on the NMR, Raman and IR results, they suggested the formation of a crystalline lithium metasilicate solid-solution having the same symmetry of MS but formed out of three different tetrahedral species. In this solid solution, four Li ions from the metasilicate structure would be statistically replaced by one Si ion and three vacancies, without changing the crystal symmetry. This replacement would be detected by a shift of the peaks in the XRD pattern related to the change in lattice parameters that was, in fact, observed by the authors and corroborated by the NMR results about the amount of BO and NBO atoms. The XPS results found in the present work for the sample crystallized at 7.7 GPa indicated two chemical environments around Si atoms, associated with Q^3species of lithium metasilicate (Li_2SiO_3) and Q^2 associated to residual SiO_2. No evidences of other Q species were observed. Maybe the reason for the differences between the consistent results found by Fuss et al. [9] and the results found in this work is related to the thermal treatment: they investigated temperatures from 608°C during 20 min up to 753°C during 60 min at 6 GPa. In this work we considered 500°C during 2 h for nucleation followed by 0.5 h at 610°C for crystal growing, both under high pressure. Probably the higher temperatures used by Fuss et al. [9] were not suitable for nucleation of the crystalline phases, resulting in a solid solution favored at higher temperatures where the average Si-O-Si bonding angle would decrease and there would be at least four different Si sites, according to the NMR results.

4. CONCLUSION

Densification of lithium disilicate at room temperature under high pressure induced large and irreversible modifications in the chemical environment of Si 2p, O 1s and Li 1s. These modifications would be related to the rearrangement of the atoms induced by high pressure due to the structural freedom of the glass structure, even at room temperature. XPS showed to be a suitable technique to detect the structural changes induced by densification at room temperature. For samples crystallized under high pressure, it was observed that a similar behavior for pressures up to 4 GPa, related to the formation of monoclinic and orthorhombic lithium disilicate. The interpretation of the XPS results for Si 2p and O 1s for the sample crystallized at 7.7 GPa was consistent with the formation of lithium metasilicate and SiO_2.

5. ACKNOWLEDGEMENTS

The authors would like to thank CNPq, CAPES and FAPERGS for the financial support and to LaMaV for the fusion of the samples.

REFERENCES

1. N. Kitamura, K. Fukumi, H. Mizoguchi, M. Makihara, A. Higuchi, N. Ohno and T. Fukunaga, "High Pressure Densification of Lithium Silicate Glasses," Journal of NonCrystalline Solids, Vol. 274, No. 3, 2000, pp. 244-248.

2. P. W. Bridgmann and I. Simon, "Effects of Very High Pressures on Glass," Journal of Applied Physics, Vol. 24, No. 4, 1953, pp. 405-413.

3. J. W. P. Schmelzer, E. D. Zanotto and V. M. Fokin, "Pressure Dependence of Viscosity," The Journal of Chemical Physics, Vol. 122, No. 7, 2005, Article ID: 074511.

4. S. Buchner, P. Soares, A. S. Pereira, E. B. Ferreira and N. M. Balzaretti, "Effect of High Pressure in the Li_2O-$2SiO_2$ Crystallization," Journal of Non-Crystalline Solids, Vol. 356, No. 52-54, 2010, pp. 3004-3308.

5. S. Buchner, C. M. Lepienski, P. C. Soares Jr. and N. M. Balzaretti, "Effect of High Pressure on the Mechanical Properties of Lithium Disilicate Glass Ceramic," Materials Science and Engineering: A, Vol. 528, No. 10-11, 2011, pp. 3921-3924.

6. S. Buchner, M. B. Pereira and N. M. Balzaretti, "Behavior of the Refractive Index of Lithium Disilicate Glass Ceramic Processed at High Pressure and High Temperature," Optical Materials, Vol. 34, No. 5, 2012, pp. 826- 831.

7. S. Buchner, "Efeito de altas pressões na Astrutura e nas Aropriedades da vitrocerâmica dissilicato de lítio," Ph.D. Thesis, Federal University of Rio Grande do Sul, São José, 2011.

8. T. Fuss, C. S. Ray, N. Kitamura, M. Makihara and D. E. Day, "Pressure Induced Nucleation in a $Li_2O{\cdot}2SiO_2$ Glass," Journal of Non-Crystalline Solids, Vol. 318, No. 1-2, 2003, pp. 157-167.

9. T. Fuss, A. Mogus-Milanković, C. S. Ray, C. E. Lesher, R. Youngman and D. E. Day, "Ex Situ XRD, TEM, IR, Raman and NMR Spectroscopy of Crystallization of Lithium Disilicate Glass at High Pressure," Journal of NonCrystalline Solids, Vol. 352, No. 38, 2006, pp. 4101- 4111.

10. T. Fuss, C. S. Ray, C. E. Lesher and D. E. Day, "In Situ Crystallization of Lithium Disilicate Glass: Effect of Pressure on Crystal Growth Rate," Journal of Non-Crystalline Solids, Vol. 352, No. 21-22, 2006, pp. 2073-2081.

11. I. Gutzow, B. Durschang and C. Russel, " Crystallization of Glass Forming Melts under Hydrostatic Pressure and Shear Stress Part I. Crystallization Catalysis under Hydrostatic Pressure: Possibilities and Limitations" Journal of Materials Science, Vol. 32, No. 20, 1997, pp. 5389- 5403.

12. J. W. Adams and B. H. W. S. De Jong, Materials Research Society 1993 Fall Meeting Symposium Proceedings, Vol. 321, 1994, p. 339.

13. P. C. Soares, P. A. P. Nascente and E. D. Zanotto, "XPS Study of Lithium Disilicate Glass Crystallisation," Physics and Chemistry of Glasses, Vol. 43, No. 3, 2002, pp. 143-146.

14. P. C. Soares, E. D. Zanotto, V. M. Fokin and H. Jain, "TEM and XRD Study of Early Crystallization of Lithium Disilicate Glasses," Journal of Non-Crystalline Solids, Vol. 331, No. 1-3, 2003, pp. 217-227.

15. H. W. Nesbitt, G. M. Bancroft, G. S. Henderson, R. Ho, K. N. Dalby, Y. Huang and Z. Yan, "Bridging, NonBridging and Free (O^{2-}) Oxygen in Na_2O-SiO_2 Glasses: An X-Ray Photoelectron Spectroscopic (XPS) and Nuclear Magnetic Resonance (NMR) Study," Journal of Non-Crystalline Solids, Vol. 357, No. 1, 2011, pp. 170- 180.

16. R. Sawyer, H. W. Nesbitt and R. A. Secco, "High Resolution X-Ray Photoelectron Spectroscopy (XPS) Study of K_2O-SiO_2 Glasses: Evidence for Three Types of O and at Least Two Types of Si," Journal of Non-Crystalline Solids, Vol. 358, No. 2, 2012, pp. 290-302.

17. M. I. Eremets, "High Pressure Experimental Methods," Oxford University Press, New York, 1996.

18. P. C. Soares, "Mecanismos de Cristalização em Vidros Disilicato de Lítio," Ph.D. Thesis, Universidade Federal de São Carlos, Brazil, 2002.

19. R. Brückner, H. U. Chun, H. Goretzki and M. Sammet, "XPS Measurements and Structural Aspects of Silicate and Phosphate

Glasses," Journal of Non-Crystalline Solids, Vol. 42, No. 1-3, 1980, pp. 49-60.

20. J. W. Adams, "Morphogenesis in Lithium Disilicate Glass: Internal Nucleation in Amorphous Structure," Ph.D. Thesis, University of Cambridge, Cambridge, 1993.

21. W. Y. Ching, R. A. Murray, D. J. Lam and B. W. Veal, "Comparative Studies of Electronic Structures of Sodium Metasilicate and α and β Phases of Sodium Disilicate," Physics Review B, Vol. 28, No. 8, 1983, pp. 4724-4735.

22. A. K. Pant and D. W. J. Cruickshank, "The Crystal Structure of α-$Na_2Si_2O_5$," Acta Crystallographica Section B, Vol. 24, No. 13-19, 1968, pp. 13-19.

23. J. Du and A. N. Cormack, "The Medium Range Structure of Sodium Silicate Glasses: a Molecular Dynamics Simulation," Journal of Non-Crystalline Solids, Vol. 349, 2004, pp. 66-79.

24. G. Mountjoy, "The Local Atomic Environment of Oxygen in Silicate Glasses from Molecular Dynamics," Journal of Non-Crystalline Solids, Vol. 353, No. 18-21, 2007, pp. 1849-1853.

Chapter 10

Fatigue Behavior of an Ultrafine-Grained Al-Mg-Si Alloy Processed by High-Pressure Torsion

Maxim Murashkin [1,2,*], Ilchat Sabirov [3], Dmitriy Prosvirnin [4,†], Ilya Ovid'ko [5], Vladimir Terentiev [4,†], Ruslan Valiev [1,2] and Sergey Dobatkin [4,†]

[1]*Institute for Physics of Advanced Materials, Ufa State Aviation Technical University, Ufa 450000, Russia*

[2]*Laboratory for Mechanics of Bulk Nanostructured Materials, Saint Petersburg State University, Saint-Petersburg 199034 Russia*

[3]*IMDEA Materials Institute, Getafe, Madrid 28906, Spain*

[4]*Baikov Institute for Metallurgy and Materials Science of the Russian Academy of Sciences, Moscow 119991, Russia*

[5]*Research Laboratory for Mechanics of New Nanomaterials, Saint-Petersburg State Polytechnical University, Saint-Petersburg 195251, Russia*

ABSTRACT

The paper presents the evaluation of the mechanical and fatigue properties of an ultrafine-grained (UFG) Al 6061 alloy processed by high-pressure torsion (HPT) at room temperature (RT). A comparison is made between the UFG state and the coarse-grained (CG) one subjected to the conventional aging treatment T6. It is shown that HPT processing leads to the formation of the UFG microstructure with an average grain size of 170 nm. It is found that yield strength ($\sigma_{0.2}$), ultimate tensile strength (σ_{UTS}) and the endurance limit (σ_f) in the UFG Al 6061 alloy are higher by a factor of 2.2, 1.8 and 2.0 compared to the CG counterpart subjected to the conventional aging treatment T6. Fatigue fracture surfaces are analyzed, and the fatigue behavior of the material in the high cycle and low cycle regimes is discussed.

Keywords: Al alloy; severe plastic deformation; high pressure torsion; ultrafine-grained microstructure; mechanical strength; fatigue properties

1. INTRODUCTION

The enhancement in the level of physical and mechanical properties in Al alloys as a result of ultrafine-grained (UFG) structure formation induced by severe plastic deformation (SPD) is currently of great interest [1,2,3,4,5]. As a rule, SPD processing leads to the formation of the microstructure with a grain/subgrain size in the range of 50 nm–1 μm in bulk Al billets [1,2,3,4,5]. Earlier, it was demonstrated that UFG Al alloys have a mechanical strength 1.2–2.0-times as much as their coarse-grained (CG) counterparts processed by a standard aging treatment [1,2,3,4,6]. Numerous studies have also shown that other properties, including ductility [1,7,8], fracture resistance [9] and electrical conductivity [1,10,11], can be enhanced. The fatigue behavior of the UFG Al alloys has been widely studied over the last decade, and a detailed overview of those studies can be found in [12,13,14,15,16,17]. The outcomes of these studies can be roughly generalized as follows. The high-cycle fatigue (HCF) properties are usually not enhanced as significantly as mechanical strength. An insignificant improvement in fatigue life observed in the UFG Al alloys in the HCF regime was related to a low increase in resistance against crack nucleation, as the fatigue life in the HCF is determined by the resistance of material to the crack nucleation [13]. The low cycle fatigue (LCF) behavior of the UFG Al alloys is very complex, since ultrafine grains have a low ability to sustain cyclic loads in the LCF regime. This was related to the limited ductility on monotonic and cycling deformation promoting early crack initiation, as well as a higher fraction of grain boundaries favorable for crack propagation [13]. Unlike CG Al alloys, their UFG counterparts often show an insignificant increase of the stress amplitude (low cyclic hardening) or even its decrease (cyclic softening) when they are cyclically deformed under a constant strain amplitude [18,19,20]. Nevertheless, intelligent microstructural design in the UFG Al alloys can significantly improve their HCF and LCF behavior [21,22].

It should be noted that the fatigue behavior of the UFG Al alloys has been studied mainly on specimens processed by equal-channel angular pressing (ECAP) [12,13,14] or by cryorolling [18,23]. It is well known that the ECAP technique or other ECAP-based methods result in the formation of a UFG microstructure with a grain/subgrain size of 300–500 nm [12,13,14], whereas a microstructure consisting mainly of cells/subgrains having a similar size is formed during cryorolling [18,23]. However, in the literature, there are no reports on the fatigue properties of the UFG Al alloys with a grain size of <200 nm, which can demonstrate record mechanical strength up to 1,000 MPa and are usually processed by HPT [24,25]. The main limitation for fatigue studies on the HPT-processed metallic materials was the small size of the processed disks. Recent progress in up-scaling of the HPT technique along with miniaturization of the specimens for fatigue testing allows one to investigate the fatigue properties of the HPT processed disks [26]. Therefore, the main objective of the present work is to study the fatigue properties and fatigue behavior of an HPT-processed Al-Mg-Si alloy with a grain size of 170 nm. The Al-Mg-Si alloy was selected due to the wide application of the material in construction, auto- and aero-space engineering [21] and as a conductor in electrical engineering [27].

2. EXPERIMENTAL SECTION

Commercial-grade Al 6061 alloy having a chemical composition (in wt%) 1.00 Mg, 0.61 Si, 0.12 Cu, 0.38 Cr, 0.60 Fe, 0.02 Ti and balance Al was selected for this investigation. Disks having a diameter of 20 mm and a thickness of 1.4 mm were machined from the as-received hot-pressed rod. The disks were solution-treated at 530 °C for 2 h and water quenched. Some disks after water quenching were subjected to the conventional treatment T6, namely artificial ageing at 160 °C for 12 h.

The disks were HPT processed for 10 rotations at RT under a load of 188 tons (e.g., a pressure of 6 GPa). The as-processed disks had a thickness of 0.9 mm. The microstructure and properties of the HPT-processed alloy were studied in the areas located at a distance of ~5 mm (half of radius) from the center of the disks.

Microstructure studies were performed using the transmission electron microscope (TEM, JEOL, Tokyo, Japan) JEM 2100 at an accelerating voltage of 200 kV. Observations were made in both the bright and the dark field imaging modes, and selected area electron diffraction (SAED) patterns were recorded from areas of interest using an aperture of a 1-µm nominal diameter. The lineal intercept method was used to estimate the grain size. At least 200 grains were analyzed to estimate the average grain size.

Specimens for tensile testing and fatigue testing were machined from the solution-treated material and HPT-processed disks. The central axis of the tensile and fatigue specimens was located at a distance of 5.2 mm from the disk center, as shown in Figure 1, where a drawing of the samples is also presented. Two specimens were machined from each disk. All machined tensile and fatigue specimens were mechanically polished to a mirror-like surface using silica colloidal solution at the final stage. Tensile tests were carried out using an Instron 8862 universal testing machine (JEOL, Tokyo, Japan). Tensile specimens were deformed to failure at room temperature with a constant cross-head speed corresponding to an initial strain rate of 10^{-3} s^{-1}. Three tensile specimens were tested for each material condition, and the results thus obtained were found to be reproducible.

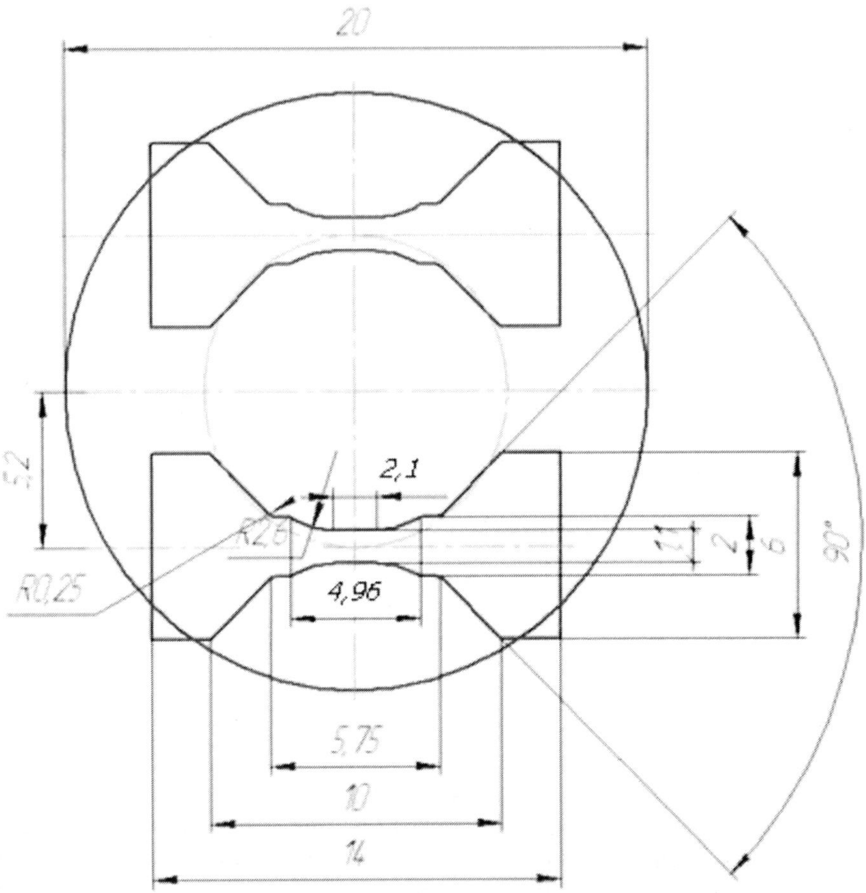

Figure 1. Schematic drawing of specimens machined from disks.

Fatigue testing was carried out using an Instron Electropuls E 3000 testing module with a frequency of 30 Hz under repeated tension conditions with a constant minimum cycle stress of σ_{min} = 100 MPa.

The fatigue fracture surfaces were studied using the scanning electron microscope (SEM, JEOL, Tokyo, Japan) JSM-6490LV operating at an accelerating voltage of 20 kV. The SEM is equipped with the INCA X-sight attachment for energy-dispersive X-ray (EDX) analysis.

3. RESULTS AND DISCUSSION

3.1. Effect of HPT Processing on Microstructure of the Al 6061 Alloy

Figure 2 illustrates the microstructure of the Al 6061 alloy before and after HPT processing. Conventional treatment T6 of the as-received alloy resulted in the formation of a CG microstructure with an average grain size of 70 μm (Figure 2a). Particles of primary excess phase forming bands oriented along the pressing direction are observed in the microstructure (Figure 2a). They have a complex chemical composition of $Al_{68}Fe_{21}Si_{6.5}Cr_{2.5}$(wt%) and are typically formed during the crystallization of slabs. These particles have a spherical shape; their average size is about 1.8 μm, and the volume fraction is about 2.4%. Such a distribution of these particles is usually observed in the hot-pressed Al-Mg-Si alloys. Second phase precipitates having a needle shape are also present in the microstructure (Figure 2b). As is well known, these are the metastable β''-Mg_2Si precipitates, which are typically formed during T6 treatment of the Al-Mg-Si alloys [28,29].

HPT processing of the solution-treated Al 6061 alloy resulted in the formation of a very homogeneous UFG microstructure consisting mainly of equiaxed grains having an average size of 170 nm (Figure 2c,d). This is also confirmed by SAED patterns (Figure 2d). No precipitates, like Guinier–Preston zones and/or second phases, were observed in the processed disks. Such a microstructure is typically formed during HPT processing of the Al-Mg-Si alloys [11,30,31]. It should be also noted that earlier studies using the atom probe tomography (APT) technique revealed segregations of Mg, Cu and Si atoms along grain boundaries in the Al 6061 alloy [30] and in the Al 6060 alloy [32] processed by HPT under similar conditions. HPT processing did not have any effect on the size and volume fraction of the primary excess phases. However, particles changed their orientation and formed a banded structure along the shear plane during HPT processing (Figure 3). It should be noted that in order to avoid the effect of the orientation of the particle bands on the properties of the alloy, specimens for tensile and fatigue testing were cut out, so that particle bands in the gauge sections were oriented along the specimen axis.

3.2. Effect of HPT Processing on Tensile and Fatigue Behavior of the Al 6061 Alloy

The results of mechanical tensile testing of the Al 6061 alloy after conventional T6 treatment and HPT processing are presented in Table 1. It is seen that HPT processing resulted in dramatic improvement of the mechanical strength of the material. The yield strength increased from 276 MPa to 605 MPa, whereas the ultimate tensile strength increased from 365 MPa to 675 MPa. No other SPD processing technique, such as ECAP [33,34], or ECAP combined with cold rolling (CR), or artificial aging (AA) [33], led to such a high mechanical strength in this alloy. Such a significant increase in mechanical strength of the HPT-processed Al 6061 alloy can be related primarily to the formation of a homogeneous UFG structure with a very small grain size, promoting the grain size hardening according to the Hall–Petch relationship [2]. At the same time, segregations of solute atoms in the Al solid solution, which were found earlier in the structure of the HPT-processed Al 6061 alloy in [30], can also provide some additional strengthening.

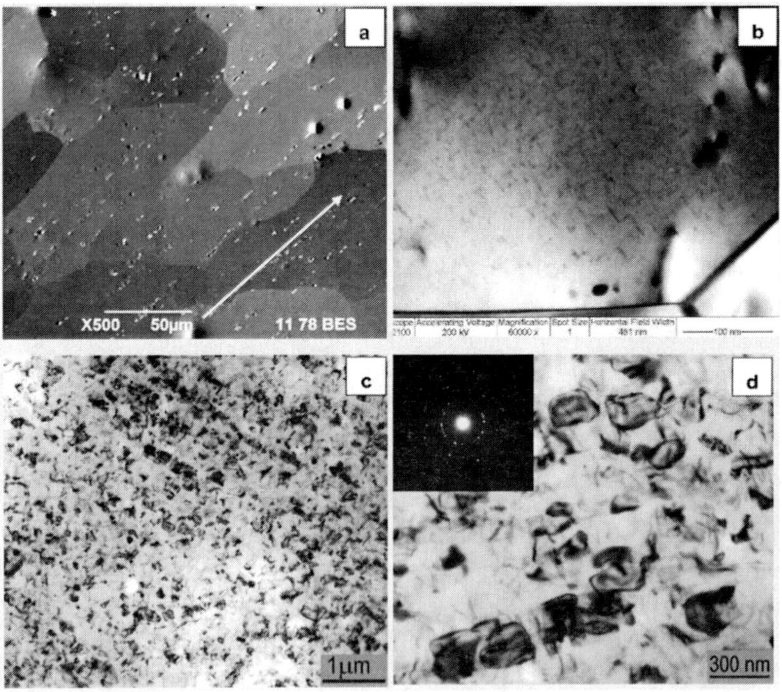

Figure 2. Microstructure of the Al 6061 alloy after conventional T6 treatment (**a**,**b**) and high-pressure torsion (HPT) processing at RT (**c**,**d**). The arrow indicates the pressing direction and the orientation of the bands formed by primary phase particles.

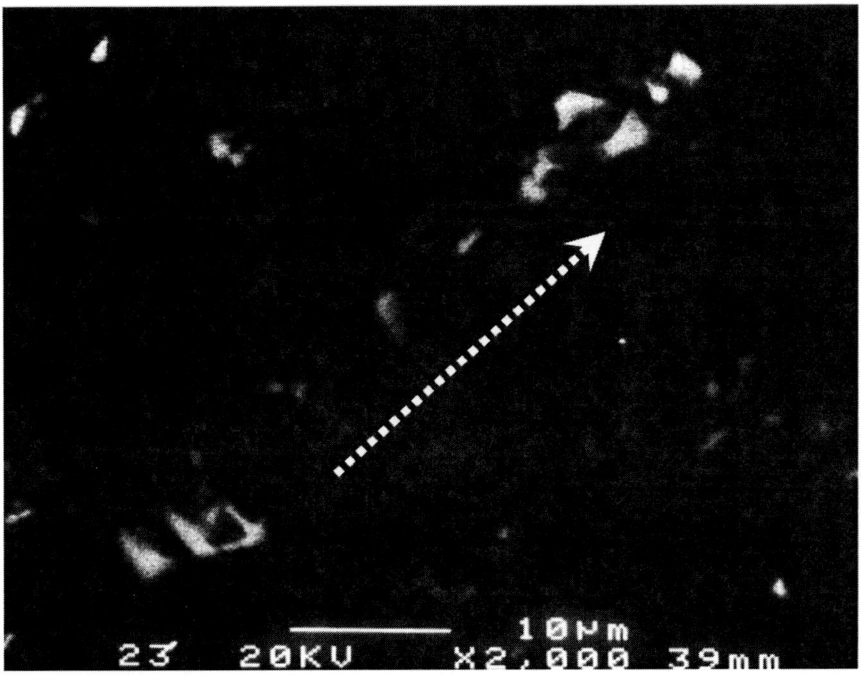

Figure 3. Distribution of the primary excess phases in the Al 6061 alloy processed by HPT at RT (the arrow indicates the shear plane and the orientation of the bands formed by primary phase particles).

Table 1. Mechanical properties of the Al 6061 alloy in the present work in comparison with the mechanical properties reported in earlier studies. ECAP, equal-channel angular pressing; AA, artificial aging; CR, cold rolling; UFG, ultrafine grained; CG, coarse grained.

Processing	State	$\sigma_{0.2}$ (MPa)	σ_{UTS} (MPa)	δ (%)	σ_f (MPa)	σ_f/σ_{UTS}	Reference
T6	CG	276	365	14	100	0.27	present work
HPT for 10 turns at RT	UFG	605	675	5.5	200	0.30	
T6	CG	275	310	12	97	0.31	[34]
ECAP at 125 °C for 1 pass	UFG	310	375	20	80	0.21	[21]
ECAP at 100 °C, 4 passes	UFG	386	434	11	-	-	[33]
ECAP + AA (130 °C for 24 h)	UFG	434	470	10	-	-	
ECAP + CR (15%)	UFG	470	500	8	-	-	

Figure 4 shows the results of fatigue testing of CG T6 treated and HPT-processed material. It is seen that HPT processing has doubled the endurance limit (σ_f) of the material at 10^7 cycles: it increased from 100 MPa in the T6-treated CG alloy to 200 MPa after HPT processing (Figure 4). It should be noted that the σ_f-values obtained after cyclic testing of sub-size specimens with the CG microstructure subjected to the conventional T6 treatment are in very good agreement with the results of the testing of standard reference specimens machined from a similar material [12,13,14,35]. For example, according to the results of fatigue testing of standard samples, the σ_f-value of the Al 6061 alloy in the T6 state is 97 MPa [13].

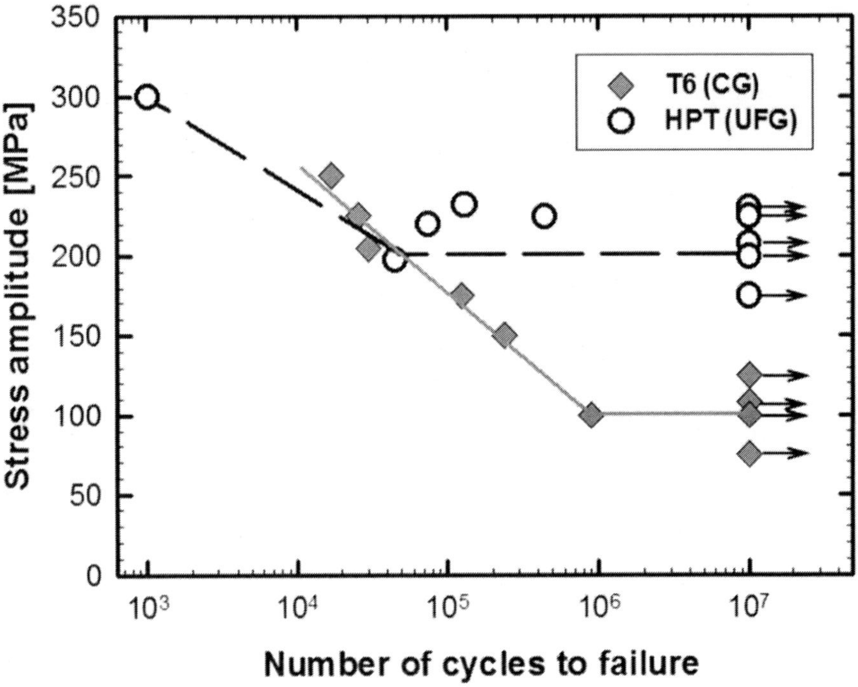

Figure 4. Fatigue curves for the studied Al 6061 alloy.

It should be also noted that the level of σ_f-value achieved in the HPT-processed Al 6061 alloy with an average grain size of 170 nm is also twice the cyclic strength of the UFG material with an average grain size of ~500 nm processed by ECAP in [34].

In Figure 5, the endurance limit is plotted *vs.* the ultimate tensile strength of the CG Al 6061 alloy processed by conventional thermal treatments (O, T4, T6) and by thermo-mechanical treatment (T91) [13,35]. The analysis of the data shows that the ratio of the fatigue limit to the ultimate tensile strength (σ_f/σ_{UTS}) for the given material decreases from ~0.50 to 0.25 with increasing σ_{UTS} (Figure 5). It should be noted that a lower σ_f/σ_{UTS} ratio of 0.20 was reported for the UFG Al 6061 alloy processed by ECAP in [21]. Such low σ_f/σ_{UTS} ratios were found also in other UFG Al-Mg-Si alloys [12,13,14]. In our work, grain refinement down to 170 nm in the Al 6061 alloy via HPT processing resulted in an increase of the σ_f/σ_{UTS} ratio from 0.27 to 0.30 and in a record value of σ_f = 200 MPa. As is well known, fatigue life in the HCF regime is determined by the resistance of the material to crack initiation [13]. Unlike UFG Al 6061 alloy produced by ECAP in [34], our UFG alloy shows a very homogeneous grain structure and, therefore, should have also a homogeneous resistance to crack nucleation all over the gauge section without any 'weakest link'. Thus, it can be concluded that further grain refinement down to the nanoscale and the generation of a very homogeneous grain structure can be proposed as a recipe for the improvement of HCF properties in the Al alloys.

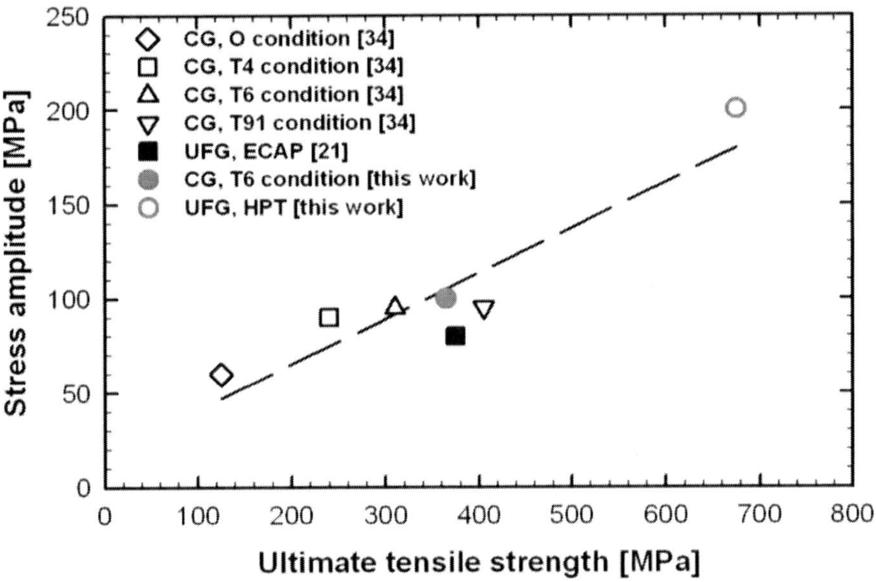

Figure 5. Ultimate tensile strength *vs.* endurance limit for the Al 6061 alloy subjected to various processing treatments.

In the LCF regime, the endurance limit of the HPT-processed Al alloy was somewhat lower than in its CG counterpart (Figure 4). This observation can be rationalized based on the low strain hardening ability of the UFG material, compared to that of the CG T6 one (Table 1), promoting early crack initiation, as well as a much higher volume fraction of grain boundaries favorable for crack propagation [13].

3.3. Analysis of Fatigue Fracture Surfaces

Figure 6 and Figure 7 present SEM images of the fatigue fracture surface of CG T6-treated and HPT-processed Al 6061 alloy, respectively. The area of fatigue crack initiation in the CG T6 material is located on the surface of specimens and is extended into the specimen by ~150 μm (Figure 6a). This is known as Stage I or the short crack growth propagation stage. The increase of the stress intensity factor K as a consequence of crack growth results in the development of slips in different planes close to the crack tip, initiating Stage II, e.g., the stage of stable fatigue crack growth (Figure 6b). An important characteristic of Stage II is the presence of surface ripples, known as "striations". The ductile striations are clearly observed on the fatigue fracture surface (Figure 6c). Their width is in the range of 1–2 μm. The average width of striations is about 3 μm. The most accepted mechanism for the formation of striations is the successive blunting and re-sharpening of the crack tip during cyclic loading [35,36]. Striations are grouped in areas of homogenous propagation separated by tear ridges in which the material fails by shearing, resulting in the apparent continuity of the fatigue advancing front. Finally, Stage III is related to unstable crack growth as K approaches K_{IC} (Figure 6e). At this stage, crack growth is controlled by static modes of failure and is very sensitive to the microstructure, load ratio and stress state. The ductile fracture surface consisting of spherical dimples is observed (Figure 6g). The size of the dimples is in the range of 2–20 μm. Figure 6g shows a fracture pattern within the transition area from a fatigue fracture to the final breakdown.

Similar stages of fatigue crack initiation and propagation can be also easily defined on the fatigue fracture surface of the HPT-processed alloy (Figure 7). The area of fatigue crack initiation is also located on the surface of specimens, though its size is smaller: ~50 μm (Figure 7a). The fracture surface relief in the areas of stable and unstable crack growth is less developed as compared to that on in the fatigue-tested CG T6 specimens. Brittle striations are observed in the area of stable crack propagation in the UFG material (Figure 7b) [36]. Unlike in the fatigue-tested CG T6 specimens, stable and unstable crack propagation is accompanied by the formation and growth of micro-cracks in the direction parallel to the applied load (Figure 7c). The length of these micro-cracks is 50–100 μm. These micro-cracks should result in energy dissipation, thus delaying the growth of the main fatigue crack [35]. The dimpled ductile fracture surface is observed in the area of unstable crack growth and the final fracture of the specimen (Figure 7d–g). However, the size of these dimples is in the range of 1–2 μm, which is much lower compared to that in the CG T6 material. A significant difference in the morphology of fatigue fracture surfaces of the CG T6-treated and UFG

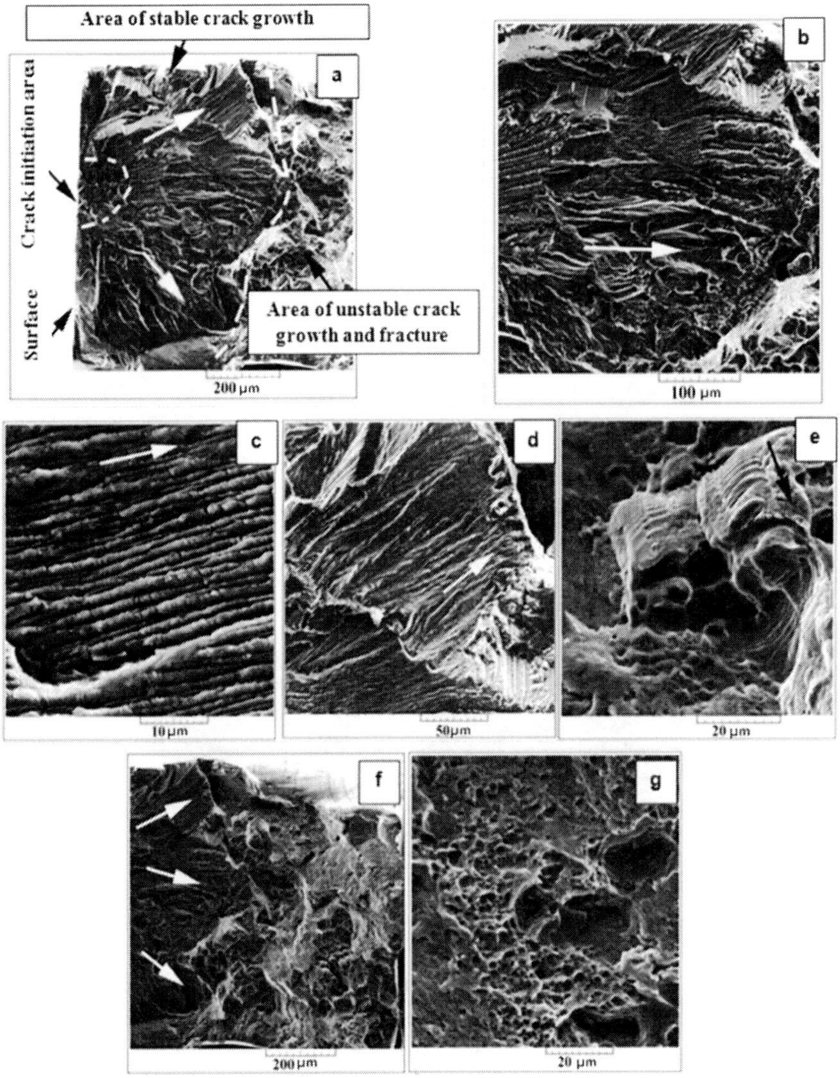

Figure 6. SEM images of the fatigue fracture surface of the CG T6 Al 6061 alloy tested under a stress amplitude of 150 MPa to 2.2×10^5 cycles) (the arrows indicate the direction of fatigue crack propagation).

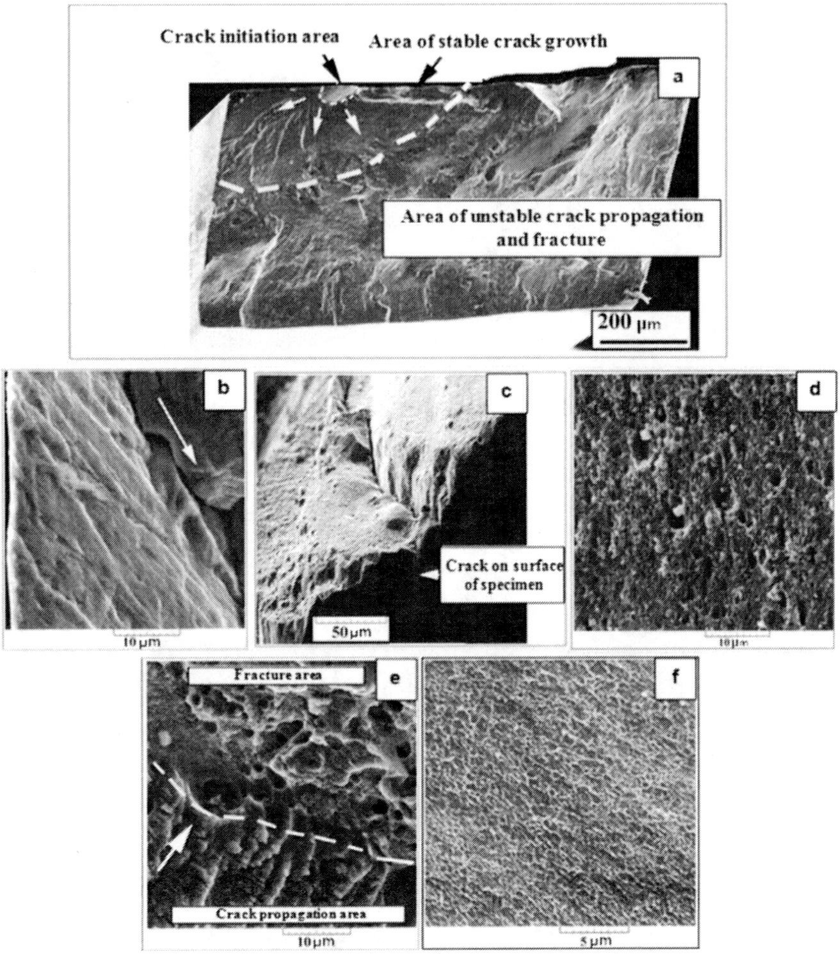

Figure 7. SEM images of the fatigue fracture surface of the UFG Al 6061 alloy: (**a**,**b**,**e**,**f**) tested under a stress amplitude of 225 MPa to 4.4×10^5cycles; (**c**,**d**) tested under a stress amplitude of 200 MPa to 4.35×10^4cycles (the arrows indicate the direction of fatigue crack propagation).

Al 6061 alloy can be related to the grain structure of these material conditions. As is well known, during the loading phase, the crack is opened by normal stress, which activates plastic slips at the tip. At the crack tip of the CG T6 material with a grain size of 70 μm, dislocations are generated in the grain interior and do not necessarily reach the grain boundaries, when the material is cyclically deformed in Stage II. Conventional transgranular fatigue crack growth followed by specimen failure is observed. In the UFG material with a grain size of 170 nm, dislocation pile-ups are definitely formed at grain boundaries near the crack tip, and intragranular slip coupled with grain boundary sliding can occur [37]. Unaccommodated grain boundary sliding might result in local microcracking or nanovoid formation at grain boundaries. Triple junction nanovoids can also appear due to unaccommodated grain boundary sliding. Both transgranular and intragranular fatigue crack growth can take place in Stage II in this case, though this cannot be clearly resolved in SEM due to the very small grain size. In Stage III, nanovoids can grow and partially relieve the constraints on a grain. Individual single crystal ligaments deform extensively and finally experience chisel-point failure. These grain boundary and triple junction voids also act as sites for nucleation of the dimples in the UFG alloy, which are significantly larger than the individual grains, and the rim of these dimples on the fracture surface (1–2 μm) is typically a magnitude larger than the grain size (170 nm) [37].

4. CONCLUSIONS

The following conclusions are outlined.

1. High-pressure torsion (HPT) of the Al 6061 alloy at room temperature leads to the formation of a very homogeneous ultrafine-grained (UFG) microstructure with an average grain size of 170 nm. The yield strength and ultimate tensile strength of the HPT processed alloy are increased as compared to the CG counterpart subjected to the conventional T6 heat treatment from 276 MPa to 605 MPa and from 365 MPa to 675 MPa, correspondingly.

2. HPT processing of the Al 6061 alloy improves its endurance limit by a factor of two (from 100 MPa after T6 treatment to 200 MPa after HPT). This is related to the formation of a very homogeneous UFG microstructure with homogeneous resistance to fatigue crack initiation in the HCF regime. In the LCF regime, the UFG alloy shows somewhat lower fatigue resistance due to its lower strain hardening ability.

3. Classical stages of fatigue crack initiation and propagation are clearly observed on the fatigue fracture surfaces of the CG T6-treated alloy and UFG alloy. Ductile striations are observed in the stage of stable crack propagation in the CG T6-treated alloy, whereas brittle striations seem to dominate on the fatigue fracture surface of the UFG alloy. The dimpled fracture surface is observed at the final stage of crack propagation in both material conditions, with dimples having a smaller size in the UFG alloy.

ACKNOWLEDGMENTS

The work has been done under the financial support of the Russian Federal Ministry for Education and Science. R.Z. Valiev and M. Yu. Murashkin gratefully acknowledge the support through Grant No. 14.B25.31.0017, and I. Ovidko acknowledges the support through Programme "5-100-2020". I. Sabirov acknowledges gratefully the Spanish Ministry of Economy and Competitiveness for financial support through the Ramon y Cajal Fellowship.

AUTHOR CONTRIBUTIONS

Maxim Murashkin formulated the idea of this work, processed the samples, conceived of the workflow and created the initial draft. Ilchat Sabirov carried out mechanical tensile tests and performed microstructural characterization of the processed material. Dmitriy Prosvirnin performed fatigue testing. Vladimir Terentiev performed the SEM studies of the fatigue fracture surfaces. Ilya Ovidko, Ruslan Valiev and Sergey Dobatkin contributed with the overall development of the main concepts presented in this paper. All authors discussed the experimental results, participated in manuscript preparation and approved the final manuscript.

CONFLICTS OF INTEREST

The authors declare no conflict of interest.

REFERENCES

1. Valiev, R.Z.; Zhilyaev, A.P.; Langdon, T.G. *Bulk Nanostructured Materials: Fundamentals and Applications*; John Wiley & Sons: Hoboken, NJ, USA, 2014; p. 456.
2. Sabirov, I.; Murashkin, M.Y.; Valiev, R.Z. Nanostructured aluminium alloys produced by severe plastic deformation: New horizons in development. *Mater. Sci. Eng. A* **2013**, *560*, 1–24.
3. Markushev, M.V.; Murashkin, M.Y. Mechanical properties of submicrocrystalline Al alloys processed by equal-channel angular pressing. *Phys. Met. Metall.* **2000**, *90*, 506–515.
4. Roven, H.J.; Nesboe, H.; Werenskiold, J.C.; Seibert, T. Mechanical properties of aluminum alloys processed by SPD: Comparison of different alloy systems and possible product areas. *Mater. Sci. Eng. A***2005**, *410–411*, 426–429.

5. Lyakishev, N.P.; Alymov, M.I.; Dobatkin, S.V. Structural bulk nanomaterials. *Russ. Metall.* **2003**, *3*, 191–202.

6. Dobatkin, S.V.; Zakharov, V.V.; Vinogradov, A.Y.; Kitagava, N.; Krasilnikov, N.A.; Rostova, T.D.; Bastrash, E.N. Nanocrystalline structure in Al-Mg-Sc alloys by severe plastic deformation. *Russ. Metall.***2006**, *6*, 533–540.

7. Zhao, Y.; Liao, X.; Cheng, S.; Ma, E.; Zhu, Y. Simultaneously increasing the ductility and strength of nanostructured alloys. *Adv. Mater.* **2006**, *18*, 2280–2283.

8. Horita, Z.; Ohashi, K.; Fujita, T.; Kaneko, K.; Langdon, T.G. Achieving high strength and high ductility in precipitation-hardened alloys. *Adv. Mater.* **2005**, *17*, 1599–1603.

9. Markushev, M.V.; Bampton, C.C.; Murashkin, M.Y.; Hardwick, D.A. Structure and properties of ultra-fine grained Al alloys produced by severe plastic deformation. *Mater. Sci. Eng. A.* **1997**, *234–236*, 927–931.

10. Murashkin, M.; Sabirov, I.; Kazykhanov, V.; Bobruk, E.; Dubravina, A.; Valiev, R.Z. Enhanced mechanical properties and electrical conductivity in ultra-fine grained Al alloy processed via ECAP-PC. *J. Mater. Sci.***2013**, *48*, 4501–4509.

11. Valiev, R.Z.; Murashkin, M.Y.; Sabirov, I. A nanostructural design to produce high-strength Al alloys with enhanced electrical conductivity. *Scr. Mater.* **2014**, *76*, 13–16.

12. Vinogradov, A.Y.; Khasimoto, S. Fatigue in ultrafine-grained materials processed by equal-channel angular pressing. *Russ. Metall.* **2004**, *1*, 42–51.

13. Estrin, Y.; Vinogradov, A. Fatigue behaviour of light alloys with ultrafine grain structure produced by severe plastic deformation: An overview. *Int. J. Fatigue* **2010**, *32*, 898–907.

14. Estrin, Y.; Vinogradov, A. Extreme grain refinement by severe plastic deformation: A wealth of challenging science. *Acta Mater.* **2013**, *61*, 782–817.

15. Höppel, W.; Kautz, M.; Murashkin, M.Y.; Xu, C.; Langdon, T.G.; Valiev, R.Z.; Mughrabi, H. An overview: Fatigue Behavior of Ultrafine-Grained Metals and Alloys. *Int. J. Fatigue* **2006**, *28*, 1001–1010.

16. Höpel, H.W.; Göken, M. Fatigue Behavior in Nanostructured Metals. In *Nanostructured Metals and Alloys: Processing, Microstructure,*

Mechanical Properties and Applications; Whang, S.H., Ed.; Woodhead Publishing Limited: Suite, PA, USA, 2011; pp. 507–541.

17. Mughrabi, H.; Höppel, H.W.; Kautz, M. Fatigue and microstructure of ultra-fine grained metals produced by severe plastic deformation. *Scr. Mater.* **2004**, *51*, 807–812.

18. Malekjani, S.; Hodgson, P.D.; Cizek, P.; Sabirov, I.; Hilditch, T.B. Cyclic deformation response of UFG 2024 Al alloy. *Int. J. Fatigue* **2011**, *33*, 700–709.

19. Canadinc, D.; Maier, H.J.; Gabor, P.; May, J. On the cyclic deformation response of ultrafine-grained Al–Mg alloys at elevated temperatures. *Mater. Sci. Eng. A* **2008**, *496*, 114–120.

20. Vinogradov, A.; Washikita, A.; Kitagawa, K.; Kopylov, V.I. Fatigue life of fine-grain Al-Mg-Sc alloys produced by equal-channel angular pressing. *Mater. Sci. Eng. A* **2003**, *349*, 318–326.

21. Chung, C.S.; Kim, J.K.; Kim, H.K.; Kim, W.J. Improvement of high-cycle fatigue life in a 6061 Al alloy produced by equal channel angular pressing. *Mater. Sci. Eng. A* **2002**, *337*, 39–44.

22. Lapovok, R.; Loader, C.; Dalla Torre, F.H.; Semiatin, S.L. Microstructure evolution and fatigue behavior of 2124 aluminum processed by ECAE with back pressure. *Mater. Sci. Eng. A* **2006**, *425*, 36–46.

23. Malekjani, S.; Hodgson, P.D.; Cizek, P.; Hilditch, T.B. Cyclic deformation response of ultra-fine pure Al.*Acta Mater.* **2011**, *59*, 5358–5367.

24. Valiev, R.Z.; Enikeev, N.A.; Murashkin, M.Y.; Kazykhanov, V.U.; Sauvage, X. On the origin of the extremely high strength of ultra-fine grained Al alloys. *Scr. Mater.* **2010**, *63*, 949–952.

25. Liddicoat, P.V.; Liao, X.Z.; Zhao, Y.; Zhu, Y.; Murashkin, M.Y.; Lavernia, E.J.; Valiev, R.Z.; Ringer, S.P. Nanostructural hierarchy increases the strength of aluminium alloys. *Nat. Comm.* **2010**, *1*, 63–70.

26. Ruffing, C.; Ivanisenko, Y.; Kerscher, E. Fatigue behavior of ultrafine grained medium Carbon steel processed by severe plastic deformation. *IOP Conf. Series: Mater. Sci. Eng.* **2014**, *63*, 012163.

27. Polmear, I.J. Light Alloys. In *From Traditional Alloys to Nanocrystals*; Butterworth-Heinemann: Oxford, UK, 2006; p. 417.

28. Mondolfo, L.F. *Aluminum Alloys: Structure and Properties*; Butterworth: Oxford, UK, 1976; p. 971.

29. Murayama, M.; Hono, K. Pre-precipitate clusters and precipitation processes in Al-Mg-Si alloys. *Acta Mater.* **1999**, *47*, 1537–1548.

30. Nurislamova, G.; Sauvage, X.; Murashkin, M.; Islamgaliev, R.; Valiev, R. Nanostructure and related mechanical properties of an Al-Mg-Si alloy processed by severe plastic deformation. *Phil. Mag. Lett.* **2008**,*88*, 459–466.

31. Moreno-Valle, E.C.; Sabirov, I.; Perez-Prado, M.T.; Murashkin, M.Y.; Bobruk, E.V.; Valiev, R.Z. Effect of the grain refinement via severe plastic deformation on strength properties and deformation behavior of an Al 6061 alloy at room and cryogenic temperatures. *Mater. Lett.* **2011**, *65*, 2917–2919.

32. Sha, G.; Tugcu, K.; Liao, X.Z.; Trimby, P.W.; Murashkin, M.Y.; Valiev, R.Z.; Ringer, S.P. Strength, grain refinement and solute nanostructures of an Al-Mg-Si alloy (AA6060) processed by high-pressure torsion.*Acta Mater.* **2014**, *63*, 169–179.

33. Murashkin, M.Y.; Markushev, M.V.; Ivanisenko, Y.V.; Valiev, R.Z. Strength of commercial aluminum alloys after equal channel angular pressing and post-ECAP processing. *Solid State Phen.* **2006**, *114*, 91–96.

34. Hatch, J.E. *Aluminium: Properties and Physical Metallurgy*; ASM International: Materials Park, OH, USA, 1984; p. 636.

35. Suresh, S. *Fatigue of Materials*; Cambridge University Press: Cambridge, UK, 1998; p. 679.

36. Milella, P.P. *Fatigue and Corrosion in Metals*; Springer: Berlin, Germany, 2013; p. 763.

37. Dalla Torre, F.; van Swygenhoven, H.; Victoria, M. Nanocrystalline electrodeposited Ni: Microstructure and tensile properties. *Acta Mater.* **2002**, *50*, 3957–3970.

Chapter 11

Suppression of Superconductivity and Structural Phase Transitions Under Pressure in Tetragonal Fes

Xiaofang Lai, Ying Liu, Xujie Lü, Sijia Zhang, Kejun Bu, Changqing Jin, Hui Zhang, Jianhua Lin & Fuqiang Huang

Beijing National Laboratory for Molecular Sciences and State Key Laboratory of Rare Earth Materials Chemistry and Applications, College of Chemistry and Molecular Engineering, Peking University, Beijing 100871, China

ABSTRACT

Pressure is a powerful tool to study iron-based superconductors. Here, we report systematic high-pressure transport and structural characterizations of the newly discovered superconductor FeS. It is found that superconductor FeS (tetragonal) partly transforms to a hexagonal structure at 0.4 GPa, and then completely transforms to an orthorhombic phase at 7.4 GPa and finally to a monoclinic phase above 9.0 GPa. The superconducting transition temperature of tetragonal FeS was gradually depressed by pressure, different from the case in tetragonal FeSe. With pressure increasing, the S-Fe-S angles only slightly change but the anion height deviates farther from 1.38 Å. This change of anion height, together with the structural instability under pressure, should be closely related to the suppression of superconductivity. We also observed an anomalous metal-semiconductor transition at 6.0 GPa and an unusual increased resistance with further compression above 9.6 GPa. The former can be ascribed to the tetragonal-orthorhombic structural phase transition, and the latter to the electronic structure changes of the high-pressure monoclinic phase. Finally, a phase diagram of tetragonal FeS as functions of pressure and temperature was mapped out for the first time, which will shed new light on understanding of the structure and physics of the superconducting FeS.

INTRODUCTION

Iron-based superconductors are of great interest and fundamental importance due to their rich structural and physical properties. Up to now, at least four structure types of iron-based superconductors have been developed, such as ZrCuSiAs-type LnOFeAs (Ln = rare earth metal) and *AeFFeAs* (*Ae* = alkaline metal), $ThCr_2Si_2$-type $AeFe_2As_2$ and $A_xFe_{2-y}Se_2$ (A = alkali metal), PbFCl-type AFeAs and anti-PbO-type FeSe1′2′3′4′5′6′7. Among them, the Fe_2X_2 (X = pnictogen or chalcogen) layers built up by edge-sharing FeX_4 tetrahedra are responsible for superconductivity. Their parent compounds usually exhibit a tetragonal to orthorhombic structural transition as well as a paramagnetic to antiferromagnetic transition at low temperatures8′9, and superconductivity emerges when the structural distortion and magnetic ordering are suppressed by chemical doping or external pressure10′11. Compared with chemical doping, pressure is a 'cleaner' way to tune the crystal structure and physical properties of iron-based superconductors without introducing impurities.

In the iron-based superconductors, the response of the superconducting transition temperature (T_c) to pressure is complex and sensitively depends on the composition of the materials and their doping level. For iron pnictides, dT_c/dP is positive for underdoped compounds and negative for the overdoped ones12′13. First-principle calculations suggest that this may be caused by the increase of orbital degeneracy in the underdoped regime and decrease of orbital degeneracy at the overdoped regions under pressure14. For $A_xFe_{2-y}Se_2$, it is interesting that a second superconducting phase suddenly reemerges above 11.5 GPa, after the T_c drops from the first maximum of 32 K at 1 GPa15. The strongest pressure effect was found in FeSe, whose T_c could be dramatically enhanced from 8 K to 36.7 K under external pressure at 8.9 GPa16. ^{77}Se NMR (nuclear magnetic resonance) measurement reveals that this may be positively related to the strongly enhanced antiferromagnetic spin fluctuations under high pressure17. Moreover, it is found the anion height (the distance of the anion from the Fe plane), rather than the geometry of $FeSe_4$ tetrahedra, is the key factor determining T_c of FeSe18′19, in contrast to the iron pnictide superconductors for which T_c attains a maximum value when the $FeAs_4$tetrahedra form a regular shape and meanwhile the anion height dependence of T_c shows a symmetric curve with a peak around 1.38 Å20′21′22. However, more iron chalcogenide superconductors are needed to verify this issue.

Recently, the discovery of superconductivity in tetragonal FeS23, obtained by a low-temperature hydrothermal route similar to that used for producing superconductor $Li_{1-x}Fe_xOHFeSe$24′25′26, has raised considerable interest. Although its T_c is only 5 K, FeS provides a new and simple platform to realize high-temperature superconductors and to study the underlying mechanism of iron-based superconductivity. According to the low-temperature X-ray diffraction investigations, superconducting FeS does not undergo any structural phase transition on cooling27, in contrast to FeSe that shows a structural phase transition from tetragonal to orthorhombic at 70~90 K28′29. Given the similar crystal and electronic structures between FeS and FeSe30, it is of fundamental importance to know whether the strong pressure effect still exists in FeS and how the pressure-tuned superconductivity of FeS relates to its structural properties at atomic level. Here, the behavior of superconductor FeS under pressure was systematically investigated using a combination of electrical resistance, Raman spectroscopy and synchrotron X-ray diffraction (XRD) measurements.

RESULTS

At ambient pressure, tetragonal FeS shows metallic behavior with a resistance that decreases with decreasing temperature; when the temperature is below 5 K, it enters into a superconducting state, consistent with the magnetization measurements (Supplementary Fig. S1). Figure 1 shows the electrical properties of tetragonal FeS under pressure. Upon increasing pressure, the T_c of tetragonal FeS gradually drops and the overall resistance decreases until the pressure reaches 6.0 GPa. The pressure-induced resistance drop below 6.0 GPa is associated with the broadening of bands, caused by the shortening and bending of bonds31′32. It should be noted that the decrease of resistance with pressure was also observed for non-superconducting tetragonal FeS in which neither metallization nor superconductivity was induced by pressure33. The low-temperature resistance at 4.5 GPa is a little higher than that at 1.9 and 2.8 GPa and the reason is not clear yet. One possibility is that there exist other phases with higher resistance, as revealed by synchrotron XRD in the following. Obviously, a metal-semiconductor transition occurs at 6.0 GPa, above which the sample becomes semiconducting. As discussed below, this should be associated with a structural phase transition. The pressure dependence of resistance at 2 K, 100 K and 273 K shows that the resistance first decreases with pressure and then increases sharply upon further compression from 6 to 12 GPa, which also implies a phase transition at ~6.0 GPa. It is unusual that, for pressure above 9.6 GPa, the resistance further increases with compression. Since no structural phase transitions occur in the pressure range, this phenomenon should be attributed to electronic structure changes. As shown in Supplementary Fig. S2, when pressure is released, the sample exhibits metallic behavior at 2.6 GPa but insulating behavior below 1.8 GPa; superconductivity is not recovered even at 0.4 GPa. These results suggest that the pressure-induced changes of electrical properties in tetragonal FeS are irreversible.

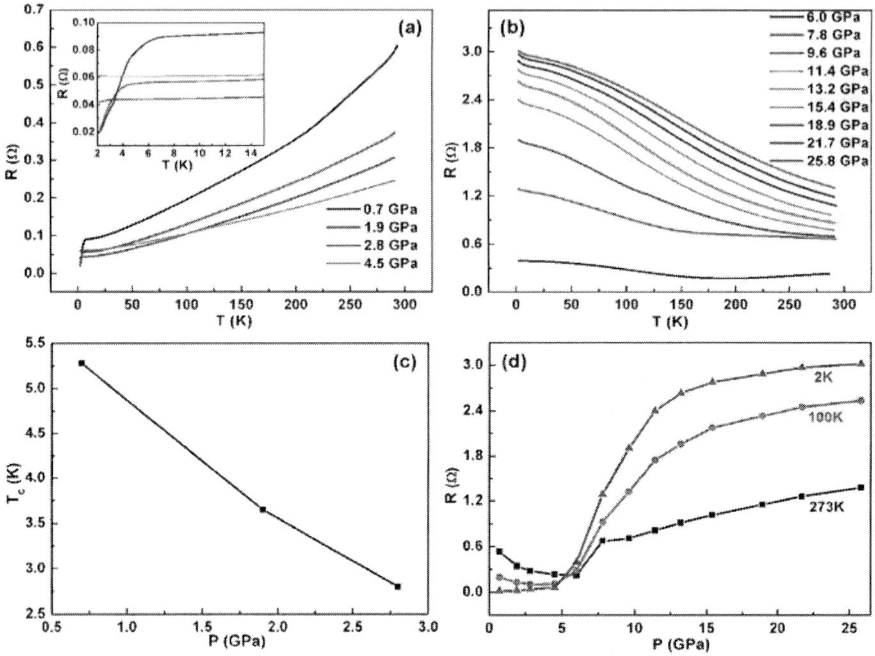

Figure 1: (**a,b**) Temperature dependence of resistance for tetragonal FeS at various pressures, demonstrating a metal-insulator transition at 6.0 GPa. (**c**) Pressure dependence of the superconducting transition temperature T_c. (**d**) Selected pressure dependence of resistance at various temperatures, showing a phase transition at around 6.0 GPa.

Raman spectra of tetragonal FeS collected up to 35.4 GPa at room temperature are shown in Fig. 2. The frequencies of the two well-resolved features seen at 0.4 GPa are 209 and 300 cm^{-1}, close to those observed at ambient pressure34′35. For clarity, we denote these two modes as M1 and M2, which may correspond respectively to one lattice mode of FeS and the symmetric stretching mode34′35. Upon increasing pressure, the positions of both modes shift to a high energy, with M1 mode hardening by 16.1 cm^{-1} and M2 mode hardening by 29.4 cm^{-1} at 5.2 GPa. At 7.2 GPa, both modes disappear. Generally speaking, vanishing of a Raman peak can be accounted for in two ways: (1) a structural transition where no Raman mode is allowed by symmetry, (2) a transition to a metallic state where the Raman signal becomes extremely weak due to the limited penetration depth of the exciting laser. In this case, our synchrotron XRD results clearly show that the vanishing of the Raman peaks is due to the absence of the tetragonal phase.

This structural change is irreversible as gradual depressurization to 1.2 GPa does not lead to reemergence of the characteristic Raman peaks of tetragonal FeS.

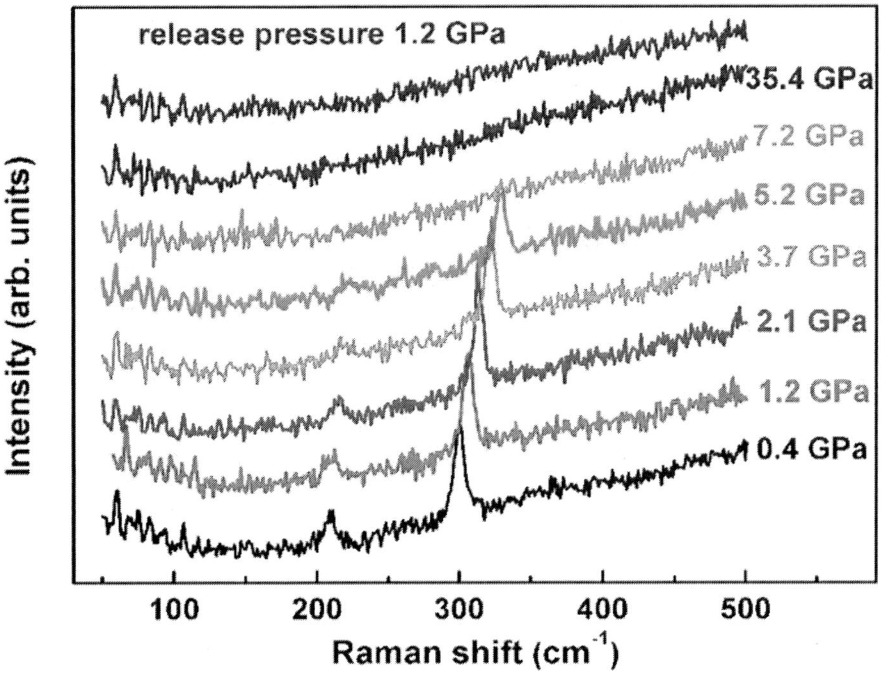

Figure 2: Raman spectra of tetragonal FeS at various pressures. It shows that tetragonal FeS is absent above 7.2 GPa.

Stoichiometric FeS can form tetragonal (space group *P4/nmm*) or hexagonal troilite-type structure (space group *P*-62*c*) at ambient pressure and room temperature. A plenty of experimental investigations on high-pressure structures of troilite have been reported, which shows that at both room temperature36'37 and 17 K38, troilite first transforms to an orthorhombic MnP-type structure (space group Pnma) at 3 GPa and then to a monoclinic structure (space group $P2_1/a$) at 7 GPa. Interestingly, semiconductor-metal-semiconductor transitions were observed for troilite under pressure, which is believed to be driven by the above successive structural phase transitions39. However, it remains controversial whether the orthorhombic MnP-type FeS is metallic throughout the whole pressure range or not, since semiconducting behavior was reported for FeS at 6 GPa40. Here we present the pressure effect on the structure of tetragonal FeS. Synchrotron XRD data of tetragonal FeS at high pressures up to 26.2 GPa at room temperature are shown in Fig. 3. All the observed diffraction reflections obtained at ambient pressure can be indexed with a tetragonal structure. With the onset of pressure, taking $P = 0.4$ GPa for example, new diffraction peaks appear, with the one in regions of 2ϑ between 10.5° and 12° the most prominent. This signals the appearance of the hexagonal structure. Actually, a pressure-induced irreversible tetragonal-hexagonal phase transition has been observed in the non-superconducting tetragonal FeS (the maximum pressure applied is 1.64 GPa) but has not caused much attention[33]. For 0 GPa < P < 4.6 GPa, the relative diffraction intensity of the tetragonal and hexagonal structures only slightly change. Then an orthorhombic phase arises at $P = 4.6$ GPa, with the characteristic diffraction peak at $2\vartheta = 5.7°$. This is in accord with that observed for nanocrystalline mackinawite in which an irreversible first-order structural phase transition from tetragonal to orthorhombic takes place at 3~3.7 GPa, depending on the particle size[41]. As can be seen, the tetragonal phase disappears at 7.4 GPa, consistent with the Raman results. A further structural phase transition to a monoclinic phase begins at 9.0 GPa, above which no more structural changes can be observed. During decompression (Supplementary Fig. S3), the monoclinic phase transforms back to orthorhombic phase at 4.0 GPa and to hexagonal phase at 0 GPa. It is worth noting that tetragonal phase is not recovered. At this point, we reexamined the transport properties shown in Fig. 1 and Supplementary Fig. S2 and found that the hexagonal phase is a semiconductor with extremely high resistance, much higher than the insulating monoclinic phase, and confirmed that the orthorhombic phase is not metallic throughout the whole pressure range. For example, the orthorhombic phase is metallic at 2.6 GPa but semiconducting at 4.6 GPa. This reminds us of the pressure-induced metal-semiconductor transition in troilite which should be closely related to the high-spin to low-spin transition rather than the orthorhombic-monoclinic transition from the point of view of their critical pressures[37,39,40,42,43]. Moreover, high-pressure-temperature studies of troilite showed that, for the high-pressure phase between 3~7 GPa, there existed an electronic transition, probably high-spin to low-spin transition, which should not be related to any structural phase transition[44,45]. Therefore, we presume that the metal-semiconductor transition of the orthorhombic phase observed in this study may be related to a pressure-induced high-spin to low-spin transition. More measurements, such as Mössbauer spectroscopy and synchrotron X-ray emission spectroscopy are needed to further clarify the nature of this transition.

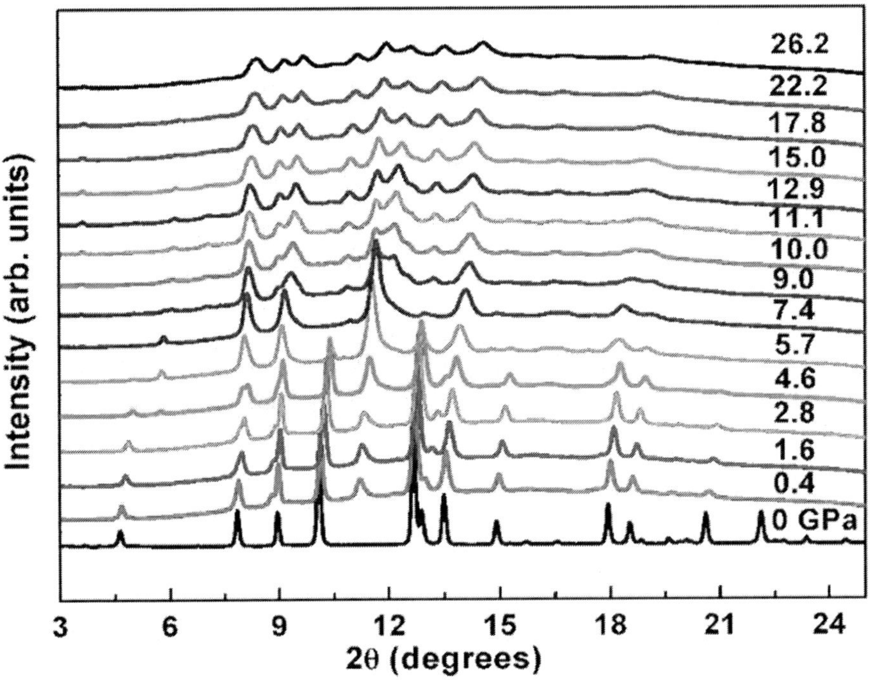

Figure 3: Room-temperature synchrotron XRD patterns of tetragonal FeS at various pressures. Tetragonal FeS first transforms to a hexagonal structure at 0.4 GPa and then to an orthorhombic structure at 4.6 GPa and finally to a monoclinic structure above 9.0 GPa.

DISCUSSION

The lattice parameters and volume of tetragonal FeS for the whole pressure range studied, determined by indexing the synchrotron XRD patterns, are shown in Supplementary Fig. S4. To gain more structural information, Rietveld refinements were performed against the diffraction data collected at ambient pressure and at pressures of 0.4, 1.6, 2.8, 7.4 and 12.9 GPa. Representative refinement profiles and refined structure parameters were displayed in Fig. 4 and Supplementary Table SI. These results confirm that the sample is a mixture of tetragonal and hexagonal phases at low pressures and the two high-pressure phases are orthorhombic and monoclinic, respectively. The evolution of the structural parameters of the tetragonal phase (superconducting phase) with pressure is of particular interest because it is closely correlated with superconductivity. Figure 5(a) shows the pressure dependence of the lattice constants of the tetragonal phase, normalized to the ambient-pressure values. The *c* axis is more compressible than *a* axis, with the former contracting by 5% and the latter contracting by

1% at 2.8 GPa. Similar anisotropic compressibility was also observed in FeSe superconductor, which can be accounted for by the anisotropy in bonding of this type of structure18. Correspondingly, a remarkable reduction in the unit-cell volume was observed, with 7% reduction at 2.8 GPa. It is well known that the maximum T_c value of iron pnictide superconductors is apparently attained when the $FeAs_4$ tetrahedron assumes a regular shape20′22. However, this rule is not applicable to FeS. On one hand, FeS_4 tetrahedron is more regular than $FeSe_4$ tetrahedron, but FeS has lower T_c than FeSe. On the other hand, upon compression, the S-Fe-S angles of FeS_4 tetrahedron only slightly change (Fig. 5c), while T_c decreases significantly. As shown in Fig. 5(d), the ambient-pressure S height (anion height) is 1.324 Å and it monotonically decreases with increasing pressure. We know that for both iron pnictides and FeSe, optimal T_c is obtained when the anion height is close to 1.38 Å18′19′21′22. Here, it seems also applicable to FeS, that is, $T_c$decreases as S height deviates farther from 1.38 Å. Therefore, our results support the proposition that the anion height, rather the geometry of FeX_4 tetrahedra, is the key factor determining T_c of iron-based superconductors containing various anions19.

The high-pressure structural phase transitions in tetragonal FeS and troilite are similar except that there exists a tetragonal phase in tetragonal FeS at low pressures. According to the low-temperature X-ray diffraction investigations, the tetragonal phase does not undergo any structural phase transition on cooling at ambient pressure27. Actually, at high pressures, the tetragonal phase would not present structural phase transitions on cooling either; otherwise, we could not observe superconductivity at low temperature. The high-pressure studies of troilite reveal that the pressure-induced structural phase transitions of troilite are the same at room temperature and 17 K36′37′38, that is, at a certain high pressure, all the hexagonal, orthorhombic and monoclinic phases would not exhibit structural phase transitions upon cooling. Based on these results, it is reasonable for us to map out a pressure-temperature (*P-T*) phase diagram shown in Fig. 6. This *P-T*phase diagram shares some similarity to that of FeSe which starts to transform to a hexagonal NiAs-type structure at 9.0 GPa and then to an orthorhombic MnP-type structure at higher pressures18. Combining with the high-pressure studies on troilite39′40, we can make a better understanding of the *P-T* phase diagram mapped out here: the tetragonal phase is metallic and the hexagonal phase is semiconducting with strong electronic correlation, both tending to have better electrical conductivity

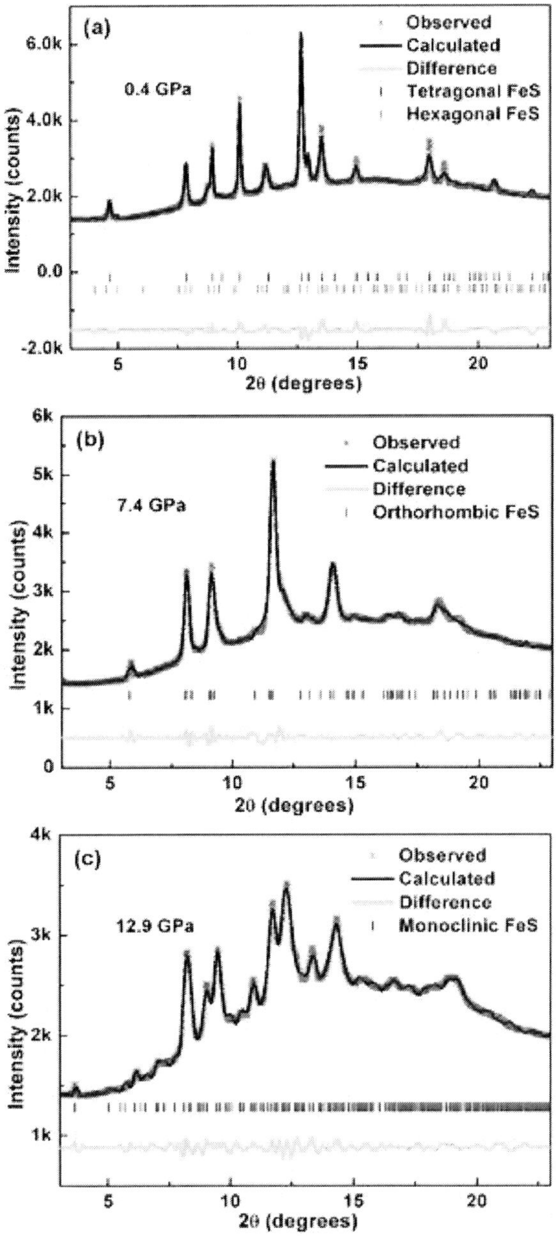

Figure 4: Rietveld refinements of the synchrotron XRD patterns of tetragonal FeS at (**a**) 0.4 GPa, (**b**) 7.4 GPa, and (**c**) 12.9 GPa. Red small points represent the experimental values, black solid line is the calculated pattern, green solid line is the difference between the experimental and calculated values, and blue or dark yellow vertical bars are the Bragg positions.

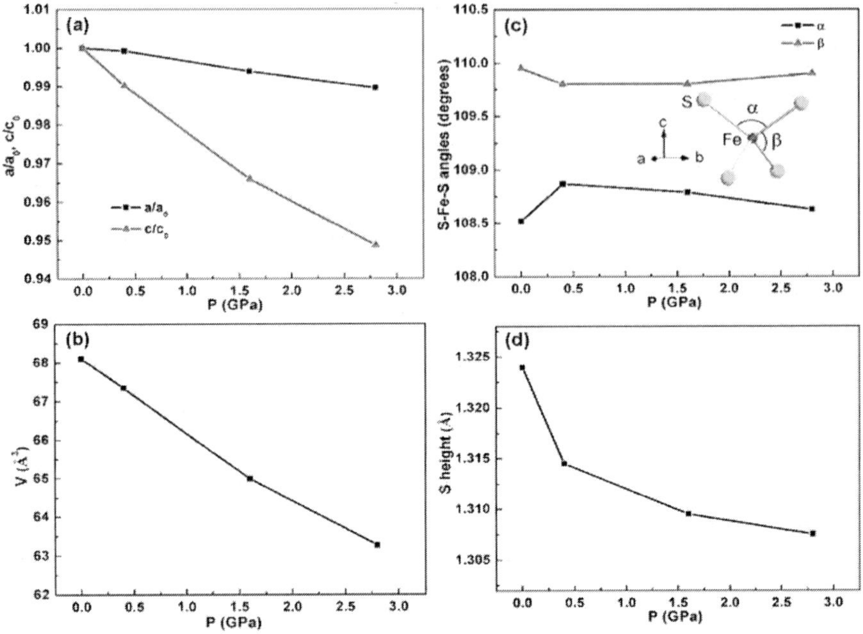

Figure 5: Structural parameters of the tetragonal phase under pressure. (**a**) Pressure dependence of the lattice constants normalized to the ambient-pressure values. (**b–d**) Pressure dependence of the unit cell volume, S-Fe-S bond angles and the S height from the iron layer, respectively.

under pressure; the orthorhombic phase indeed shows metallic behavior at low pressures as reported in the literature but it becomes semiconducting at high pressures; the monoclinic phase is a band insulator whose band gap opens up between the nonbonding and antibonding bands, and because the gap increases more rapidly with pressure than the bandwidth, it becomes more insulating with increasing pressure. Based on this *P-T* phase diagram, we can ascribe the observed metal-semiconductor transition in Fig. 1 to the tetragonal-orthorhombic structural phase transition and the increased resistance with pressure above 9.6 GPa to the electronic structure changes of the monoclinic phase.

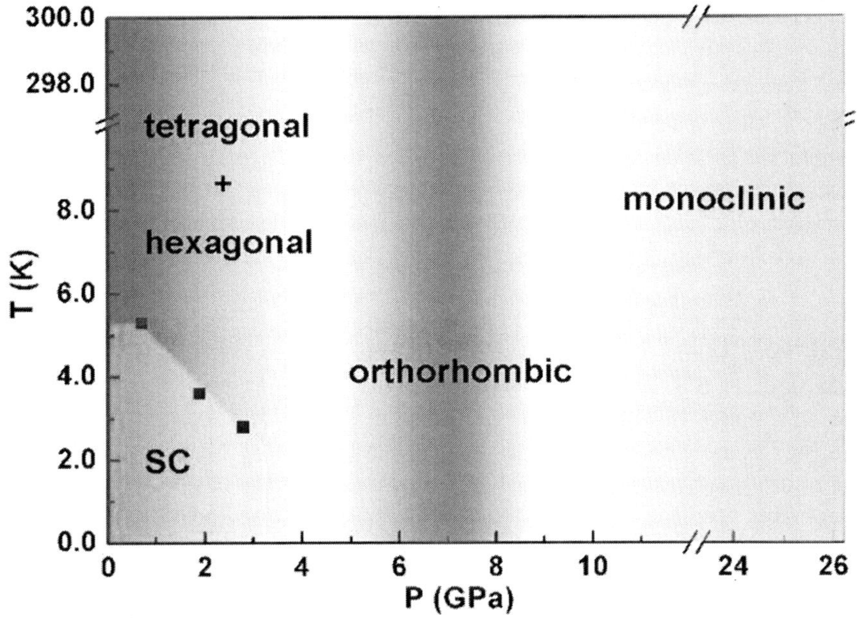

Figure 6: Phase diagram of tetragonal FeS as functions of pressure and temperature. At ambient pressure, the sample is tetragonal (*P*4/*nmm*); once the pressure is onset, the sample becomes a mixture of tetragonal (*P*4/*nmm*) and hexagonal (*P*-62*c*) phases; at intermediate pressure region, the sample is orthorhombic (*Pnma*); at high pressure region, the sample is monoclinic ($P2_1/a$). The solid squares represent T_c extracted from electrical resistance measurements.

Comparatively speaking, tetragonal FeS is less stable under pressure than the other iron-based superconductors. Although pressure-induced structural phase transitions were also observed in iron pnictide superconductors, the phase transitions are commonly tetragonal to collapsed tetragonal and isostructural in nature. For example, isostructural phase transitions were reported in ZrCuSiAs-type $Nd(O_{0.88}F_{0.12})FeAs$, $ThCr_2Si_2$-type $CaFe_2As_2$ and $EuFe_2As_2$, and PbFCl-type $Na_{1-x}FeAs$22′46′47′48. In these cases, the high-pressure tetragonal phase also possesses superconductivity and provides additional information to study the relationship between superconductivity and atomic crystal structure. For FeSe and $FeSe_{0.57}Te_{0.43}$, the pressure onset of structural phase transition is much higher than FeS, 9.0 GPa and 3.0 GPa, respectively18′49. It is suggested that much higher T_c should be realized by suppressing both the pressure-induced competing tetragonal to hexagonal and orthorhombic to monoclinic structural instabilities in FeSe and $FeSe_{0.57}Te_{0.43}$49. So apart from the anion height change, the complex structural instabilities of FeS may also be one of the possible reasons responsible for the suppression of superconductivity under pressure.

In conclusion, we have presented systematic high-pressure study on the superconductor FeS. Once the pressure is onset, part of tetragonal FeS will transform to a hexagonal structure. At higher pressures, another two successive phase transitions take place, with the sample being solely orthorhombic at 7.4 GPa and solely monoclinic above 9.0 GPa. The superconductivity of tetragonal FeS is gradually suppressed by pressure, probably due to the S height becoming farther from 1.38 Å and the structural instability under pressure. Thereafter, a metal-semiconductor transition, driven by the tetragonal-orthorhombic phase transition, was observed at 6.0 GPa. The unusual increase of electrical resistance with pressure above 9.6 GPa was attributed to the electronic structure changes of the high-pressure monoclinic phase39. Both the structural transitions and the anion height dependence of superconductivity under pressure resemble those observed for FeSe superconductor18′19, emphasizing the importance of anion height and structural stability, rather than the geometry of FeX_4 tetrahedra, on superconductivity. Our results will shed new light on understanding of the superconducting mechanism and structure-property relationship of iron chalcogenide superconductors.

METHODS

Sample synthesis and characterization

Tetragonal FeS was synthesized by the hydrothermal reaction of iron powder with sulfide solution, as described in ref. 23. The ambient-pressure superconducting properties were studied by electrical resistance and magnetic susceptibility measurements in a physical property measurement system (PPMS) of Quantum Design.

High-pressure electrical resistance measurements

High-pressure electrical resistance measurements were carried out in a diamond anvil cell (DAC) made of CuBe alloy50′51′52′53′54. A T301 stainless steel gasket was preindented from a thickness of 250 μm to 40 μm, and a hole was drilled at the center with a diameter of about 200 μm. Fine cubic boron nitride (cBN) powder was used to cover the gasket to protect the electrode leads from the metallic gasket. The cBN powder was pressed and further drilled to give a center chamber of 100 μm in diameter. FeS flake with a dimension of 80 μm × 80 μm × 10 μm was loaded with soft NaCl fine powder surrounding it as pressure transmitting medium that can provide a good quasi-hydrostatic pressure environment. We used slim Au wire of 18 μm in diameter as electrodes. Pressure was measured via the ruby fluorescence method55at room temperature before and after each cooling. The DAC was placed inside a MagLab system to perform the experiments with an automatic temperature control. A thermometer located near the diamond of the DAC is used for monitoring the sample temperature.

High-pressure laser Raman spectroscopy

High-pressure Raman experiments were conducted on FeS flake using Renishaw in Via Raman microscope with laser wavelength of 532 nm and spectral resolution of ~0.2 cm^{-1}. The gasket was T301 stainless steel, which was preindented from a thickness of 250 μm to 40 μm and drilled to give a center hole of 120 μm in diameter. Ruby spheres were pressure monitors.

High-pressure synchrotron XRD experiments

High-pressure angle-dispersive XRD experiments were carried out at the 16 ID-B station of the High Pressure Collaborative Access Team (HPCAT) at the Advanced Photon Source (APS), Argonne National Laboratory (ANL) using a wavelength of 0.406626 Å. A symmetrical diamond anvil cell was employed to generate high pressure. A rhenium gasket was preindented to 40 μm in thickness followed by laser-drilling the central part to form a 200 μm diameter hole to serve as the sample chamber. The FeS sample and two small ruby balls were loaded into the chamber. Neon was used as the pressure-transmitting medium and the pressure was measured using a ruby fluorescence calibrate55. XRD images were collected with a Pilatus detector and were integrated using the Fit2D program56. The integrated XRD patterns were further analyzed using the Fullprof program package57 for the Rietveld refinements.

ADDITIONAL INFORMATION

How to cite this article: Lai, X. *et al*. Suppression of superconductivity and structural phase transitions under pressure in tetragonal FeS. *Sci. Rep.* **6**, 31077; doi: 10.1038/srep31077 (2016).

ACKNOWLEDGEMENTS

This work was financially supported by Innovation Program and "Strategic Priority Research Program (B)" of the Chinese Academy of Sciences (Grants

KJCX2-EW-W11 and XDB04040200), National Natural Science Foundation of China (Grants 91122034, 51125006, 51202279, 61376056, 21201012, and 11275012), Science and Technology Commission of Shanghai (Grant 12XD1406800). Works at IOPCAS are supported by NSF & MOST & CAS of China through research projects. HPCAT operations are supported by US-DOE-NNSA under Award DE-NA0001974 and DOE-BES under Award DE-FG02-99ER45775, with partial instrumentation funding by NSF. The gas loading was performed at GeoSoilEnviroCARS, APS, ANL, supported by EAR-1128799 and DE-FG02-94ER14466.

CONTRIBUTIONS

C.J., J.L. and F.H. conceived & designed the research; X.L. synthesized the sample and performed preliminary characterizations with the help of K.B.; Y.L. conducted the high-pressure transport and Raman spectroscopy measurements with assistance from S.Z.; X.L., Y.L., X.J.L., C.J. and F.H. analyzed the data with the help of H.Z. and wrote the paper. All authors reviewed the manuscript.

COMPETING INTERESTS

The authors declare no competing financial interests.

REFERENCES

1. Kamihara, Y., Watanabe, T., Hirano, M. & Hosono, H. Iron-based layered superconductor La[$O_{1-x}F_x$]FeAs (x = 0.05–0.12) with T_c = 26 K. *J. Am. Chem. Soc.* **130**, 3296–3297 (2008).
2. Matsuishi, S., Inoue, Y., Nomura, T., Yanagi, H., Hirano, M. &Hosono, H. Superconductivity induced by Co-doping in quaternary fluoroarsenide CaFeAsF. *J. Am. Chem. Soc.* **130**, 14428–14429 (2008).
3. Rotter, M., Tegel, M. & Johrendt, D. Superconductivity at 38 K in the iron arsenide $(Ba_{1-x}K_x)Fe_2As_2$. *Phys. Rev. Lett.* **101**, 107006 (2008).
4. Guo, J. G. *et al.* Superconductivity in the iron selenide $K_xFe_2Se_2$ ($0 \le x \le 1.0$). *Phys. Rev. B* **79**, 180520(R) (2010).
5. Wang, X. C. *et al.* The superconductivity at 18 K in LiFeAs system.*Solid State Commun.* **148**, 538–540 (2008).
6. Tapp, J. H., Tang, Z. J., Lv, B., Sasmal, K., Lorenz, B., Chu, P. C. W. &Guloy, A. M. LiFeAs: an intrinsic FeAs-based superconductor withT_c = 18 K. *Phys. Rev. B* **78**, 060505(R) (2008).

7. Hsu, F. C. *et al.* Superconductivity in the PbO-type structure α-FeSe. *Proc. Natl. Acad. Sci.* **105**, 14262–14264 (2008).
8. de la Cruz, C. *et al.* Magnetic order close to superconductivity in the iron-based layered $LaO_{1-x}F_xFeAs$ systems. *Nature* **453**, 899–902 (2008).
9. Huang, Q. Z. *et al.* Neutron-Diffraction measurement of magnetic order and a structural transition in the parent $BaFe_2As_2$compound of FeAs-Based high temperature superconductors. *Phys. Rev. Lett.* **101**, 257003 (2008).
10. Paglione, J. & Green, R. L. *High-temperature superconductivity in iron-based materials. Nat. Phys.* **6**, 645–658 (2010).
11. Stewart, G. R. Superconductivity in iron compounds. *Rev. Mod. Phys.* **83**, 1589–1652 (2011).
12. Chu, C. W. & Lorenz, B. High pressure studies on Fe-pnictide superconductors. *Physica C* **469**, 385–395 (2009).
13. Takabayashi, Y. *et al.* Doping dependence of the pressure response of T_c in the $SmO_{1-x}F_xFeAs$ Superconductors. *J. Am. Chem. Soc.* **130**, 9242–9243 (2008).
14. Zhang, H. J., Xu, G., Dai, X. & Fang, Z. Enhanced orbital degeneracy in momentum space for LaOFeAs. *Chin. Phys. Lett.* **26**, 017401 (2009).
15. Sun, L. L. *et al.* Re-emerging superconductivity at 48 kelvin in iron chalcogenides. *Nature* **483**, 67–67 (2012).
16. Medvedev, S. *et al.* Electronic and magnetic phase diagram of β-$Fe_{1.01}Se$ with superconductivity at 36.7 K under pressure. *Nature Mater.* **8**, 630 (2009).
17. Imai, T., Ahilan, K., Ning, F. L., McQueen, T. M. & Cava, R. J. Why does undoped FeSe become a high-T_c superconductor under pressure? *Phys. Rev. Lett.* **102**, 177005 (2009).
18. Margadonna, S. *et al.* Pressure evolution of the low-temperature crystal structure and bonding of the superconductor FeSe (T_c = 37 K). *Phys. Rev. B* **80**, 064506 (2009).
19. Okabe, H., Takeshita, N., Horigane, K., Muranaka, T. & Akimitsu, J.Pressure-induced high-T_c superconducting phase in FeSe: correlation between anion height and T_c. *Phys. Rev. B* **81**, 205119 (2010).
20. Lee, C. H. *et al.* Effect of structural parameters on superconductivity in fluorine-free LnFeAsO_{1-y} (*Ln* = La, Nd). *J. Phys. Soc. Jpn.* **77**, 083704 (2008).
21. Mizuguchi, Y. *et al.* Anion height dependence of T_c for the Fe-based superconductor. *Supercond. Sci. Technol.* **23**, 054013 (2010).
22. Liu, Q. Q. *et al.* Pressure-induced isostructural phase transition and correlation of FeAs coordination with the superconducting

properties of 111-type $Na_{1-x}FeAs$. *J. Am. Chem. Soc.* **133**, 7892–7896 (2011).

23. Lai, X. F. *et al.* Observation of superconductivity in tetragonal FeS.*J. Am. Chem. Soc.* **137**, 10148–10151 (2015).
24. Lu, X. F. *et al.* Superconductivity in $LiFeO_2Fe_2Se_2$ with anti-PbO-type spacer layers. *Phys. Rev. B* **89**, 020507(2014).
25. Lu, X. F. *et al.* Coexistence of superconductivity and antiferromagnetism in $(Li_{0.8}Fe_{0.2})OHFeSe$. *Nature Mater.* **14**, 325 (2015).
26. Dong, X. L. *et al.* Phase diagram of $(Li_{1-x}Fe_x)OHFeSe$: a bridge between iron selenide and arsenide superconductors. *J. Am. Chem. Soc.* **137**, 66–69 (2015).
27. Pachmayr, U., Fehn, N. & Johrendt, D. Structural transition and superconductivity in hydrothermally synthesized Fe*X* (*X* = S, Se).*Chem. Commun.* **52**, 194–197 (2016).
28. Margadonna, S. *et al.* Crystal structure of the new $FeSe_{1-x}$superconductor. *Chem. Commun.* 5607–5609 (2008).
29. McQueen, T. M. *et al.* Tetragonal-to-orthorhombic structural phase transition at 90 K in the superconductor $Fe_{1.01}Se$. *Phys. Rev. Lett.* **103**, 057002–057004 (2009).
30. Subedi, A., Zhang, L. J., Singh, D. J. & Du, M. H. Density functional study of FeS, FeSe, and FeTe: electronic structure, magnetism, phonons, and superconductivity. *Phys. Rev. B* **78**, 134514 (2008).
31. Lü, X. J. *et al.* Pressure-induced amorphization in single-crystal Ta_2O_5 nanowires: a kinetic mechanism and improved electrical conductivity. *J. Am. Chem. Soc.* **135**, 13947–13953 (2013).
32. Lü, X. J. *et al.* Enhanced electron transport in Nb-doped TiO_2nanoparticles via pressure-induced phase transitions. *J. Am. Chem. Soc.* **136**, 419–426 (2014).
33. Denholme, S. J. *et al.* Evidence for non-metallic behaviour in tetragonal FeS (mackinawite). *Mater. Chem. Phys.* **147**, 50–56 (2014).
34. Bourdoiseau, J. A., Jeannin, M., Rémazeilles, C., Sabot, R. & Refait, Ph. The transformation of mackinawite into greigite studied by Raman spectroscopy. *J. Raman Spectrosc.* **42**, 496–504 (2011).
35. Bourdoiseau, J. A., Jeannin, M., Sabot, R. & Rémazeilles, C. & Refait, Ph. Characterisation of mackinawite by Raman spectroscopy: effects of crystallisation, drying and oxidation.*Corros. Sci.* **50**, 3247–3255 (2008).
36. Kusaba, K., Research, Y., Kikegawa, T. & Shimomura, O. Structure of FeS under high pressure. *J. Phys. Chem. Solids* **58**, 241–246 (1997).
37. Nelmes, R. J., McMahon, M. I., McMahon, M. I. & Parise, J. B.Structure of the high-pressure phase III of iron sulfide. *Phys. Rev. B* **59**, 9048 (1999).

38. Kobayashi, H., Kamimura, T., Ohishi, Y., Takeshita, N. & Ocirc, N.Structural and electrical properties of stoichiometric FeS compounds under high pressure at low temperature. *Phys. Rev. B*71, 014110 (2005).
39. Kobayashi, H., Takeshita, N., Môri, N., Takahashi, H. & Kamimura, T. Pressure-induced semiconductor-metal-semiconductor transitions in FeS. *Phys. Rev. B* **63**, 115203 (2001).
40. Takele, S. & Hearne, G. R. Electrical transport, magnetism, and spin-state configurations of high-pressure phases of FeS. *Phys. Rev. B* **60**, 4401 (1999).
41. Ehm, L. *et al.* Structural changes in nanocrystalline mackinawite (FeS) at high pressure. *J. Appl. Crystallogr.* **42**, 15–21 (2009).
42. Marshall, W. G. *et al.* High-pressure neutron-diffraction study of FeS. *Phys. Rev. B* **61**, 11201 (2000).
43. Rueff, J.-P. *et al.* Pressure-induced high-spin to low-spin transition in FeS evidenced by X-ray emission spectroscopy. *Phys. Rev. Lett.* **82**, 3284 (1999).
44. Fei, Y., Prewitt, C. T., Mao, H.-k. & Bertka, C. M. Structure and density of FeS at high pressure and high temperature and the internal structure of Mars. *Science* **268**, 1892 (1995).
45. Kusaba, K., Syono, Y., Kikegawa, T. & Shimomura, O. High pressure and temperature behavior of FeS. *Journal of Physics and Chemistry of Solids* **59**, 945–950 (1998).
46. Zhao, J. G. *et al.* Structure stability and compressibility of iron-based superconductor $Nd(O_{0.88}F_{0.12})FeAs$ under high pressure. *J. Am. Chem. Soc.* **130**, 13828–13829 (2008).
47. Kreyssig, A. *et al.* Pressure-induced volume-collapsed tetragonal phase of $CaFe_2As_2$ as seen via neutron scattering. *Phys. Rev. B* **78**, 184517 (2008).
48. Yu, Z. *et al.* Conventional empirical law reverses in the phase transitions of 122-type iron-based superconductors. *Sci. Rep.* **4**, 7172 (2014).
49. Gresty, N. C. *et al.* Structural phase transitions and superconductivity in $Fe_{1+\delta}Se_{0.57}Te_{0.43}$ at ambient and elevated pressures. *J. Am. Chem. Soc.* **131**, 16944–16952 (2009).
50. Zhang, J. L. *et al.* Pressure-induced superconductivity in topological parent compound Bi_2Te_3. *Proc. Natl. Acad. Sci.* **108**, 24 (2011).
51. Zhang, S. J. *et al.* The comprehensive phase evolution for Bi_2Te_3topological compound as function of pressure. *J. Appl. Phys.* **111**, 112630 (2012).
52. Zhu, J. *et al.* Superconductivity in topological insulator Sb_2Te_3induced by pressure. *Sci. Rep.* **3**, 2016 (2013).
53. Kong, P. P. *et al.* Superconductivity of the topological insulator Bi_2Se_3 at high pressure. *J. Phys.: Condens. Matter* **25**, 362204 (2013).

54. Kong, P. P. *et al.* Superconductivity in strong spin orbital coupling compound Sb_2Se_3. *Sci. Rep.* **4**, 6679 (2014).
55. Mao, H. K., Xu, J. & Bell, P. M. Calibration of the ruby pressure gauge to 800 kbar under quasi-hydrostatic conditions. *J. Geophys. Res.* **91**, 4673 (1986).
56. Hammersley, A. P., Svensson, S. O., Hanfland, M., Fitch, A. N. &Hausermann, D. Two-dimensional detector software: from real detector to idealised image or twotheta scan. *High Pres. Res.* **14**, 235 (1996).
57. Carvajal, J. R. Recent advances in magnetic structure determination by neutron powder diffraction. *Physica B* **192**, 55–69 (1993).

Chapter 12

Pressure Dependence of ^{15}N Chemical Shifts in Model Peptides Ac-Gly-Gly-X-Ala-NH_2

Joerg Koehler [1], Markus Beck Erlach [1], Edson Crusca Jr. [2], Werner Kremer [1], Claudia E. Munte [2] and Hans Robert Kalbitzer [1,*]

[1]*Institute of Biophysics and Physical Biochemistry and Centre of Magnetic Resonance in Chemistry and Biomedicine, University of Regensburg, Regensburg 93040, Germany*

[2]*Physics Institute of São Carlos, University of São Paulo, São Carlos 13566-590, Brazil*

ABSTRACT

High pressure NMR spectroscopy has developed into an important tool for studying conformational equilibria of proteins in solution. We have studied the amide proton and nitrogen chemical shifts of the 20 canonical amino acids X in the random-coil model peptide Ac-Gly-Gly-X-Ala-NH_2, in a pressure range from 0.1 to 200 MPa, at a proton resonance frequency of 800 MHz. The obtained data allowed the determination of first and second order pressure coefficients with high accuracy at 283 K and pH 6.7. The mean first and second order pressure coefficients $\langle B_1^{15N} \rangle$ and $\langle B_2^{15N} \rangle$ for nitrogen are 2.91 ppm/GPa and −2.32 ppm/GPa^2, respectively. The corresponding values $\langle B_1^{1H} \rangle$ and $\langle B_2^{1H} \rangle$ for the amide protons are 0.52 ppm/GPa and −0.41 ppm/GPa^2. Residual dependent $^1J_{1H15N}$-coupling constants are shown.

Keywords: tetrapeptide; high pressure; NMR spectroscopy; random-coil; chemical shift; J-coupling; nitrogen; amide group, backbone

1. INTRODUCTION

High pressure nuclear magnetic resonance spectroscopy (HP-NMR) is a powerful tool for investigating thermodynamic behavior of macromolecules in solution at atomic resolution (for a review see e.g., [1]). Application of pressure has direct effects on molecular properties such as bond lengths in hydrogen bonds but more importantly usually shifts conformational equilibria in the structural ensemble of biological macromolecules such as proteins by its contribution to the free energies of the system. These structural changes influence all NMR parameters but notably chemical shifts are very precisely observable. Since proteins have a highly anisotropic compressibility the observed chemical shift changes show large variations from atom to atom. In rigid proteins such as the basic pancreatic trypsin inhibitor (BPTI), the pressure-induced changes of chemical shifts can be fitted well by a linear dependence for almost all atoms [2]. In contrast, in proteins that exist in more than one conformational state in solution, usually non-linear pressure responses are found. The first example was the histidine containing protein (HPr) from *S. carnosus* that shows strong deviations from the linearity in the pressure response. Usually, these shifts are observed in regions of the protein where interactions with small ligands or other proteins occur. In HPr it is the active-centre loop around His15 (interaction site with enzyme I and enzyme II) and in the regulatory helix B (interaction site with the HPr-kinase/phosphatase) [3]. These regions should also show a higher internal mobility [4] and compressibility since structural transitions should be possible.

At a given temperature T_0 and pressure p the chemical shift δ^k of a given nucleus k in a multistate system depends on the exchange correlation times τ_{ij}^k and the chemical shift differences $\Delta\omega_{ij}^k$ of the states i and j. In many cases one can assume that fast exchange prevails for all states i and j with

$$|\Delta\omega_{ij}^k \ \tau_{ij}^k| \quad << 1 \tag{1}$$

The fast exchange condition should also hold for the substates assigned to an expectation value $<\delta_i^k>$ of a state i.

The pressure dependence of the chemical δ^k is than given by

$$\delta^k(T_0,p) = \frac{1}{Z}\sum_{i=1}^{N} <\delta_i^k> p'_i = \frac{1}{Z}\sum_{i=1}^{N} <\delta_i^k> \exp\left(-\frac{1}{RT}\left(G_i^0 + V_i^0(p-p_0) + \frac{1}{2}\beta_i^0(p-p_0)^2\right)\right) \tag{2}$$

with p'_i the population of state i, Z the states sum, G_i^0 the free energy at (p_0, T_0), V_i^0 the partial molar volume at (p_0, T_0), β_i^0 the compressibility factor at (p_0, T_0) and R the gas constant. For the interpretation of the data it is important to separate trivial chemical shift changes as they occur also in unfolded peptides from structurally or functionally important pressure induced changes. The simplest way to do such a correction is just subtracting the relatively small pressure dependent chemical shifts of model peptides from the experimental data. The small pressure dependent changes of the chemical shift δ *(T, p)* can be fitted sufficiently well by a second order Taylor expansion as

$$\delta(T_0, p) = \delta_0(T_0, p_0) + B_1(p - p_0) + B_2(p - p_0)^2 \qquad (3)$$

with T the temperature, p the pressure and B_1 and B_2 the first and second order pressure coefficients, respectively [5].

Such a correction has been applied earlier to the pressure response of the human prion protein [6] and helped to identify stretches with residual structures in the disordered N-terminal tail of the protein. However, only ^{1}H pressure coefficients were reported by Arnold *et al.* [5]. In the present study we will report ^{15}N shifts and refined ^{1}H chemical shifts for the amide groups of the 20 canonical amino acids X in the tetrapeptide Ac-Gly-Gly-X-Ala-NH_2. The N- and C-terminally unprotected tetrapeptide Gly-Gly-X-Ala has been introduced as random coil model by Wüthrich and Bundi [7].

2. RESULTS AND DISCUSSION

2.1. Pressure Dependence of the Backbone ^{15}N Chemical Shifts

Generally, the largest pressure dependent shift changes in protein backbones are expected for the amide groups since they are involved in hydrogen bonding. Figure 1 shows a typical pressure dependence of amide chemical shifts in an ^{1}H–^{15}N HSQC spectrum. Since the amino acid X in the tetrapeptide was enriched in ^{15}N as well as ^{13}C, ^{1}H and ^{13}C decoupling was performed in the indirect dimension. In the direct dimension broadband decoupling was not possible because the long acquisition time was required for the necessary resolution. All ^{15}N resonances shift downfield with increasing pressure, most probably because of the expected shortening of the hydrogen bonds with pressure [8]. Such an effect has been described earlier by Kalbitzer *et al.* [3].

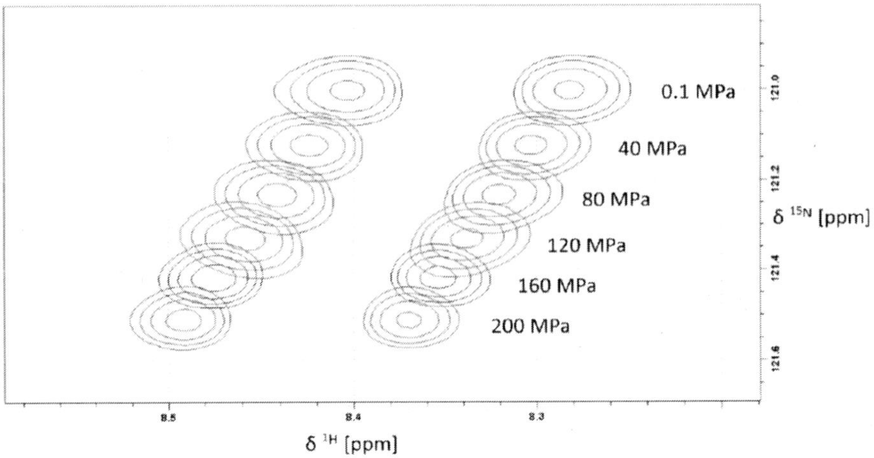

Figure 1. Pressure dependent ^{1}H-^{15}N-HSQC spectra of backbone amide region of arginine. The samples contained 5 mM Ac-Gly-Gly-Arg-Ala-NH_2 in 20 mM Tris-HCl in 90% H_2O/10% D_2O, pH 6.7. Measurements were performed at 283 K. Data processing was done with Gaussian filter in both directions using negative line broadenings of −10 Hz and −3 Hz for ^{15}N and ^{1}H, respectively.

In first approximation the pressure dependence of backbone ^{15}N resonances is linear with a mean first order pressure coefficient $\langle B_1^{15N} \rangle$ value of 2.91 ppm/GPa (Table 1). Only small deviations from the linearity are observed, corresponding to a mean second order pressure coefficient $\langle B_2^{15N} \rangle$ of −2.32 ppm/GPa^2.

With 1.36 ppm/GPa cis-proline shows the smallest ^{15}N first order pressure response of all amino acids but an average pressure response in the trans-configuration (Figure 2, Table 1). This is in agreement with the supposed hydrogen bonding mechanism that should be negligible in the cis-configuration since the water interaction with peptide nitrogen is largely blocked. The largest pressure induced ^{15}N chemical shift changes are observed for glycine residues with a first order pressure coefficient of 3.79 ppm/GPa (Table 1). Concerning the second order coefficients again cis-proline shows the smallest non-linearity of the pressure dependent ^{15}N chemical shifts (−0.38 ppm/GPa^2) and tyrosine the largest one (−4.52 ppm/GPa^2) (Table 1). Figure 3 shows that these differences in non-linearity are clearly significant. For nitrogen backbone resonances all B_2 values are negative suggesting an asymptotic behavior at very high pressures.

Table 1. Pressure dependence of backbone ^{15}N chemical shifts. The samples contained 5 mM Ac-Gly-Gly-X-Ala-NH_2 in 20 mM Tris-HCl in 90% H_2O/10% D_2O, pH 6.7. HSQC measurements were performed at 283 K in a pressure range from 0.1 MPa to 200 MPa. δ_0 corresponds to chemical shift at atmospheric pressure, B_1 and B_2 were obtained by fitting the pressure dependence of the chemical shift δ to Equation (3). The errors given are standard deviations from the fitting algorithm. The errors of the δ_0 and $^1J_{1H15N}$ can be estimated as 0.01 ppm and 0.1 Hz, respectively.

	δ_0^{15N} (ppm)	B_1^{15N} (ppm/GPa)	B_2^{15N} (ppm/GPa^2)	$^1J_{1H15N}$ (Hz)
Ala	124.39	2.74 ± 0.03	−1.49 ± 0.16	−97.5
Arg	121.00	3.01 ± 0.04	−2.44 ± 0.21	−95.5
Asn	119.12	2.98 ± 0.17	−1.73 ± 0.82	−95.9
Asp	120.94	2.94 ± 0.07	−2.58 ± 0.36	−96.0
Cys	119.21	2.75 ± 0.08	−1.87 ± 0.41	−96.0
Gln	120.25	2.87 ± 0.06	−2.15 ± 0.29	−93.5
Glu	120.95	2.78 ± 0.17	−1.66 ± 0.83	−99.5
Gly	109.22	3.79 ± 0.08	−2.16 ± 0.37	−96.9
His pH 4.0	118.34	3.42 ± 0.08	−2.86 ± 0.38	−98.2
His pH 8.5*	120.33	3.15 ± 0.10	−3.53 ± 0.50	–
Ile	120.39	2.35 ± 0.03	−1.89 ± 0.17	−92.9
Leu	122.07	2.65 ± 0.10	−2.92 ± 0.48	−96.2
Lys	121.29	3.00 ± 0.03	−2.24 ± 0.13	−94.7
Met	120.17	2.76 ± 0.05	−1.89 ± 0.22	−92.7
Phe	120.62	3.01 ± 0.04	−3.65 ± 0.19	−95.5
Pro trans*	134.78	2.45 ± 0.02	−1.76 ± 0.10	–
Pro cis*	135.03	1.36 ± 0.02	−0.38 ± 0.10	–
Ser	116.27	3.41 ± 0.15	−2.46 ± 0.70	−95.7
Thr	114.14	3.28 ± 0.05	−2.36 ± 0.24	−93.8
Trp	121.43	3.35 ± 0.03	−2.90 ± 0.17	−95.5
Tyr	120.33	3.56 ± 0.07	−4.52 ± 0.32	−93.9
Val	119.41	2.32 ± 0.02	−1.70 ± 0.12	−95.0

Note: * Data acquisition performed via HA(CA)N experiment.

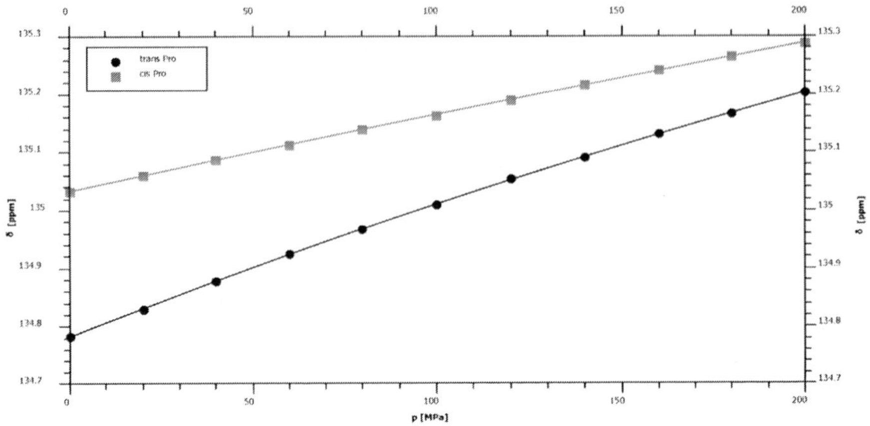

Figure 2. Pressure dependence of backbone ^{15}N chemical shift resonances of proline. The samples contained 5 mM Ac-Gly-Gly-Pro-Ala-NH_2 in 20 mM Tris-HCl in 90% H_2O/10% D_2O, pH 6.7. Measurements were performed at 283 K. The chemical shift δ was fitted as a function of pressure p to Equation (3) with the parameters given in Table 1.

In random coil peptides histidine residues are in equilibrium between protonated and deprotonated states of the imidazole ring. In Gly-Gly-His-Ala the pK_a is 7.0 [7]. The protonation/deprotonation equilibrium of the ring may also influence the pressure induced shifts at the backbone atoms. Therefore, the pressure dependence of the nitrogen chemical shift was measured at two different pH-values, at pH 4.0 where the ring should be fully protonated and pH 8.5 where the ring is fully deprotonated. Since at high pH the amide proton exchange is too fast for a polarization transfer from the amide proton to the nitrogen, an HA(CA)N experiment was performed where the primary polarization transfer occurs from the H^{α}-proton. The backbone amide resonance is shifted upfield at lower pH and the first order pressure coefficient increases slightly (Figure 4, Table 1).

2.2. Measurement of the One-Bond Amide Coupling Constants

In our tetrapeptides a residue dependent variation of the $^{1}J_{1H15N}$-coupling constant of nearly 7 Hz is shown inTable 1. Significant changes of the $^{1}J_{1H15N}$-coupling constant were observed in HPr from *S. carnosus* with a pressure dependent increase of the coupling constant up to 20 Hz [3].

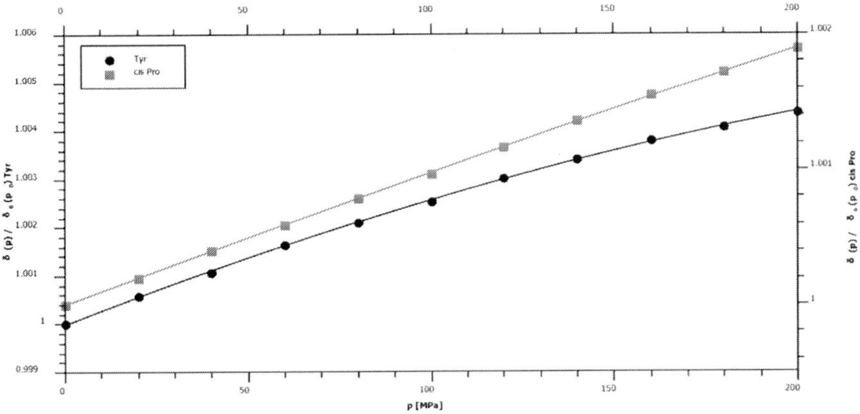

(a)

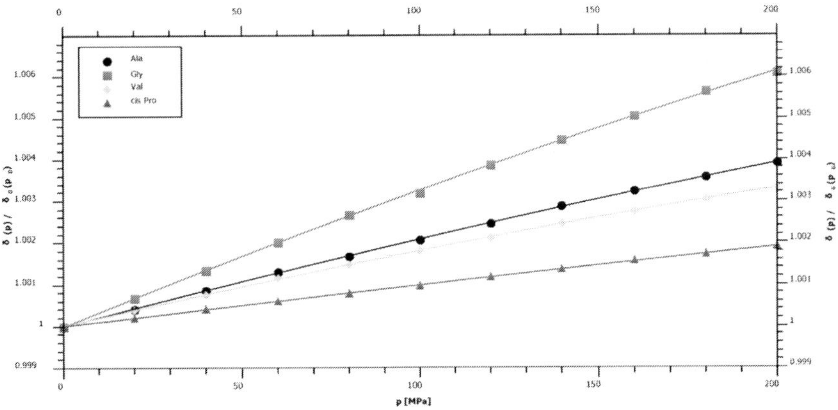

(b)

Figure 3. Pressure dependent ^{15}N chemical shift changes of selected residues. The chemical shifts δ are plotted as a function of pressure p for some residues X. For a better comparison the inverse of the initial value at ambient pressure was used as normalization factor for each data series. (**a**) Non-linearity of the pressure dependence of cis-proline and tyrosine residues. (**b**) The pressure dependence of chemical shifts of selected residues.

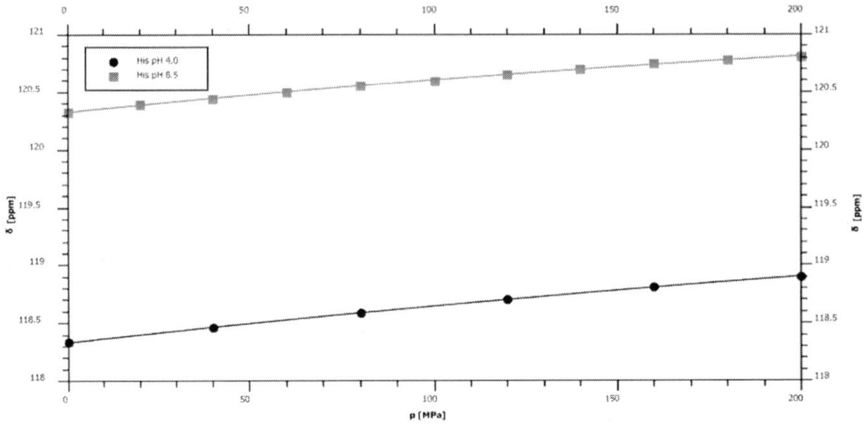

Figure 4. Pressure dependence of backbone ^{15}N chemical shift resonances of histidine Ac-Gly-Gly-His-Ala-NH_2. Sample composition, experimental conditions and data fitting are described in Table 1. The sample was measured at pH 4.0 and pH 8.5.

2.3. Pressure Dependence of the Amide ^1H Chemical Shifts

Pressure dependent chemical shifts were measured earlier for Gly-Gly-X-Ala by Arnold *et al.* [5] at a proton resonance frequency of 500 MHz in the pressure range from 0.1 to 200 MPa. They were determined in 50 mM phosphate buffer at pH 5.0 and 5.4 at 305 K. Since our data were recorded at different experimental conditions (20 mM Tris-HCl, pH 6.7, 283 K) it seems to be worthwhile to compare them with the older data for obtaining information about the pH and temperature effects. With Ac-Gly-Gly-X-Ala-NH_2 we also used an N-terminally and C-terminally protected peptide where the N- and C-terminus remain uncharged independent of the pH used. In addition they were measured at 800 MHz and thus promise a higher accuracy of the obtained pressure coefficients (Table 2). The lower temperature of 283 K used in our study is expected to lead to a mean downfield shift of 0.17 ppm [9] of the amide resonance frequencies relative to the values reported by Arnold *et al.* [5] for 305 K. In contrast to this expectation only a very small shift deviation could be observed (0.005 ± 0.05 ppm). This is probably due to the protection of the N- and C-terminal charged groups. In addition a pH-dependent upfield shift might partially compensate the temperature dependent effect for some residues [5]. The chemical shift for histidine was measured at pH 4.0 (Table 2) since at high pH the amide proton exhanges too fast with the bulk water to be observable. Since the pK of histidine is 7.0 in the tetrapeptide [7] the value given here corresponds to a positively charged, fully protonated imidazole side chain. At neutral pH one would expect an upfield shift of the amide proton resonance and a downfield shift of the nitrogen resonance [9]. The pH dependence of the histidine nitrogen chemical shifts reported here are in aggreement to the expectation (Table 1).

Table 2. Pressure dependence of backbone amide H^N chemical shifts. The samples contained 5 mM Ac-Gly-Gly-X-Ala-NH_2 in 20 mM Tris-HCl in 90% H_2O/10% D_2O, pH 6.7. Measurements were performed at 283 K in a pressure range from 0.1 MPa to 200 MPa. δ_0 corresponds to chemical shift at atmospheric pressure, B_1 and B_2 were obtained by fitting the pressure dependence of the chemical shift δ to Equation (3). The errors given are standard deviations from the fitting algorithm. The error of the δ_0 can be estimated as 0.001 ppm.

	δ_0^{HN} (ppm)	B_1^{HN} (ppm/GPa)	B_2^{HN} (ppm/GPa^2)
Ala	8.368	0.45 ± 0.02	−0.15 ± 0.08
Arg	8.342	0.48 ± 0.02	−0.22 ± 0.10
Asn	8.503	0.50 ± 0.03	−0.42 ± 0.14
Asp	8.433	0.56 ± 0.02	−0.48 ± 0.09
Cys	8.418	0.55 ± 0.01	−0.49 ± 0.05
Gln	8.433	0.44 ± 0.03	−0.17 ± 0.13
Glu	8.539	0.25 ± 0.02	−0.03 ± 0.08

The average first and second order shift coefficients $<B_1^{1H}>$ and $<B_2^{1H}>$ are 0.52 ppm/GPa and −0.41 ppm/GPa^2 in Ac-Gly-Gly-X-Ala-NH_2. In Gly-Gly-X-Ala they are with 0.42 ppm/GPa and −0.18 ppm/GPa^{2}somewhat smaller. The first order pressure coefficients are positive in both peptides, that is a downfield shift of the^{1}H amide resonances with pressure is observed. The only exception described by Arnold *et al.* [5] is glutamate where a negative first order pressure coefficient and a large second order coefficient had been observed. However, these unusual pressure coefficients are due to an interaction with the negative charge of the C-terminal carboxyl group since its methylation [10] or amination leads to a "normal" behavior of the amide resonance (Figure 5).

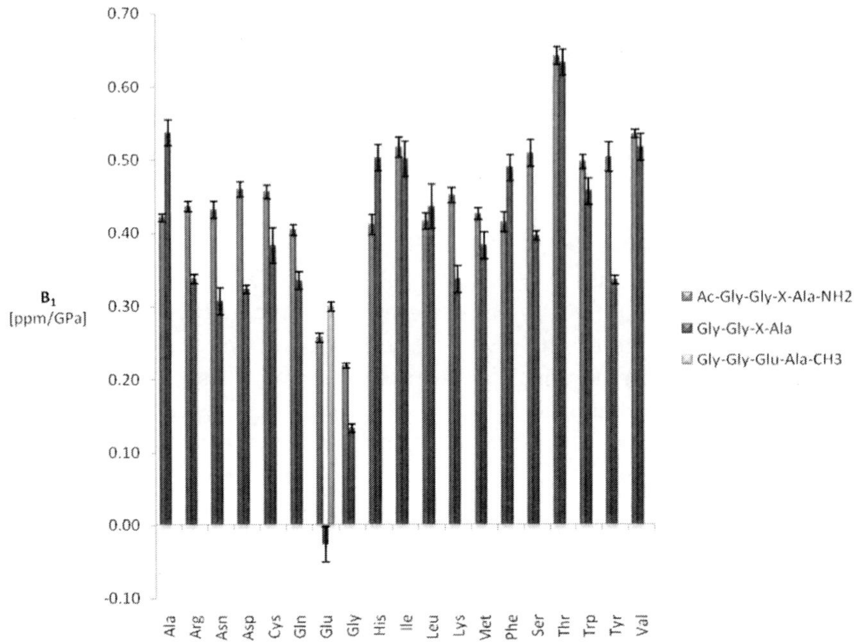

Figure 5. First order amide proton pressure coefficients in Ac-Gly-Gly-X-Ala-NH_2 and Gly-Gly-X-Ala. Sample composition, experimental conditions and data fitting for Ac-Gly-Gly-X-Ala-NH_2 (blue) are described in Table 2. The values for Gly-Gly-X-Ala (red) were taken from Arnold *et al.* [5] and were recorded for 5 mM tetrapeptide in 50 mM phosphate buffer, pH 5.4 at 305 K. The glutamate pressure coefficients of Gly-Gly-Glu-Ala-CH_3 (green) [10] were recorded under the same experimental conditions as those of Gly.Gly-Glu-Ala.

For the first order pressure coefficients the largest differences between the two peptides are observed for glutamate (0.28 ppm/GPa), followed by tyrosine and asparagine with 0.17 ppm/GPa and 0.14 ppm/GPa, respectively. As already observed for the amide nitrogen resonances the second order coefficients for protons are negative for all amino acids leading to a maximum in the pressure dependence at high pressures but probably are only an approximation for an asymptotic behavior at high pressures. In the data published by Arnold *et al.* [5] the sign of the second order pressure coefficients vary from amino acid to amino acid. This is most probably due to the lower NMR frequency used in the earlier study and thus a potentially larger error of measurements.

2.4. pH-Dependence of the Pressure Response

Amide proton shifts in random-coil peptides can be influenced by temperature and pH (see e.g., [9,11]). At ambient pressure the temperature dependence of the amide chemical shifts can be predicted sufficiently well from the amino acid specific temperature coefficients [9]. However, since the pK values of charged groups of the buffer as well as of the peptides are pressure dependent, pressure may induce a pH shift in the solution that also may influence the observed chemical shifts. Tris-HCl shows only a relatively small change of its pK-value with pressure compared to other buffers such as phosphate buffers [12]. This is the reason why it is often used as buffer in high-pressure NMR experiments on proteins. For having similar conditions we added Tris-HCl also for the study of the model peptides, although its buffer capacity is very low at pH 6.7. Here it can be used as pH-probe since the chemical shift dependence of the Tris-signal on pressure was studied in detail by Huberth [13]. At 200 MPa the pH increases in average by 0.01 pH units. The study by Arnold *et al.* [5] showed that the pressure coefficients of model peptides are only significantly influenced by pH when the side chain pK value is close to pH-value of the experiments. This is not the case in our study, only the pK-value of histidine is close enough to pH 6.7 but here the experiments were performed in the fully protonated and deprotonated state at pH 4.0 and 8.5 (Table 1).

3. EXPERIMENTAL SECTION

3.1. Synthesis of Peptides

Uniformly ^{15}N and ^{13}C enriched and N-terminally Fmoc protected amino acids were purchased from Sigma Aldrich (St. Louis, MO, USA) with an isotope enrichment larger than 98%. Side chains were protected by Pbf (Arg), Trt (Asn, Gln, His), Boc (Lys, Trp), tBu (Ser, Thr, Tyr), and OtBu (Asp, Glu). Solvents and reagents were obtained from Merck (Darmstadt, Germany). The tetrapeptides Ac-Gly-Gly-X-Ala-NH_2 were synthesized by solid phase synthesis using a Rink Amide resin and Fmoc (N-(9-fluorenyl)methoxycarbonyl) N-terminal group protection of amino acids. Coupling of all amino acids was achieved using three equivalents of HBTU (O-Benzotriazole-N.N.N'.N'-tetramethyl-uronium-hexafluoro-phosphate) and two equivalents of DIEA (N.N-diisopropylethylamine) for each equivalent of Fmoc amino acid. When the peptide chain was finished, a mixture of 10% acetic anhydride in DMF (dimethylformamide) was used for N-terminal acetylation. The peptides were cleaved from the resin and the side-chain protecting groups were removed simultaneously with appropriate TFA/scavengers cocktails. The identity and purity of all peptides was confirmed by ESI-MS (Bruker, Billerica, MA, USA) and RP-HPLC (Waters, Milford, MA, USA).

3.2. Sample Preparation

The peptide concentration was 5 mM in aqueous solution of 90% H_20 and 10% D_2O. Additionally 20 mM perdeuterated Tris-HCl (tris(hydroxymethyl)aminomethane hydrochloride) and 0.5 mM DSS (4.4-dimethyl-4-silapentane-sulfonic acid) were added. In general, the pH value was adjusted to 6.7 with a Hamilton Spintrode attached to a Beckman Coulter pH-meter. Only histidine was measured at pH 4.0 and pH 8.5. The measured pH-values have not been corrected for the deuterium isotope effect.

3.3. High Pressure System

A homebuilt online-pressure system according to Yamada-method [16] was used. Pressure produced by a homemade manually operated piston compressor was transmitted via a high pressure line (High Pressure Equipment Company, Linden, PA, USA) by methylcyclohexane to the high pressure ceramic cell (with an outer diameter of 5 mm and an inner diameter of 3 mm) from Daedalus Innovations LLC (Aston. PA. USA). A PET (polyethylene terephthalate) membrane acts as a flexible separator between the pressure fluid and the aqueous sample. To reduce the volume of the ceramic cell a cylindric PEEK (polyether-ether-ketone) displacement body was used. A full-metal autoclave connects the ceramic cell with the closed pressure line (Figure 6). It is similar to the original autoclave provided by Daedalus Innovations [14] but is produced from titanium and contains a safety valve [15] that closes rapidly when the cell brakes.

3.4. NMR Spectroscopy

All NMR experiments were carried out on an 800 MHz BRUKER Avance spectrometer with a QXI probe at 283 K and the magnetic field was locked to D_2O. Temperature calibration was done before each sample-change via the difference of the resonance lines of the hydroxyl- and methyl-group protons in 100% methanol according to [16]. ^{15}N resonances of amide-groups were detected indirectly via ^{1}H–^{15}N-HSQC (Hetero-Single-Quantum-Coherence) without ^{15}N decoupling published by Davis *et al.* [18]. The ^{15}N resonances of proline and histidine (pH 8.5) were detected by a HA(CA)N pulse sequence according to [19,20], which uses a polarization transfer from $^{1}H^{\alpha}$ proton to backbone ^{15}N via $^{13}C^{\alpha}$ and return. In 2D spectra the digital resolution in ^{15}N

dimension was 0.16 Hz and in ^{1}H 0.12 Hz. Proton resonances were additionally measured with the PURGE (Presaturation utilizing relaxation gradients and echoes) sequence [21] with a digital resolution of 0.02 Hz and were referenced to DSS used as internal standard. ^{15}N chemical shifts were referenced indirectly to DSS using a ^{15}N/^{1}H -ratio of 0.101329118 [22].

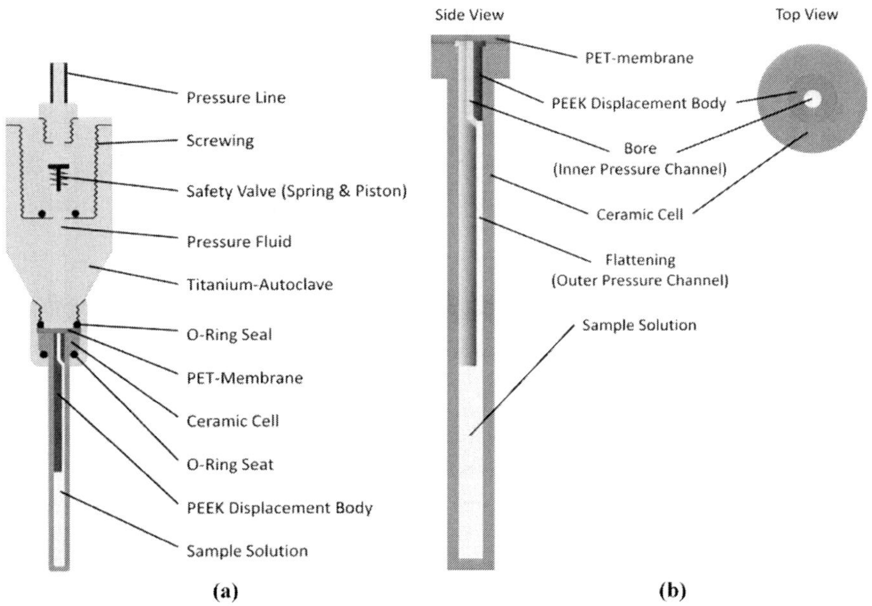

Figure 6. High pressure autoclave. (**a**) The high pressure autoclave holding the ceramic cell is similar to that provided by Daedalus Corporation [14], but has additionally a safety valve analogous to that described earlier by Beck Erlach *et al.* [15]; (**b**) A PET membrane was posed on top of the ceramic cell, and a bored and flattened cylindrical displacement body was inserted into the tube.

3.5. Data Evaluation

Data processing, spectral analysis and peak picking was performed with BRUKER Topspin 2.1. The pressure dependence of chemical shift δ was fitted to Equation (3) with QtiPlot 0.9.8 using a Scaled Levenberg-Marquardt algorithm.

4. CONCLUSIONS

We have presented here random-coil values for both the amide proton and nitrogen that can be used to correct for unspecific pressure effects in proteins. It is also important to show residual structure in disordered polypeptides (see e.g., [23]). The use of a high magnetic field proved to be essential to obtain second order pressure coefficients with sufficient accuracy.

ACKNOWLEDGEMENTS

This work was supported by the DFG (Deutsche Forschungsgemeinschaft), the Bayerische Forschungsstiftung and the FAPESP (Fundação de Amparo à Pesquisa do Estado de São Paulo). The authors thank Eduardo Maffud Cilli for generously supplying peptide synthesis laboratory (IQ-Unesp/CAr).

REFERENCES

1. Akasaka, K. Probing conformational fluctuations by pressure perturbations. *Chem. Rev.* **2006**, *106*, 1814–1835.
2. Li, H.; Yamada, H.; Akasaka, K. Effect of pressure on individual hydrogen bonds in proteins. Basic pancreatic trypsin inhibitor. *Biochemistry* **1998**, *37*, 1167–1173.
3. Kalbitzer, H.R.; Görler, A.; Li, H. ^{15}N and ^{1}H NMR study of Histidine containing protein (HPr) from*Staphylococcus carnosus* at high-pressure. *Protein Sci.* **2000**, *9*, 693–703.
4. Baldwin, A.J.; Kay, L.E. NMR spectroscopy brings invisible protein states into focus. *Nat. Chem. Biol.***2009**, *11*, 808–814.
5. Arnold, M.A.; Kremer, W.; Luedemann, H.D.; Kalbitzer, H.R. ^{1}H-NMR parameters of common amino acid residues measured in aqueous solutions of the linear tetrapeptides Gly-Gly-X-Ala at pressures between 0.1 and 200 MPa. *Biophys. Chem.* **2002**, *96*, 129–140.
6. Kremer, W.; Kachel, N.; Kuwata, K.; Akasaka, K.; Kalbitzer, H.R. Species specific differences in the intermediate states of human and Syrian hamster prion protein detected by high pressure NMR spectroscopy. *J. Biol. Chem.* **2007**, *282*, 22689–22698.
7. Bundi, A.; Wuethrich, K. ^{1}H-NMR parameters of the common amino acid residues measured in aqueous solutions of the linear tetrapeptides H-Gly-Gly-X-L-Ala-OH. *Biopolymers* **1979**, *18*, 285–297.

8. Asakawa, N.; Kameda, T.; Kuroki, S.; Kurosu, H.; Ando, S.; Shoji, A. Structural studies of hydrogen-bonded peptides and polypeptides by solid state. *Ann. Reports NMR Spectr.* **1998**, *35*, 56–137.

9. Kjaergaard, M.; Brander, S.; Poulsen, F.M. Random coil chemical shift for intrinsically disordered proteins: Effects of temperature and pH. *J. Biomol. NMR* **2011**, *49*, 139–149.

10. Kremer, W. High-pressure NMR studies in proteins. *Ann. Reports NMR Spectr.* **2006**, *57*, 177–203.

11. Bundi, A.; Wuethrich, K. Use of amide ^{1}H-NMR titration shifts for studies of polypeptide conformation.*Biopolymers* **1979**, *18*, 299–311.

12. Quinlan, R.J.; Reinhart, G.D. Baroresistant buffer mixtures for biochemical analyses. *Anal. Biochem.* **2005**,*341*, 69–76.

13. Huberth, A. Hochdruck-NMR am Ras-Protein. Diploma thesis in Physics, University of Regensburg, July 2011.

14. Peterson, R.W.; Wand, A.J. Self-contained high-pressure cell, apparatus, and procedure for the preparation of encapsulated proteins dissolved in low viscosity fluids for nuclear magnetic resonance spectroscopy.*Rev. Sci. Instrum.* **2005**, *76*, 094101:1–094101:7.

15. Erlach, M.B.; Munte, C.E.; Kremer, W.; Hartl, R.; Rochelt, D.; Niesner, D.; Kalbitzer, H.R. Ceramic cells for high pressure NMR spectroscopy on proteins. *J. Magn. Reson.* **2010**, *204*, 196–199.

16. Yamada, H. Pressure-resisting glass cell for high pressure, high resolution NMR measurement. *Rev. Sci. Instrum.* **1974**, *45*, 640–642.

17. Raiford, D.S.; Fisk, C.L.; Becker, E.D. Calibration of methanol and ethylene glycol nuclear magnetic resonance thermometers. *Anal. Chem.* **1979**, *51*, 2050–2051.

18. Davis, A.L.; Keeler, J.; Laue, E.D.; Moskau, D. Experiments for recording pure-absorption heteronuclear correlation spectra using pulsed field gradients. *J. Magn. Reson.* **1992**, *98*, 207–216.

19. Kanelisa, V.; Donaldson, L.; Muhandiram, D.R.; Rotin, D.; Forman-Kay, J.D.; Kay, L.E. Sequential assignment of proline-rich regions in proteins: Application to modular binding domain complexes. *J. Biomol. NMR* **2000**, *16*, 253–259.

20. Wang, A.C.; Grzesiek, S.; Tschudin, R.; Lodi, P.J.; Bax, A. Sequential backbone assignment of isotopically enriched proteins in D_2O by

deuterium-decoupled HA(CA)N and HA(CACO)N. *J. Biomol. NMR* **1995**, *5*, 376–382.

21. Simpson, A.J.; Brown, S.A. Purge NMR: Effective and easy solvent suppression. *J. Magn. Reson.* **2005**,*175*, 340–346.

22. Wishart, D.S.; Bigam, C.G.; Yao, J.; Abildgaard, F.; Dyson, H.J.; Oldfield, E.; Markley, J.L.; Sykes, B.D. ^{1}H, ^{13}C and ^{15}N chemical shift referencing in biomolecular NMR. *J. Biomol. NMR* **1995**, *6*, 135–140.

23. Kachel, N.; Kremer, W.; Zahn, R.; Kalbitzer, H.R. Observation of intermediate states of the human prion protein by high pressure NMR spectroscopy. *BMC Struct. Biol.* **2006**, *6*, 16:1–16:18.

Chapter 13

Stability of Rhombohedral Phases in Vanadium at High-Pressure and High-Temperature: First-Principles Investigations

Yi X. Wang, Q. Wu, Xiang R. Chen & Hua Y. Geng

National Key Laboratory of Shock Wave and Detonation Physics, Institute of Fluid Physics, CAEP, P.O. Box 919-102, Mianyang 621900, Sichuan, People's Republic of China

ABSTRACT

The pressure-induced transition of vanadium from BCC to rhombohedral structures is unique and intriguing among transition metals. In this work, the stability of these phases is revisited by using density functional theory. At finite temperatures, a novel transition of rhombohedral phases back to BCC phase induced by thermal electrons is discovered. This reentrant transition is found not driven by phonons, instead it is the electronic entropy that stabilizes the latter phase, which is totally out of expectation. Parallel to this transition, we find a peculiar and strong increase of the shear modulus C_{44} with increasing temperature. It is counter-intuitive in the sense that it suggests an unusual harding mechanism of vanadium by temperature. With these stability analyses, the high-pressure and finite-temperature phase diagram of vanadium is proposed. Furthermore, the dependence of the stability of RH phases on the Fermi energy and chemical environment is investigated. The results demonstrate that the position of the Fermi level has a significant impact on the phase stability, and follows the band-filling argument. Besides the Fermi surface nesting, we find that the localization/delocalization of the *d* orbitals also contributes to the instability of rhombohedral distortions in vanadium.

INTRODUCTION

Revealing and elucidating the trend of structural transformations and the underlying mechanisms in elemental metals is a fundamental topic in condensed matter physics. In recent years, the transition metal vanadium has attracted much experimental[1,2,3] and theoretical[4,5,6,7] attention because of its important applications and the puzzling softening in the strength and the subsequent transition from the body-centered cubic (BCC) phase to the low-symmetry rhombohedral (RH) structures driven by pressure.

The first direct indication of a phase transition in vanadium perhaps came from the theoretical observation of a strong softening in the transverse acoustic phonon mode along the **Γ-H** direction at high pressures, which eventually becomes imaginary when beyond 130 GPa[8]. This softening is corroborated by the calculated elastic constants where the shear modulus C_{44} continuously decreases to zero and becomes negative, suggesting instability of the BCC structure[9,10]. This bizarre behavior was believed due to Fermi surface nesting and the Kohn anomalies, but the resultant crystalline structure was not proposed at that time. In fact, there had an early experimental study of vanadium up to 154 GPa, but no phase transition was reported[11]. Later, Ding *et al.*[12] conducted delicate x-ray diffraction experiments using diamond anvil cell (DAC) up to 150 GPa, and found a novel RH phase, which is a slight distortion of BCC structure and appears from about 63–69 GPa. It was soon confirmed by Lee *et al.* using static lattice density functional theory (DFT) calculations, and showed that the RH phase was the ground state when beyond 84 GPa[13,14]. They also predicted other two phase transformations that were not detected in Ding's experiment, i.e., a transformation to another RH structure at 120 GPa and the third transformation back to the BCC structure at 280 GPa. Other theoretical studies also reported qualitatively similar findings[15,16]. Lattice dynamics calculations also supported the argument that the RH phase should transform back to BCC structure when above 250 GPa[17]. On the other hand, Jenei *et al.*[18] reported a BCC → RH transition at 30 GPa when no pressure medium was used, whereas it was at about 60 GPa if in the Ne pressure medium. It seems non-hydrostatic condition has a strong impact on the transition pressure. Nonetheless, in Ding *et al.*'s experiments, the transition pressure P_c is at 69 GPa if without a pressure medium, and P_c = 63 GPa if in He pressure medium[12]. Therefore it is possible that the deviation in P_c might not be due to the non-hydrostatic condition, other factors such as chemical environment might also have some contributions. In addition, the energetics stability of the RH phase of vanadium with respect to high symmetry candidate structures, such as FCC, HCP, simple cubic, and simple hexagonal structures has been reported in literature15'19'20, and the RH phases are the most stable structures within our studied pressure range. For this reason, we will not discuss these structures below.

It has established a theoretical consensus on the phase transition sequence of BCC → RH_1 → RH_2 → BCC in vanadium under pressure. But there still are discrepancies on the exact values of the transition pressure. One of the most remarkable difference was reported by Qiu *et al.*16, in which the first transition pressure was evaluated to be 32 GPa, much smaller than all other estimations. Qiu *et al.*'s calculations were under hydrostatic condition, thus are irrelevant to Jenei *et al.*'s argument18. Except the value reported by Qiu *et al.*, all other theoretical transition pressures of BCC → RH_1 were located between 60–84 GPa. By contrast, the predicted RH_1 → RH_2 transition pressures scattered between 115–160 GPa, and the predicted P_c of RH_2 → BCC was between 240–297 GPa. The reason underlying this large uncertainty is unknown. To clarify this theoretical deviation is important for understanding the discrepancy in experimental transition pressures. In this work, we will revisit the phase transition and structural stability in vanadium with highly accurate *ab initio* calculations, in attempt to shed new light on this open issue.

On the other hand, most investigations on this problem reported in literature mainly focused on the ground state. Few study was devoted to finite temperatures. Recently, Landa *et al.*21 studied the stability of vanadium metal and vanadium-chromium alloys in BCC phase at high temperatures and pressures by calculating the anharmonic phonon dispersions. Their results showed that lattice vibrations slightly weaken the stability of RH phases; but RH_1 is stable up to the static melting curve, and RH_2 is stable beyond the melting point estimated by shock wave experiments (for vanadium there is a large discrepancy between the melting curves determined by static compression techniques such as DAC22 and dynamic shock wave method)23. Nevertheless, as revealed in previous investigations, the anomalous softening and phase transition in vanadium originate in electronic structure, and closely relate to the Fermi surface nesting, hence the contribution by lattice dynamics might be secondary. By intuition, thermo-electrons should have greater contributions, which to our best knowledge have not been studied yet. Whether the electronic temperature has significant influence on the shear modulus and phase stability of vanadium is still unknown. For this reason, a thorough and comprehensive investigation on thermo-electron effects will also be performed in this work. Furthermore, we will elucidate the possible impact of the Fermi level position and charge transfer on the softening behavior and phase stability of vanadium. This will shed light on how the compression behavior of vanadium can be changed by surrounding chemical environment, which might be helpful in understanding the large discrepancy in experimental transition pressures.

RESULTS AND DISCUSSION

Phase transition at zero Kelvin

Since Jenei *et al.* observed a low transition pressure of P_c ~30 GPa under non-hydrostatic condition18, and Qiu *et al.* also reported a similar hydrostatic P_c with theoretical calculations16, it becomes necessary to revisit the BCC → RH → BCC transition problem with careful treatment. Qiu *et al.* attributed their discrepancy with respect to other estimates in P_c to the constraints imposed by the hypothetical transition pathway that was employed in those works13'14'24. Jenei *et al.* also speculated that the thermodynamic equilibrium P_c of the first transition might be at ~30 GPa, and concluded that Ding *et al.*'s value of ~60 GPa could be due to a kinetic effect where a large energy barrier hinders the transition. In order to figure out the possible reason of such a big scattering in theoretical calculations, a comparative investigation will be performed in this work, including both the methods to locate the transition pressure and the accuracy of DFT calculations.

We first optimize the structures at zero Kelvin with all parameters being fully relaxed. We choose RH_1 and RH_2 as the initial structures, and optimize them at different pressures without any symmetry constraint being imposed. The calculated enthalpy difference with respect to the BCC phase, together with the variation of the structural parameter (angle α) in RH_1 and RH_2 structures as a function of pressure, are shown in Fig. 1. These fully relaxed results clearly reveal that the equilibrium transition of BCC → RH_1 is not at 30 GPa, where RH_1 is dynamically instable and spontaneously collapses to BCC phase. As shown in the figure, the metastable region of RH_1 extends down to 50 GPa. At lower pressures it becomes highly unstable. Nonetheless, its relaxation back to the BCC phase is not perfect. The residual angle α is about 109.51° at 20–40 GPa. This reflects a fact that the potential well of BCC is flat, and non-hydrostatic loading can easily drive vanadium towards RH-like deformations. It is necessary to point out that the angle obtained in Jenei *et al.*'s experiment is $\alpha = 109.61°$, which is far less than Lee *et al.*'s theoretical value of $\alpha = 110.25°$, but close to the imperfect BCC that collapsed from RH_1 as shown in Fig. 1. Also note that the RH_2 phase leads to a similar distorted BCC structure with $\alpha = 109.39°$ when below 110 GPa. In our calculations, the RH_1 phase becomes the ground state when beyond 98 GPa. It has a bigger angle $\alpha = 110.17°$, compared to $\alpha = 109.47°$ of the perfect BCC structure. Transition to another RH phase (RH_2) occurs at ~128 GPa, which attains the maximum stability at ~210 GPa, with $\alpha = 108.23°$. RH_2 is the ground state up to ~284 GPa. As the pressure increases further, it becomes metastable and eventually collapses back to the BCC structure at ~300 GPa. On the other hand, the metastability of RH_1 phase extends up to ~247 GPa, where it also automatically transforms back to BCC. Our results are in good agreement with ref. 13, except a small numeric difference on the phase boundaries. This indicates that Lee *et al.*'s method is compatible with the full structural relaxation.

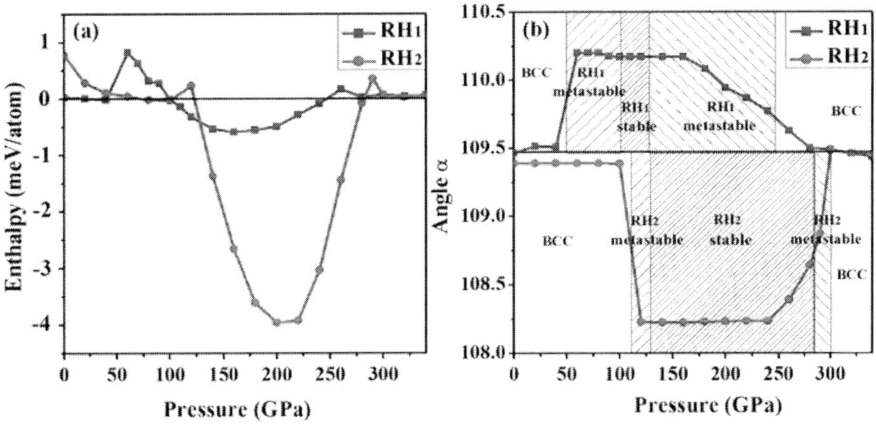

Figure 1: (**a**) Enthalpy difference of vanadium in RH_1 and RH_2 structures at zero Kelvin with respect to the BCC phase as a function of pressure. (**b**) Variation of angle α in RH_1 and RH_2 structures as a function of pressure at zero Kelvin. Note that $\alpha = 109.47°$ corresponds to the perfect BCC structure.

We also study the phase transition using the same method as Lee *et al.*13(i.e., the method II in Sec. Methods). This comparative calculation provides a consistent benchmark for the transition pressures. In this method, the unit cell volume is conserved. Qiu *et al.* argued that this treatment would give rise to a higher transition pressure in Lee *et al.* and others' calculations16. Nonetheless, our results with the full relaxation discussed above suggested that this argument might not be true. The error introduced by fixing the volume at V_0 can be corrected using the formula

$$H(\delta, P_0) \approx U(\delta, V_0) + P_0 V_0 - \frac{1}{2B(\delta, V_0)} \Delta P(\delta, V_0)^2 V_0. \qquad (1)$$

It corrects the enthalpy along the deformation path. In Eq. (1), $\Delta P(\delta,V_0)$ is the change of the pressure at a given volume V_0 and a rhombohedral deformation magnitude δ, B is the bulk modulus. In Lee *et al.*'s evaluation, only the first term was used13. We carefully tested and found that the correction from the third term is indeed very small. This supports Lee *et al.*'s assessment that the internal energy $U(\delta,V_0)$ is sufficient when studying the relative phase stability, and the contribution from volume relaxation can be safely ignored.

Our enthalpy differences as a function of the rhombohedral distortion δat several pressures calculated with method II are plotted in Fig. 2. It clearly shows that the pressure-induced transformations of BCC → RH_1, RH_1 → RH_2, and RH_2 → BCC occur in sequence with increasing pressure. A small energy barrier between equilibrium transitions is evident. Therefore the transitions should be first-order rather than continuous. In this method, the BCC is the only stable structure at low pressures. It changes to RH_1 at about 103 GPa. The stability of RH_2 structure gets enhanced with increased pressure, and becomes the ground state at ~126 GPa. The RH_2 structure attains the maximum stability at about 211 GPa, beyond which the enthalpy well becomes shallow. When above 300 GPa, the RH_2 phase disappears and the BCC becomes the only stable structure again. It should be noted that these results are in good agreement with those obtained by the method I. Also Fig. 2 clearly illustrates that there is no local enthalpy minimum when below 60 GPa, and thus no metastable region for RH_1 phase. This observation disproves Jenei *et al.*'s conjecture that the equilibrium P_c of BCC → $RH_1$is at ~30 GPa.

In order to acquire a comprehensive understanding of these results, we summarize the critical pressures in Table 1. For comparison the theoretical results by Lee *et al.*13, Qiu *et al.*16, and Verma *et al.*15 are also listed. It is evident that both our results calculated with method I and method II are consistent very well with each other. For this reason, Qiu*et al.*'s comment on Lee *et al.*'s results is not pertinent16. Compared with Lee *et al.*, our P_c of BCC → RH_1, RH_1 → RH_2, and RH_2 → BCC are slightly higher by about 14, 9, and 4 GPa, respectively. Verma *et al.*15 gave a BCC → RH_1 transition pressure of 60 GPa, which is close to Ding *et al.*'s experimental value, but their transition pressures of RH_1 → RH_2 and RH_2 → BCC are far away from others' estimate. Landa *et al.*24 predicted a similar BCC → RH_1 transition pressure, but no other two transition pressures were reported, therefore we do not discuss their results here.

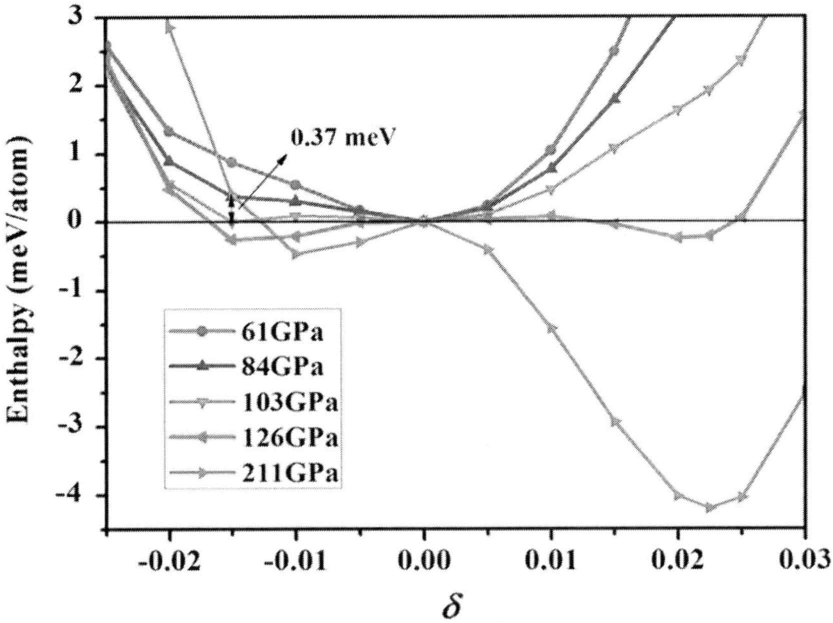

Figure 2: Variation of the calculated enthalpy difference as a function of the rhombohedral deformation parameter δ at selected pressures. The negative δ on the left side corresponds to the RH_1 phase, and the positive δ on the right side is for the RH_2 phase.

Table 1: Calculated critical pressures of vanadium at zero Kelvin, in which P_i denotes the lower bound of the metastable pressure, P_c is the thermodynamically equilibrium transition pressure, and P_m is the pressure where RH_2 phase attains the greatest stability, respectively.

Critical pressure	Method I (GPa)	Method II (GPa)	Lee *et al.*[13] (GPa)	Qiu *et al.*[16] (GPa)	Verma *et al.*[15] (GPa)
P_i of RH_1	50	—	73	19	—
P_c of BCC → RH_1	98	103	84	32	60
P_i of RH_2	110	—	103–112	65	—
P_c of RH_1 → RH_2	128	126	119	115	160
P_m of RH_2	210	211	187	—	—
P_c of RH_2 → BCC	284	278	280	297	>240

In order to understand the discrepancy in these theoretical data, we notice that for the latter two transitions, the deviation in P_c among Qiu*et al.*, Lee *et al.*, and our results is small. It is just 11 and 19 GPa for $RH_1 \rightarrow RH_2$ and $RH_2 \rightarrow$ BCC transition, respectively. In contrast, the deviation in BCC $\rightarrow RH_1$ is as big as 71 GPa. By inspecting Figs 1 and 2, we find that the enthalpy difference of RH_1 is much smaller than RH_2 phase. The small enthalpy difference requires very accurate calculation in order to obtain a reliable estimate of the transition pressure. For example, though our P_c of BCC $\rightarrow RH_1$ is about 103 GPa and larger than Lee *et al.*'s 84 GPa, the enthalpy difference between RH_1 and BCC at 84 GPa is less than 0.37 meV per atom. This is almost the accuracy limit of DFT based on pseudopotential method. We notice that both ours and Lee *et al.*'s calculation already achieved the absolute computation convergency, but we used a different PAW pseudopotential newly released with VASP. This could be the reason for the minor numeric difference. Except this, our results are in good agreement with theirs.

The shallow depth of the enthalpy well in RH_1 implies that its transition pressure is sensitive to the computation accuracy. In our calculations, we find that a P_c of 84 GPa for BCC $\rightarrow RH_1$ could be obtained if reduced the *k*-point mesh to $30 \times 30 \times 30$. A lower P_c would be obtained if further degrade the computational accuracy. By comparison, Verma *et al.*15 used a coarse *k*-point grid of $18 \times 18 \times 18$, and obtained a low P_c for BCC $\rightarrow RH_1$ (60 GPa) and a high P_c for $RH_1 \rightarrow RH_2$ (160 GPa) transition. Qiu *et al.*16 employed FLAPW method as implemented in WIEN2k_07 package. They predicted a BCC $\rightarrow RH_1$ transition pressure of ~32 GPa, which is much smaller than Ding *et al.*'s experimental value12 and all other theoretical studies13'14'15. As noted above, they ascribed this discrepancy to the constraints imposed by the transition pathway employed in calculations. But our above analysis showed that this is not the case. The true reason might be the computation precision that was not clearly specified in ref. 16.

By concluding this subsection, we emphasize that the scattering in the theoretical data of P_c mainly comes from computational accuracy. We found that there is no local enthalpy minimum and metastable region for RH_1 phase at around 30 GPa if fully convergent DFT calculations were performed. The non-hydrostatic effects reported by Jenei *et al.* might be due to unexpected exploration of angle α away from the ideal value of 109.47° by shear deformations, rather than transformation into the true RH_1 phase. We thus suggest to reinvestigate this problem experimentally by releasing pressure from the stable RH_1 phase with $\alpha = 110.17°$. This would reveal the distinct collapse of RH_1 back to BCC structure and give a lower bound of its metastability.

THERMO-ELECTRON EFFECT

The proposed Fermi surface nesting mechanism for the instability of BCC structure of vanadium10 implies that the transformation between BCC and RH phases should sensitively depend on subtle features of the Fermi surface. Usually there are three factors that will affect the Fermi surface structure: (i) lattice dynamics might modify the Fermi surface via electron-phonon interaction; (ii) thermo-electron excitations will blur and smear the subtle features in the Fermi surface; (iii) shift the Fermi level will also change the features of the Fermi surface. It is known that increasing the temperature leads to excitation of lattice vibrations and thermo-electrons. Recently, Landa *et al.*21 studied phonon effects on the phase stability of vanadium metal and vanadium-chromium alloys at high temperatures and pressures. Their results indicated that phonon has little impact on the stability of RH phases. Based on their conclusion, here we will mainly discuss the thermo-electron effects by using the finite temperature DFT method25. The effects of Fermi level position will be discussed in next subsection.

Different from lattice dynamics, we find that thermo-electrons have a very strong impact on the stability of RH phases. Our calculated results for various temperatures at a pressure of 211 GPa are plotted in Fig. 3, at this pressure the RH_2 phase attains its greatest stability. It is evident from this figure that the electronic temperature strikingly weakens the RH stability. The RH_2 phase transforms back to BCC structure when the temperature is above ~1915 K at 211 GPa. In comparison, it would be stable up to 8000 K at 182 GPa if including only the phonon contribution, as Landa *et al.* reported21. The thermo-electron effect is much more important than phonons. A simple analysis shows that including both thermo-electronic and phonon corrections will further reduce the transition temperature by ~260 K at 200 GPa (see Supplementary information). This phenomenon is compatible with the Fermi surface nesting mechanism of the BCC → RH transition, in which the temperature smears the Fermi surface and thus stabilizes the BCC phase. Nonetheless, as will be shown below, the Fermi surface nesting might not be the only mechanism for the BCC stabilization at high temperatures.

In order to deepen our understanding about this temperature-induced reentrant transition, we compute the temperature dependent electronic density of states (DOS) of RH_2 phase at 211 GPa, and show them in Fig. 4. The DOS of BCC phase at 580 K is also plotted for comparison. We can see that the BCC phase has more localized electrons (having *d* orbital character,

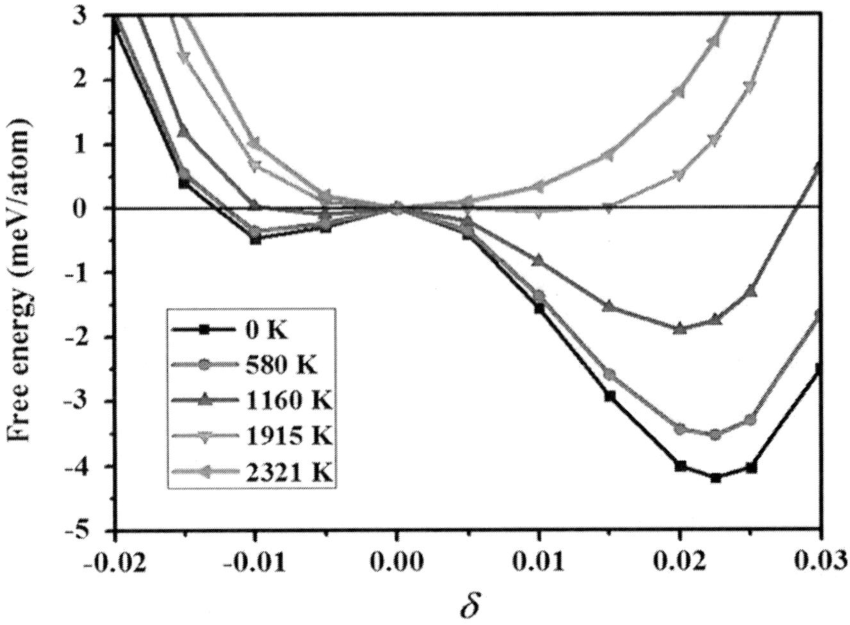

Figure 3: Effect of thermo-electrons on the phase stability of vanadium in the RH_1, RH_2, and BCC structures at 211 GPa.

at just below the Fermi level), whereas in RH phases the electrons are more delocalized. As the temperature increases, the valence electrons delocalize further and give a smooth DOS. This is in line with the expectation. The Fermi-Dirac distribution function $f(\varepsilon, T)$ is also plotted as an inset in Fig. 4, from which we can see that only a narrow region near the Fermi level (±0.5 eV) is directly affected by quantum statistics. However, its indirect influence on the whole DOS is remarkable and extends down to lower energy, mainly due to the electron-electron correlation that is automatically included in the self-consistent field procedure of the finite temperature DFT. For example, a noticeable modification on DOS still can be observed at −4 eV (measured from the Fermi level). The indication is that the classic argument based on perturbation theory, i.e., the temperature effects on electronic properties is usually restricted to an energy scale of $k_B T$ around the Fermi level26, is invalid. There are strong electron-electron correlations involved in vanadium. From this perspective, the stability of RH phases with respect to BCC might not solely be due to the Fermi surface nesting. Note that both Fermi surface nesting and band Jahn-Teller effect will result in a structure with a low DOS at the Fermi level. But at a temperature of 2321 K where BCC phase is favored, our calculations show that the DOS of the BCC phase at the

Fermi level is higher than that of RH phases. This is at odds with the proposed Fermi surface nesting mechanism. Furthermore, with increasing temperature, the DOS of RH phases at the Fermi level decreases, but its stability becomes weakened instead, which is also incompatible with the picture of Fermi surface nesting or the Jahn-Teller mechanism. As shown in Fig. 4, the thermo-electrons effects extend down to a low energy region, therefore it is natural to suspect that the transition back to BCC at high temperatures is caused by the states on the Fermi surface.

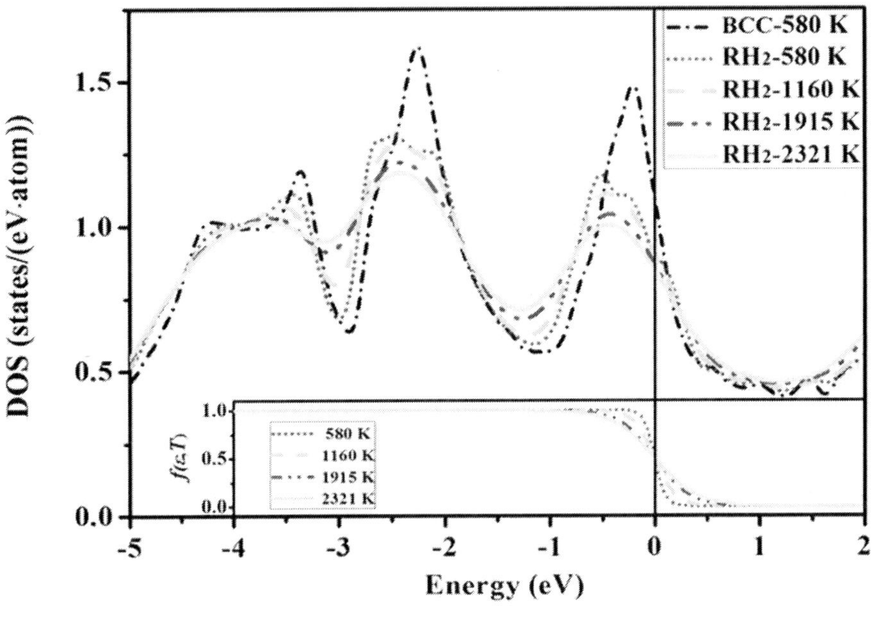

Figure 4: Comparison of the electronic density of states of vanadium at 211 GPa for BCC phase at 580 K and RH_2 phase at various temperatures.

The inset shows the Fermi-Dirac distribution function at the given temperatures. The vertical line denotes the Fermi level.

To elucidate that, we calculate the band energy and thermo-electronic entropy difference between BCC and RH phases at elevated temperatures. The results for RH_2 at a pressure of 211 GPa are given inFig. 5. Unexpectedly, we find that the band energy contribution is almost independent of the temperature, even though the modification on the DOS by temperature is remarkable. Instead, it is the *electronic entropy* that shifts the free energy difference up and finally stabilizes the BCC structure. Therefore we can conclude that the BCC phase at high temperatures is favored by electronic entropy, rather than by changes to the Fermi surface. A subsequent question is why entropy prefers BCC instead of RH? The answer is at Eq. (11) (in Sec. Methods), from which one can see that the entropy maximizes if a large DOS presents at or near the Fermi level. In other words, the more the electronic orbitals localize to the Fermi level, the greater the electronic entropy becomes. As depicted in Fig. 4, BCC structure has more localized *d* states and higher DOS near the Fermi level, hence it has larger electronic entropy and finally becomes the most thermodynamically stable phase at high temperatures. This mechanism is completely different from the band Jahn-Teller effect, Fermi surface nesting, or electronic topological transition that were proposed for the pressure-induced BCC → RH transition.

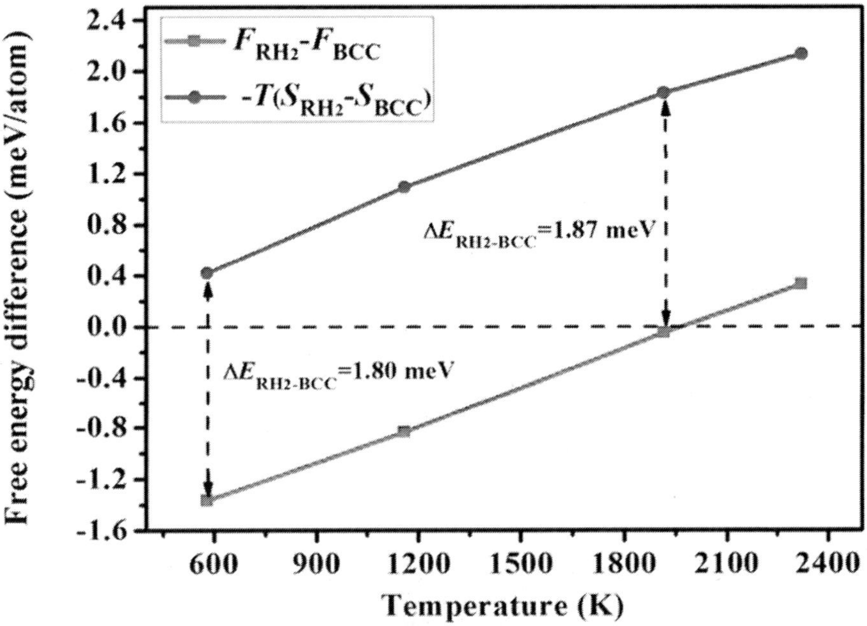

Figure 5: Variation of the free energy difference and its entropy contribution (−*TS*) between RH_2 (with δ = 0.01) and BCC phase of vanadium as a function of temperature at a pressure of 211 GPa. The dashed lines with arrowheads denote the difference of internal energy.

Phase diagram and elastic constants

Based on the calculated stability of BCC, RH_1, and RH_2 phases at high pressure and temperature, we construct a phase diagram for vanadium. This is the first time that such a comprehensive diagram of this metal is proposed. As shown in Fig. 6, BCC is the only stable solid phase up to 98 GPa (estimated with our BCC → RH_1 transition pressure) and stands up to the melting point. With increasing pressure and temperature, RH_1phase becomes stable when below 1440 K and 140 GPa. The stable region of RH_2 phase locates between 126–280 GPa, with the maximal transition temperature of 1915 K at 211 GPa. As the temperature increases, both RH_1 and RH_2 phases become unstable and transform back to BCC, which then melts to the liquid phase. This picture drastically changes our previous understanding that RH_1 could stand up to the static melting curve22 and RH_2 is stable up to the shock wave melting curve23. It is interesting to note that at the low pressure side, the phase boundary slope of BCC → RH_1 and RH_1 → RH_2 is positive. This indicates that RH_1 is denser than BCC phase and RH_2 is denser than RH_1phase, according to the Clausius-Clapeyron equation. When the pressure is higher than 211 GPa, the slope of the RH_2 → BCC phase boundary becomes negative, suggesting that under this thermodynamic condition BCC becomes denser than RH_2 phase. The BCC-RH_1-RH_2triple point is determined at 1440 K and 140 GPa, where spectacular mechanic properties could be expected. The phase diagram as shown inFig. 6 is unique. As discussed above, it is a consequence of competition between two different transition mechanisms: along the compression direction Fermi surface nesting destabilizes BCC phase, whereas along the temperature dimension, the BCC phase is favored by electronic entropy.

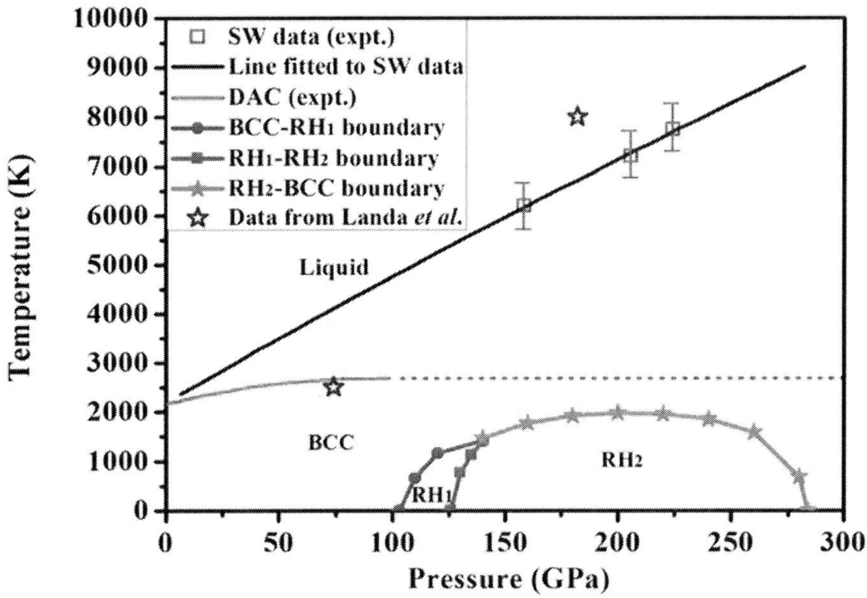

Figure 6: Phase diagram of vanadium at high pressure and finite temperature.

The DAC data of the melting line and its extrapolation are taken from ref. 22. The shock wave (SW) data on the melting curve are from ref. 23. The stability bounds of RH phases estimated by Landa *et al.* are from ref. 21.

The competition of the Fermi surface nesting and electronic entropy implies that the strength of vanadium should increase with increasing temperature. This is an astonishing prediction and at odds with widely accepted (though empirical) experience that temperature always *softens*metals. To verity this, we also calculate the elastic constants of vanadium at elevated temperatures and pressures.

The calculated elastic moduli C_{44} and C' as a function of pressure at zero Kelvin are shown in Fig. 7. For comparison the results of Landa *et al.*10 and the experimental data at ambient conditions26 are also included. It is evident that our results are in good agreement with the experimental data and Landa *et al.*'s full-potential linear muffin-tin orbitals (FPLMTO) results. We find C_{44} becomes negative between 125 and 260 GPa, at the same time C' is anomalously softened within the same pressure range (similar results were observed by using the exact muffin-tin orbitals (EMTO) calculations)10. This is in sharp contrast to the theoretical C_{44} values reported by Qiu *et al.*16 (not shown here), which are far from ours and Landa *et al.*'s results. Qiu *et al.* argued that the discrepancy is due to the pressure correction neglected in Landa *et al.*'s work. However, here we used the same formula as Qiu *et al.* and included the pressure correction explicitly when calculating the elastic moduli $C_{44}(p)$ and $C'(p)$. The perfect match between ours and Landa *et al.*'s results unequivocally demonstrates that the pressure correction is not the main reason of their discrepancy. Rather, as already mentioned in previous subsection, the computation convergence quality might be the true cause.

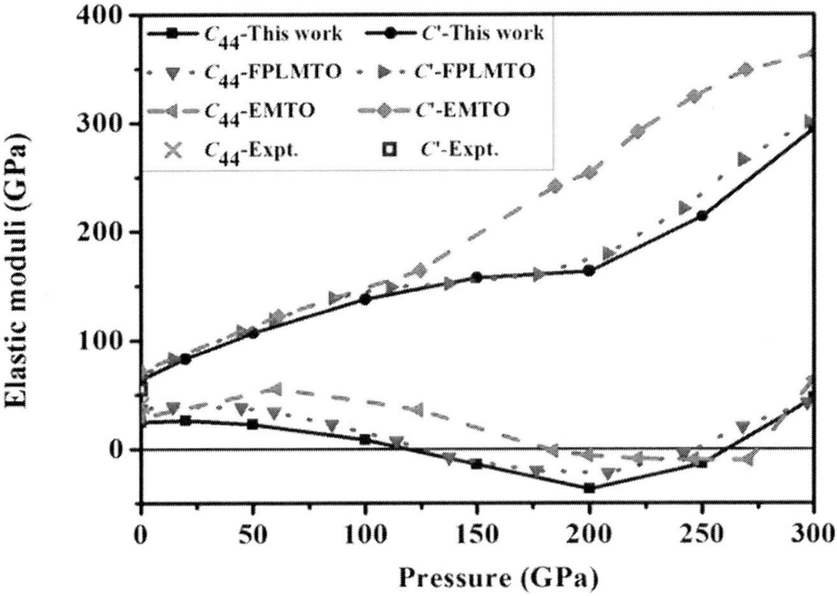

Figure 7: Elastic moduli of vanadium in BCC phase as a function of pressure.

The results of C_{44}-FPLMTO, C'-FPLMTO, C_{44}-EMTO, and C'-EMTO are taken from ref. 10. The experimental data at ambient conditions are from ref. 26.

The electronic temperature effects on the shear modulus C_{44} of BCC phase as a function of pressure at different temperatures are plotted in Fig. 8. As we envisioned above, in the whole pressure range we studied, C_{44} indeed increases with the temperature, and stabilizes the BCC against RH phases. This phenomenon originates in electronic structures and is against our empirical intuition. It should be noted that when temperature is above ~1900 K, the mechanically instable region (with negative C_{44}) of BCC phase disappears completely. This is consistent with the phase diagram as drawn in Fig. 6. Therefore we discover a novel temperature-induced *hardening* phenomenon in vanadium, which is very rare (if any) to our knowledge. At ambient pressure, when temperature increases from room temperature to 2000 K, C_{44} increases by 13.18%. At 50 GPa and 300 GPa where BCC is always stable, when temperature is increased from room temperature to 3000 K, C_{44} increases by 75.43% and 53.21%, respectively, as shown in the inset of Fig. 8. At high enough temperatures, especially when near the melting point, the thermal motion of nuclei will soften the metal via thermal activation mechanism. Therefore we predict that the strength and shear modulus of vanadium will increase to a maximum, and then drop down to zero with increasing temperature.

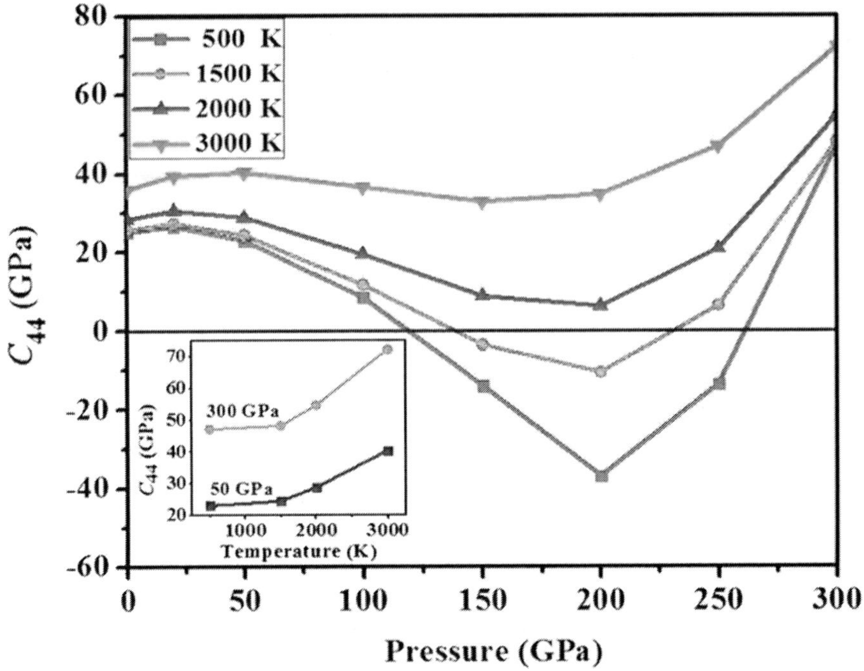

Figure 8: Calculated shear elastic constant C_{44} of vanadium in BCC phase as a function of pressure at different electronic temperatures. Inset: Variation of C_{44} as a function of temperature at given pressures of 50 and 300 GPa, respectively.

Effect of Fermi level and charge transfer

This subsection is devoted to the charge transfer induced by chemical environment and the subsequent shift of the Fermi level. Both Fermi surface nesting and band Jahn-Teller mechanism depends sensitively on the position of the Fermi energy or the electronic chemical potential E_F. By using the partial jellium model briefed in Sec. Methods, we adjust the position of E_F slightly and investigate its impact on phase stability of vanadium. In Fig. 9, the enthalpy difference as a function of the rhombohedral deformation parameter δ are shown for three shifted Fermi levels at a pressure of 211 GPa. Shifting the Fermi level means to change the orbital occupations at the valence band top or conduction band bottom by charge transfer. At 211 GPa, the RH_2 phase attains the maximum stability. If the RH structures are stabilized by $s \rightarrow d$ electronic transition or band Jahn-Teller effect, it should have the optimum orbital occupation at this pressure. Therefore, lifting (adding electrons) or descending (removing electrons) the Fermi level moves the system away from the optimum occupation, thus both should weaken the stability of RH phases. Nonetheless, our calculations show that at 211 GPa shifting down the Fermi level stabilizes the RH phases further, whereas shifting it up destabilizes RH greatly. This is in line with Landa*et al.*'s band-filling argument when alloying vanadium with the same transition series24, but is incompatible with Jahn-Teller mechanism. However, this observation is not contrary to the Fermi surface nesting, which should depend sensitively on the subtle Fermi surface structure, and thus its position.

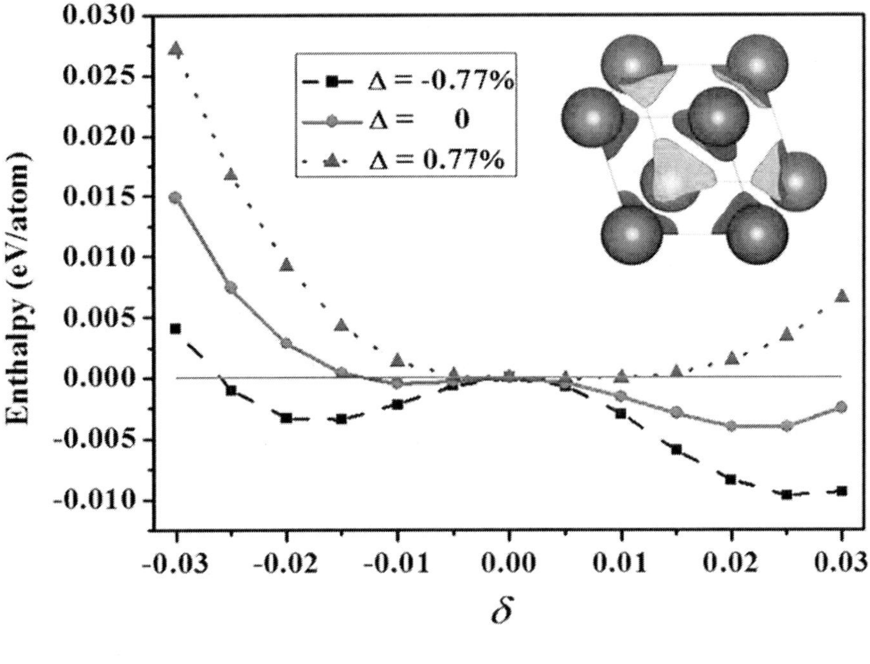

Figure 9: Variation of the enthalpy difference with respect to BCC phase as a function of the rhombohedral deformation parameter $\delta$when the Fermi level being shifted up or down at a pressure of 211 GPa. Inset: Calculated differential charge density between Δ = −0.77% and Δ = 0. Here Δ is the percentage of the total charge that are removed from (or added to) the system for the purpose to shift the Fermi level.

Furthermore, our calculation indicates that the maximal stability of RH_2phase at this pressure is attained when Δ = −2.15%. Here Δ is defined as the total charge percentage being added/removed to the system. Further shifting down the Fermi level destabilizes the RH phases. When Δ < −4.85%, the BCC phase becomes favorable again. On the other hand, BCC also becomes stable when Δ > 0.77%. The large stable range of Δ for RH_2 phase implies that the Fermi surface nesting alone can not be the distortion mechanism. In the inset of Fig. 9, we plot the calculated differential charge density between $\Delta = -0.77\%$ and $\Delta = 0$. It is evident that the removed electrons (or added holes) are well localized around the nuclei. Analysis indicates that they are mainly *d* character. The variation of enthalpy difference with Δ shows that within this pressure range the RH_2 phase of vanadium dislikes electrons, and has a low electronegativity with respect to the BCC phase. Such behavior should be a consequence of *d* orbitals delocalization. Therefore localization/delocalization of the *d* electrons also has an important role in RH stability even at low temperatures.

In Fig. 10, the variation of the electronic DOS of RH_2 phase for three shifted Fermi levels at 211 GPa are displayed. It can be seen that the modification on DOS by Fermi level shifting is very small. Based on Figs 9 and 10, it seems that the RH_2 phase becomes more stable when the DOS at Fermi level is higher. This is at odds with the band Jahn-Teller mechanism and the Fermi surface nesting. However, when compared to the BCC phase, the relative DOS difference at the Fermi level for $\Delta = -0.77\%$, $\Delta = 0$, and $\Delta = 0.77\%$ are 0.34, 0.22, and 0.05, respectively. In other words, when the Fermi level shifts up, the relative DOS difference at the Fermi level along the rhombohedral deformation path decreases. At the same time, the stability of RH_2 phase continuously weakens, and eventually transforms back to the BCC phase.

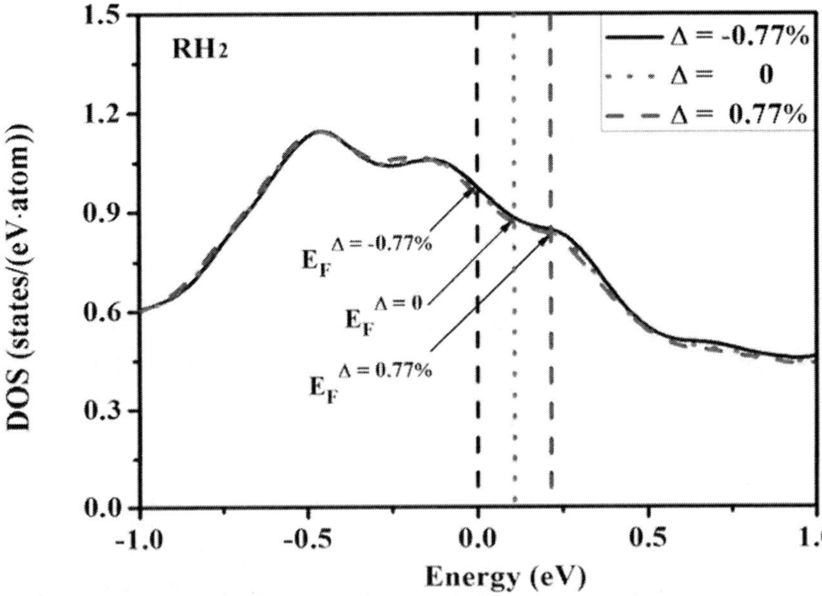

Figure 10: Variation of the electronic density of states and the position of the Fermi level when additional electrons or holes are added to vanadium in RH_2 phase at 211 GPa.

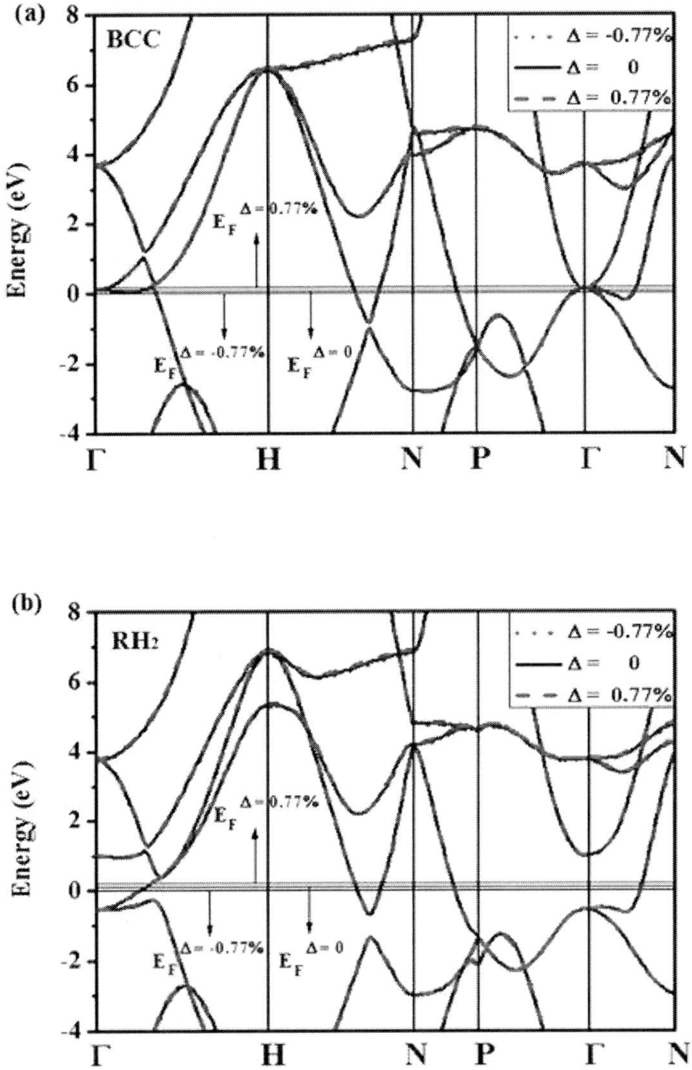

Figure 11: Band structures of vanadium at 211 GPa when the Fermi level is shifted: (**a**) BCC phase, (**b**) RH_2 phase.

Consider now how the band structure responses when the Fermi level being shifted. In Fig. 11, the band structures of vanadium in RH_2 and BCC phases at 211 GPa are plotted for three shifted Fermi levels. It can be seen that the band structure is just slightly modified when the Fermi level being changed, and is consistent with the DOS shown in Fig. 10. When comparing the band structure of RH_2 to BCC phase, we find that the rhombohedral distortion splits the levels with t_{2g} symmetry at **Γ** and **H** points, which is consistent with the results reported by Landa *et al.*10and Ohta *et al.*27 The interesting finding here is that though all Fermi levels are well positioned within the pseudo-gap opened by splitting of the t_{2g} states at **Γ** point, the RH_2 stability is changed differently. This intriguing phenomenon suggests that other factors besides the splitting of the t_{2g} might be involved in BCC → RH phase transformation.

From above analyses, it is clear that the Fermi level position, as well as accompanying charge transfer, has a strong effect on the phase stability of vanadium. Since electron topological transition has been excluded as the main driving force for the BCC → RH phase transitions28, our calculations suggest that the Fermi surface nesting, the distortion induced band splitting, and localization/delocalization of the *d* orbitals might be responsible for these unique phase transitions in vanadium, as well as the unique mechanical properties.

CONCLUSIONS

In summary, a thorough and comprehensive theoretical study of the phase transition in vanadium at high pressure and high temperature has been carried out with first-principles calculations based on density functional theory. It was found that the scattering in the theoretical pressure of the first-order phase transitions (BCC → RH_1, RH_1 → $RH_2$and RH_2 → BCC) is mainly due to computation convergence quality. With high enough precision computations, the transition pressures were pinned to 98, 128, and 284 GPa, respectively. Our calculations also predicted no local minimum and metastable region for RH_1 at low pressures, and suggested that the transition pressure of 30 GPa reported by Jenei *et al.* might not be due to BCC → RH_1 transition, rather it could arise from shear deformations. Considering the complication arisen from non-hydrostatic effects, we suggest to investigate the stability of RH_1 along the pressure releasing path, by which the abrupt collapse of RH_1 to BCC at ~60 GPa might be more distinct to observe.

The thermo-electron effect on the stability of vanadium in different structures was studied by using the finite temperature DFT. We observed that both phonon (see ref. 19) and thermo-electronic effects reduce the stability of RH phases, but the effects of thermal electrons are more important. Both RH_1 and RH_2 phases transform back to BCC structure when the temperature is above 1440 K (at 140 GPa) and 1915 K (at 211 GPa), respectively. By comparing the free energy difference, we determined the BCC → RH_1 → RH_2 → BCC boundaries and the BCC-RH_1-RH_2 triple point, which finally led us to construct a high-pressure and finite-temperature phase diagram for vanadium. The unexpected stabilization of the BCC phase by temperature was determined due to electronic entropy, which also leads to unusual hardening of the shear modulus C_{44} by temperature, a very rare and interesting phenomenon.

By using the partial jellium model and intendedly adjusting the Fermi level position, we found that decreasing of the electronic chemical potential would further stabilizes the RH_2 phase at 211 GPa, whereas destabilizes it otherwise. By inspecting the changes in the density of states and band structures of vanadium when the Fermi level being shifted, we concluded that the most possible transition mechanism is a combination of the Fermi surface nesting, band splitting due to lattice distortions, and *d* orbitals localization. The former two affect the phase stability along the compression direction, while the latter favors BCC via the electronic entropy along the temperature dimension.

METHODS

Most calculations are performed in the primitive cell of the BCC or RH phases, and the Vienna *ab initio* simulation package (VASP) is used29, which is based on first-principles density functional theory (DFT)30 and the projector augmented-wave (PAW) method31. The pseudopotential contains 13 valence electrons (including $3s^2$, $3p^6$, $3d^3$, and $4s^2$ states). The Perdew-Burke-Ernzerhof (PBE) generalized gradient approximation (GGA)32 for the electronic exchange-correlation functional is used. We speculate that the scattering in the previous theoretical results might relate to the calculation accuracy, therefore a high enough cutoff energy for the plane wave basis of 900 eV is used, as well as a 60 × 60 × 60 shifted uniform mesh for the *k*-point sampling. This set includes the Γ point and results in 5200 and 18910 *k*-points in the irreducible Brillouin zone of the BCC and RH lattices, respectively. The smearing parameters are also well tested. This parameter setting is carefully checked by increasing the cutoff energy and *k*-points to higher values and to ensure that it gives an absolutely converged total energy and pressure (in the sense of computer simulations).

In order to deliver a reliable energetics assessment, we exploit two different methods to evaluate the phase transitions and the (meta-)stability region. The first one is a conventional method (method I), in which we fully optimize the structure directly and then calculate the enthalpy of the resultant phases as a function of pressure. The transition pressures are obtained by comparing the enthalpy difference with respect to BCC phase. With this method, both the thermodynamic and mechanical (meta)stability are obtained. In the second method, we explore the structure transformation as that employed in ref. 13, namely, distorting the BCC structure along a predefined pathway (method II). The instability and phase transitions are then deduced by inspecting the resultant energy curve as a function of the distortion magnitude. Different from the first one where structure is fully relaxed, in this method the explored geometry is highly constrained by the predefined pathway. Hence the absolute stability of these phases and their transition pressures are not guaranteed in principle. For this reason, Qiu *et al.* attributed the large discrepancy of their transition pressure from that of Lee *et al.* to the limitation of this method16. A cross-check of these two methods is necessary and will be helpful to secure a reliable and consistent theoretical result.

According to ref. 13, the volume-conserved BCC → RH transformation matrix (or the deformation gradient) $T(\delta)$ is defined as

$$T(\delta) = \begin{pmatrix} k & \delta & \delta \\ \delta & k & \delta \\ \delta & \delta & k \end{pmatrix}, \qquad (2)$$

in which k is determined from the real positive solution of det $(T) = 1$, to ensure a volume-conserving transformation. The small displacement δrepresents the amount of rhombohedral deformation of the BCC crystal: a positive δ corresponds to a decrease in the angle α from the BCC value of $\alpha_0 = 109.47°$. This pathway, however, is not the one for the pure shear deformation. Therefore a different transformation matrix will be used when calculating C_{44}.

To evaluate the elastic moduli C_{44} and $C' = (C_{11} - C_{12})/2$ of the BCC phase, we use a conventional 2-atom cubic unit cell to calculate the total energy as a function of volume and its variation along the shear strains. The C_{44} and C' are then obtained from the second derivative of the total energy with respect to the deformation magnitude δ, which is defined by the strain matrices33

$$\varepsilon C_{44} = \begin{pmatrix} 0 & \delta & 0 \\ \delta & 0 & 0 \\ 0 & 0 & \delta^2/(1-\delta^2) \end{pmatrix}, \qquad (3)$$

$$\varepsilon C' = \begin{pmatrix} \delta & 0 & 0 \\ 0 & -\delta & 0 \\ 0 & 0 & \delta^2/(1-\delta^2) \end{pmatrix}. \qquad (4)$$

The corresponding strain energy is then given by

$$E(\delta) = E(0) + 2C_{44}V\delta^2 + O(\delta^4), \qquad (5)$$
$$E(\delta) = E(0) + 2C'V\delta^2 + O(\delta^4). \qquad (6)$$

As noted by Qiu *et al.*16, these formulae include the pressure corrections because of the δ^2 term for ε_3, which they took as the main cause for the big deviation of their C_{44} from that of Landa *et al.*10.

At finite temperatures, one should use the free energy rather than the internal energy to derive the structure stability and physical properties. In general, the free energy as a function of temperature and density (or atomic volume) can be expressed as

$$F(V,T) = E_c(V) + F_e(V,T) + F_{vib}(V,T). \qquad (7)$$

Here E_c is the cold static lattice energy with atoms being clamped at their equilibrium positions, F_e is the thermal free energy contributed by electronic excitations, and F_{vib} is the vibrational free energy of phonons. Landa *et al.* calculated lattice dynamics and showed that F_{vib} has little impact on the stability of RH phases21. For this reason we will mainly focus on the thermo-electrons effect in this work. After discarding the phonon term, the free energy becomes

$$F(V,T) = E_c(V) + F_e(V,T). \qquad (8)$$

Within one-electron approximation, the thermo-electronic free energy$F_e(V,T)$ can be constructed from the groundstate density of states $n(\varepsilon)$ as

$$F_e(V,T) = E_e(V,T) - TS_e(V,T), \qquad (9)$$

in which the internal energy due to electronic excitations is given by34

$$E_e(V,T) = \int_{-\infty}^{+\infty} \varepsilon n(\varepsilon,V) f(\varepsilon,T)\,d\varepsilon - \int_{-\infty}^{\varepsilon_F} \varepsilon n(\varepsilon,V)\,d\varepsilon, \qquad (10)$$

and the electronic entropy is

$$S_e(V,T) = -k_B \int_{-\infty}^{+\infty} n(\varepsilon,V)\{f(\varepsilon,T)\ln f(\varepsilon,T) + [1-f(\varepsilon,T)]\ln[1-f(\varepsilon,T)]\}\,d\varepsilon, \qquad (11)$$

where $f(\varepsilon, T)$ is the Fermi-Dirac distribution function, $n(\varepsilon, V)$ is the electronic density of states (DOS) at the energy eigenvalue ε, ε_F is the Fermi energy, and k_B is the Boltzmann's constant. This formalism can be devised and solved self-consistently with Mermin's finite temperature DFT25 For a given volume and temperature, from the variational principle of the free energy with respect to electron density, one first solves the standard Kohn-Sham equations30 using a trial density. This produces one-electron eigenstates and the Fermi level. Then one reconstructs the charge density by populating electrons onto these states according to Fermi-Dirac distribution. Mixing this density with the initial one and recalculating the Kohn-Sham equations, then repeating the whole process until the convergence is achieved, one then obtains the self-consistent free energy of electrons at finite temperatures29. With this free energy, we can evaluate the thermo-electronic effect on phase stability and elastic constants.

Chemical environment or alloying will affect materials behavior by donating (or accepting) electrons. This phenomenon in vanadium has been explored in Landa *et al.*'s pilot works24'35. They found that the band-filling argument applies when alloying vanadium with its neighbors within the same transition series, whereas it is complicated by the Madelung energies in other cases. This precluded them from drawing a conclusion on the effects of charge transfer and electron chemical potential. In order to circumvent this difficulty and obtain an unambiguous understanding, we will study this problem with an alternative method, i.e., focusing on the effects solely due to the electron chemical potential. The electron chemical potential μ, which is also the Fermi energy ε_F in a metal, is defined as

$$\mu = \left(\frac{\partial F}{\partial N_e}\right)_{V,T}, \qquad (12)$$

where N_e is the total number of electrons. At the level of free electron gas model, its value at zero Kelvin is given by $\mu(0) = \hbar^2/2m_e(3\pi^2 N_e/V)^{2/3}$. Therefore one can adjust the Fermi-level position by adding or removing electrons from the system. In this work, we employ the partial jellium model to achieve this purpose by compensating the charged simulation cell with a homogeneous positive/negative background charge density. By allowing the added (fictitious) charges interacting with the nuclei and other electrons, the system relaxes and leads to a set of self-consistent eigenstates of the Kohn-Sham equations, from which the Fermi energy can be derived. With this method, the relaxation effect due to adding/removing electrons is automatically included. This method is widely used when studying charged defects in semiconductors26. Here we exploit it to investigate how the band-filling and Fermi-level position affect the high pressure behavior of vanadium. This will benefit us in understanding the phase transition mechanism, since different mechanisms usually have different response when the Fermi-level shifts.

ADDITIONAL INFORMATION

How to cite this article: Wang, Y. X. *et al*. Stability of rhombohedral phases in vanadium at high-pressure and high-temperature: first-principles investigations. *Sci. Rep.* **6**, 32419; doi: 10.1038/srep32419 (2016).

ACKNOWLEDGEMENTS

This work is supported by the National Natural Science Foundation of China under Grant Nos 11274281 and 11174214, the CAEP Research Projects under Grant Nos 2102A0101001 and 2015B0101005, and the NSAF under Grant No. U1430117.

CONTRIBUTIONS

H.Y.G. and Y.X.W. conceived the research. Y.X.W. performed atomic and electronic structure calculations. Y.X.W., Q.W., X.R.C. and H.Y.G. analyzed the numerical results. Y.X.W. and H.Y.G. wrote the manuscript and all the authors commented on it.

COMPETING INTERESTS

The authors declare no competing financial interests.

REFERENCES

1. Ishizuka, M., Iketani M. & Endo, S. Pressure effect on superconductivity of vanadium at megabar pressures. *Phys. Rev. B*61, 3823 (2000).
2. Hu, W. Y., Xiao, S. F., Yang, J. Y. & Zhang, Z. Melting evolution and diffusion behavior of vanadium nanoparticles. *Eur. Phys. J. B* **45**, 547 (2005).
3. Bosak, A., Hoesch, M., Antonangeli, D. *et al.* Lattice dynamics of vanadium: Inelastic x-ray scattering measurements. *Phys. Rev. B*78, 020301 (2008).
4. Rudd, R. E. & Klepeis, J. E. Multiphase improved Steinberg-Guinan model for vanadium. *J. Appl. Phys.* **104**, 093528 (2008).
5. Lee, B., Rudd, R. E. & Klepeis, J. E. Using alloying to promote the subtle rhombohedral phase transition in vanadium. *J. Phys.: Condens. Matter* **22**, 465503 (2010).
6. Krasilnikov, O. M., Vekilov, Y. K., Isaev, E. I. & Bondarenko, N. G.Theory of elastic phase transitions in metals at high pressures. Application to vanadium. *J. Exp. Thero. Phys.* **112**, 240 (2011).
7. Krasilnikov, O. M., Vekilov, Y. K., Mosyagin, I. Y. *et al.* Elastic phase transitions in metals at high pressures. *J. Phys.: Condens. Matter*24, 195402 (2012).

8. Suzuki, N. & Otani, M. Theoretical study on the lattice dynamics and electron-phonon interaction of vanadium under high pressures. *J. Phys.: Condens. Matter* **14**, 10869 (2002).
9. Landa, A., Klepeis, J., Söderlind, P. *et al.* Fermi surface nesting and pre-martensitic softening in V and Nb at high pressures. *J. Phys.: Condens. Matter* **18**, 5079 (2006).
10. Landa, A., Klepeis, J., Söderlind, P. *et al. Ab initio* calculations of elastic constants of the bcc V-Nb system at high pressures. *J. Phys. Chem. Solids* **67**, 2056 (2006).
11. Manghnani, M. H., Nellis, W. J. & Nicol, M. F. *Science and technology of high pressure: proceedings of the International Conference on High Pressure Sciene and Technology* (AIRAPT-17), Honolulu, Hawaii, 25–30 July, 1999. Vol. 1 (University Press, Hyderabad, India, 2000).
12. Ding, Y., Ahuja, R., Shu, J. *et al.* Structural phase transition of vanadium at 69 GPa. *Phys. Rev. Lett.* **98**, 085502 (2007).
13. Lee, B., Rudd, R. E., Klepeis, J. E., Söderlind, P. & Landa, A.Theoretical confirmation of a high-pressure rhombohedral phase in vanadium metal. *Phys. Rev. B* **75**, 180101(R) (2007).
14. Lee, B., Rudd, R. E., Klepeis, J. E. & Becker, R. Elastic constants and volume changes associated with two high-pressure rhombohedral phase transformations in vanadium. *Phys. Rev. B* **77**, 134105 (2008).
15. Verma, A. K. & Modak, P. Structural phase transitions in vanadium under high pressure. *Europhys. Lett.* **81**, 37003 (2008).
16. Qui, S. L. & Marcus, P. M. Phases of vanadium under pressure investigated from first principles. *J. Phys.: Condens. Matter* **20**, 275218 (2008).
17. Luo, W., Ahuja, R., Ding, Y. & Mao, H. K. Unusual lattice dynamics of vanadium under high pressure. *Proc. Natl. Acad. Sci. USA* **104**, 16428 (2007).
18. Jenei, Z., Liermann, H. P., Cynn, H. *et al.* Structural phase transition in vanadium at high pressure and high temperature: Influence of nonhydrostatic conditions. *Phys. Rev. B* **83**, 054101 (2011).
19. Louis, C. N. & Iyakutti, K. Electronic phase transition and superconductivity of vanadium under high pressure. *Phys. Rev. B***67**, 094509 (2003).
20. Nagasako, N., Jahnátek, M., Asahi, R. & Hafner, J. Anomalies in the response of V, Nb, and Ta to tensile and shear loading: *Ab initio*density functional theory calculations. *Phys. Rev. B* **81**, 094108 (2010).
21. Landa, A., Söderlind, P. & Yang, L. H. Ab initio phase stability at high temperatures and pressures in the V-Cr system. *Phys. Rev. B***89**, 020101(R) (2014).

22. Errandonea, D., Schwager, B., Ditz, R. *et al.* Systematics of transition-metal melting. *Phys. Rev. B* **63**, 132104 (2001).
23. Dai, C. D., Jin, X. G., Zhou, X. M. *et al.* Sound velocity variations and melting of vanadium under shock compression. *J. Phys. D: Appl. Phys.* **34**, 3064 (2001).
24. Landa, A., Söderlind, P., Ruban, A. V. *et al.* Stability in BCC transition metals: Madelung and band-energy effects due to alloying. *Phys. Rev. Lett.* **103**, 235501 (2009).
25. Mermin, N. D. Thermal properties of the inhomogeneous electron gas. *Phys. Rev.* **137**, 1441 (1965).
26. Ashcroft, N. W. & Mermin, N. D. *Solid State Physics* (Holt, Rinehart and Winston, New York, 1976).
27. Ohta, Y. & Shimizu, M. Influence of shear strains on the electronic structure of vanadium. *J. Phys. F: Met. Phys.* **13**, 761 (1983).
28. Kobi, L., Ma, Y., Oganov, A. R. *et al.* Elasticity of the superconducting metals V, Nb, Ta, Mo, and W at high pressure.*Phys. Rev. B* **77**, 214101 (2008).
29. Kresse, G. & Furthmüller, J. Efficiency of ab-initio total energy calculations for metals and semiconductors using a plane-wave basis set. Comp. *Mater. Sci.* **6**, 15 (1996).
30. Kohn, W. & Sham, L. J. Self-consistent equations including exchange and correlation effects. *Phys. Rev.* **140**, 1133 (1965).
31. Blöchl, P. E. Projector augmented-wave method. *Phys. Rev. B* **50**, 17953 (1994).
32. Perdew, J. P., Burke, K. & Ernzerhof, M. Generalized gradient approximation made simple. *Phys. Rev. Lett.* **77**, 3865 (1996).
33. Liu, Z. H. & Shang, J. X. First principles calculations of electronic properties and mechanical properties of bcc molybdenum and niobium. *Rare Metals* **30**, 354 (2011).
34. Geng, H. Y., Sluiter, M. H. F. & Chen, N. X. Cluster expansion of electronic excitations: Application to fcc Ni-Al alloys. *J. Chem. Phys.* **122**, 214706 (2005).
35. Landa, A., Söderlind, P., Velikokhatnyi, O. I. *et al.* Alloying-driven phase stability in group-VB transition metals under compression.*Phys. Rev. B* **82**, 144114 (2010).

Chapter 14

Pressure-Induced Zigzag Phosphorus Chain and Superconductivity in Boron Monophosphide

Xinyu Zhang, Jiaqian Qin, Hanyu Liu, Shiliang Zhang, Mingzhen Ma, Wei Luo, Riping Liu & Rajeev Ahuja

State Key Laboratory of Metastable Materials Science and Technology, Yanshan University, Qinhuangdao 066004, P.R. China

ABSTRACT

We report on the prediction of the zinc-blende structure BP into a novel *C2/m* phase from 113 to 208 GPa which possesses zigzag phosphorus chain structure, followed by another $P4_2/mnm$ structure above 208 GPa above using the particle-swarm search method. Strong electron-phonon coupling λ in compressed BP is found, in particular for*C2/m* phase with the zigzag phosphorus chain, which has the highest λ (0.56–0.61) value among them, leading to its high superconducting critical temperature T_c (9.4 K–11.5 K), which is comparable with the 4.5 K to 13 K value of black phosphorus phase I (orthorhombic, *Cmca*). This is the first system in the boron phosphides which shows superconductivity from the present theoretical calculations. Our results show that pressure-induced zigzag phosphorus chain in BP exhibit higher superconducting temperature T_C, opening a new route to search and design new superconductor materials with zigzag phosphorus chains.

INTRODUCTION

The crucial thermodynamic parameter-high pressure has been emerging as a powerful tool to investigate physical and chemical behaviours of materials, especially to synthesize or design materials with excellent properties such as superhardness and superconductivity1′2′3′4′5′6. Recently, B and P have been studied extensively in physical, chemical, and material science fields due to their interesting structural properties when pressure is applied4′5′7′8′9. Boron has always been recognized as a complex element, both structurally and electronically: its crystalline phases are numerous and inevitably complicated, which is related to its electron deficiency and thus the tendency to form multicenter bonds5′7. Phosphorus, on the other hand, has six known phases at room temperature under high pressure10′11′12′13′14′15. Black phosphorus phase I (orthorhombic, *Cmca*) is known to be most stable under ambient temperature and pressure15. The experiment at low temperature and high pressure showed that the superconducting temperature of phase I increased from 4.5 K to 13 K under pressure14. These peculiar physical properties of compressed B and P solids have motivated our attention on high-pressure study of BP. The binary semiconductor boron phosphide (BP) is the zinc-blende crystal structure (space group *F*-43*m*) with lattice parameter a = 4.537 Å16, which has extraordinary properties, such as high hardness, high temperature stability, resistance to chemical corrosion, high thermoelectric powers for direct energy conversion, and is regarded as an important candidate for electronic, optical, and other engineering applications17′18′19′20′21. In order to understand in detail the pressure-induced structural behaviour, including mainly the phase transitions of BP, further high-pressure studies are essential.

Here, on the basis of comprehensive density functional theory (DFT) computations, we report the design of two new high pressure structures of BP. We performed variable-cell structure prediction simulations using CALYPSO22′23 approach for BP at 0, 50, 100, 150, 200, 250, and 300 GPa, respectively. For comparison, the structures reported in boron nitride (*P-3m1*, $P6_3mc$, $P6_3/mmc$) experimentally and theoretically, are also considered in our calculations. Our calculations demonstrate that, *F*-43*m* is the most stable phase at ambient conditions, two new high pressure phases *C*2/*m* and $P4_2/mnm$ can be predicted during compression up to 300 GPa. We further elucidate the energetic, mechanical, electronic and superconducting properties of the obtained novel phases, confirming that *F*-43*m* is semiconductor, and discovering two high-pressure superconducting phases *C*2/*m* and$P4_2/mnm$.

RESULTS

The analysis of the predicted structures gives us a list of candidate structures with space groups *F-43m, C2/m, P4$_2$/mnm, P6$_3$mc, Cmmm, Immm, P63/mmc-I, P63/mmc-II*, and the previously reported high pressure rock salt structure *Fm-3m*20 are depicted in Figure S1 (Supporting Information). The relative enthalpies per chemical formula unit vs. pressure curves of selected structures are plotted in Figure 1(a). Considering that *F-43m* and *P6$_3$mc* structures hold very close enthalpy, the enthalpy difference vs. pressure curves for the two phases are specially inserted in Figure 1(a), and it indicates that *F-43m* has lower enthalpy, which is consistent with the experimental results. In order to investigate the phase transition pressures clearly, the enthalpies vs. pressure curves for *F-43m, Immm, P4$_2$/mnm, C2/m,* and *P-3m*1 are given in Figure 1(b). From Figure 1, one can see that, for compressed BP, the zinc-blende crystal structure *F-43m* is the most stable structure below 113 GPa, a *C2/m* phase takes over the pressure range from 113 to 208 GPa, followed by another *P4$_2$/mnm* structure above 208 GPa. The pressure evolution of the unit cell volume of BP in the structures *F-43m, C2/m,* and *P4$_2$/mnm* is inserted in Figure 1(b). The experimental equation of state data24 for single-crystal BP with *F-43m* structure up to 55 GPa were listed for comparison. It can be seen that our calculated data are in good agreement with the experimental results. The zinc-blende crystal structure *F-43m* is stable on compression to 113 GPa, and then complete change to *C2/m* is observed. Comparisons with the *F-43m* cell at 113 GPa shows a 14.5% reduction in volume. Upon further compression into the region of *P4$_2$/mnm* above 208 GPa, the volume difference peaks at only 1.2%

The optimized structural parameters of zinc-blende crystal structure *F-43m* are listed in Table S1. For this phase, the equilibrium lattice constants are 4.547 Å and 4.492 Å, for PBE and LDA calculation, respectively. The present results are in good agreement with the previous results21'25'26'27. In this structure, four boron atoms lie in the Wyckoff 4*a* site and four phosphorus atoms occupy the 4*c* site, in which three-dimensional (3D) boron and phosphorus network in this structure is formed, as shown in Figure S1. The optimized structural parameters of another two phases *C2/m* and *P4$_2$/mnm* at high pressure are presented in Table 1. For monoclinic *C2/m* phase, the equilibrium lattice constants at 120 GPa are a = 4.275 Å, b = 3.681 Å, c = 8.357 Å and β= 122.627 degree. The structure of *C2/m* is shown in Figure 2(a) andFigure 2(b), it can be seen that the 1D-zigzag phosphorus chain is formed, which is connected by boron atoms. For hexagonal *P4$_2$/mnm*phase (Figure 2(c)), the equilibrium lattice constants at 210 GPa are a = 3.631 Å, c = 3.640 Å. The B-P, B-B, and P-P bonds are formed in this structure, but P-P bonds are not continuous (Figure 2(d)).

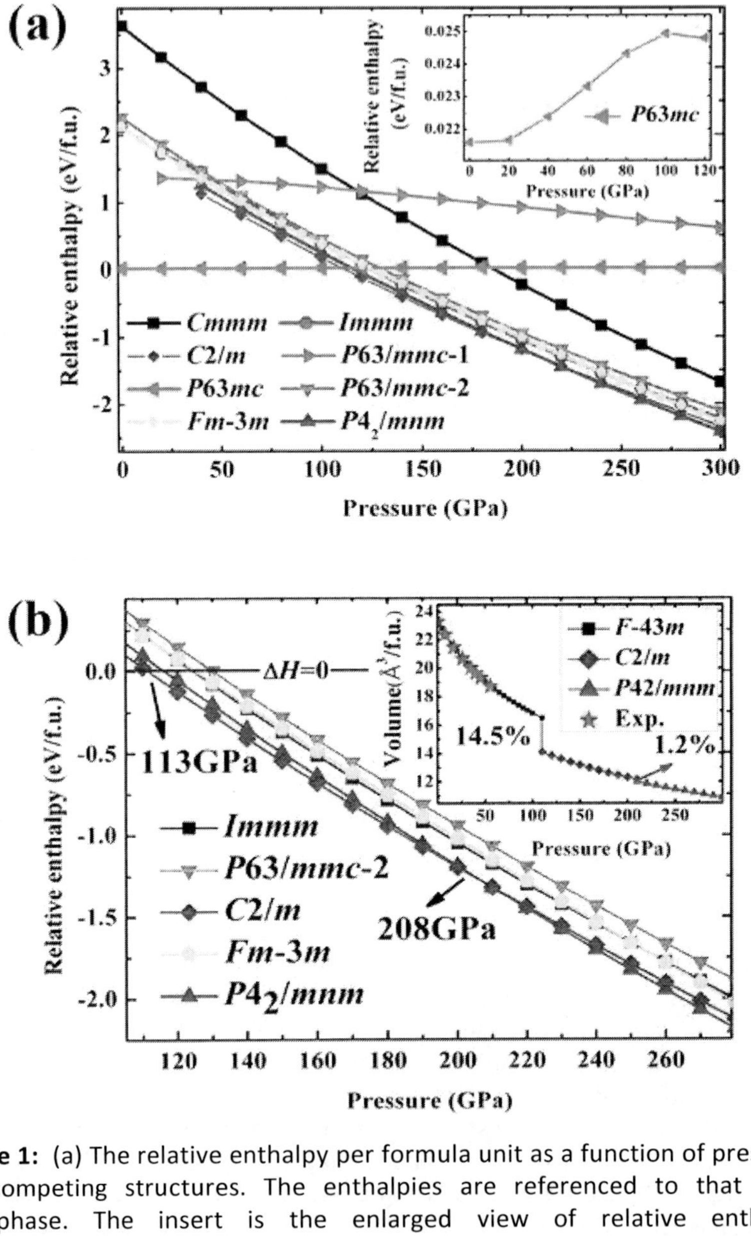

Figure 1: (a) The relative enthalpy per formula unit as a function of pressure for competing structures. The enthalpies are referenced to that of *F-43m* phase. The insert is the enlarged view of relative enthalpy of *P63mc* phase as a function of pressure. (b) The relative enthalpy per f.u. as a function of pressure with the region from 105 GPa to 280 GPa around the phase transition pressure. The insert is the pressure dependence of volume for *F-43m*, *C2/m* and*P42/mnm* phases of BP.

Table 1: Optimized structural parameters for two high-pressure structures of BP

Space group	Pressure (GPa)	Lattice parameters (Å or degree)	Atomic coordinates (fractional)
$C2/m$	120	$a = 4.275$, $b = 3.681$, $c = 8.357$ $\alpha = \gamma = 90$, $\beta = 122.627$	B1(4i)(1.48905, 0.00000, 0.89375) B2(4i)(1.05110, 0.00000, 0.65390) P1(4i)(0.99081, 0.00000, 0.13299) P2(4i)(0.56959, 0.00000, 0.39201)
$P4_2/mnm$	210	$a = 3.631$, $c = 3.640$	B (4g)(0.30702, 0.69298, 0.00000) P (4f) (0.70572, 0.29428, 0.5000

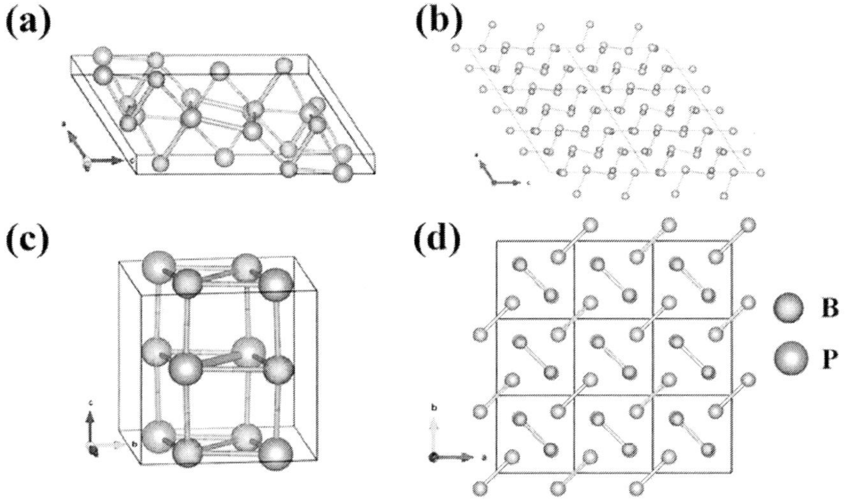

Figure 2: Predicted structures of BP at ambient and high pressures. The green and pink balls represent B and P atoms, respectively. (a) and (b) The high-pressure phase *C2/m*. (c) and (d) The high-pressure phase *P42/mnm*. One-dimensional phosphorus chain is formed in the structure of *C2/m*.

The calculated elastic constants for three BP phases with low enthalpy are presented in Table 2. From Table 2, the elastic constants reveal that all BP phases satisfy the mechanical stable criterion at the corresponding pressures. Furthermore, the hexagonal phase $P4_2/mnm$can remain stable at ambient pressure, but monoclinic *C2/m* does not match the mechanical stable criterion at ambient pressure, which only keeps stable at the high pressure range. Also, the calculated negative formation enthalpies of *C2/m* and $P4_2/mnm$ phases at the pressure ranges indicate the stability against decomposition of BP into the elements boron and phosphorus7. Phonon calculations show that BP phases (*F-43m*, *C2/m* and $P4_2/mnm$) are dynamically stable at their thermodynamically stable pressure regions. Phonon dispersions and partial phonon density of states (PPHDOS) of BP phases at their stable pressure ranges are shown in Figure 3. The maximum optical branch frequencies are 803.5 cm^{-1} for *F-43m* phase at atmospheric pressure, 983.7 cm^{-1} for *C2/m* phase at 120 GPa, and 1175.2 cm^{-1} for $P4_2/mnm$phase at 210 GPa. One can see that the maximum optical branch frequencies increase with increasing pressure due to an obvious structural difference. From PPHDOS, the lower frequency region is associated with phosphorus atoms, whereas boron atom contributes to the high-frequency region for high-pressure phases (*C2/m* and$P4_2/mnm$) at their corresponding pressure regions. But in the *F-43m*phase, the boron atom dominates the high-frequency region, whereas phosphorus atom contributes almost equally to the high and low frequency region.

To investigate the possibility of pressure induced metallization and superconducting transition, band structures for *F-43m*, *C2/m* and$P4_2/mnm$ phases have been calculated to examine the electronic properties of these polymorphs of BP at their pressure ranges as shown in Figure 3. Electronic structure calculations indicate that *F-43m* phase is a semiconductor, whereas *C2/m* and $P4_2/mnm$ phases are metallic. From projected density of states (PDOS) plotted in Figure 3, it can be observed that all phases of BP have consistent electronic distributions. B-2*p* and P-3*p* electrons dominate in the wide energy range and form the strong covalent bonds as suggested by the matching B-2*p* and P-3*p*curve shapes. The contour plots of charge density on the chosen planes are also shown in Figure 4. The strong covalent B-P bonding within in the 3D B-P network is evidenced (Figures 4(a), 4(b) and 4(c)). Moreover, according to Mulliken population analysis, P-P bonding does not exist in*F-43m* phase, but it exists in *C2/m* and $P4_2/mnm$ phases. Interestingly, P-P zigzag chains in phase *C2/m* transfer to discontinuous bonding in phase $P4_2/mnm$. This conclusion is further substantiated by the electron localization function (ELF) shown in Figures 4(d), 4(e) and 4(f), where the calculated ELF patterns have isosurface values of 0.8, 0.65, and 0.7 for *F-43m*, *C2/m*, and $P4_2/mnm$, respectively.

Table 2: Independent elastic constants C_{ij}, bulk, shear, and Young's moduli (*B*, *G* and *E* all in GPa), Poisson's ratio *v* and *G/B* of the stable phases of BP

Space group	*P*	C_{11}	C_{22}	C_{33}	C_{44}	C_{55}	C_{66}	C_{12}	C_{13}	C_{15}	C_{23}	C_{25}	C_{35}	C_{46}	*B*	*G*	*E*	*v*	*G/B*
F-43m	0	340			189			73							162	165	369	0.120	1.016
C2/m	120	761	1327	830	390	268	449	313	451	−10	355	6	−35	−15	563	319	805	0.261	0.567
$P4_2/mnm$	210	1326		2025	575		78	670	350						821	354	929	0.311	0.431

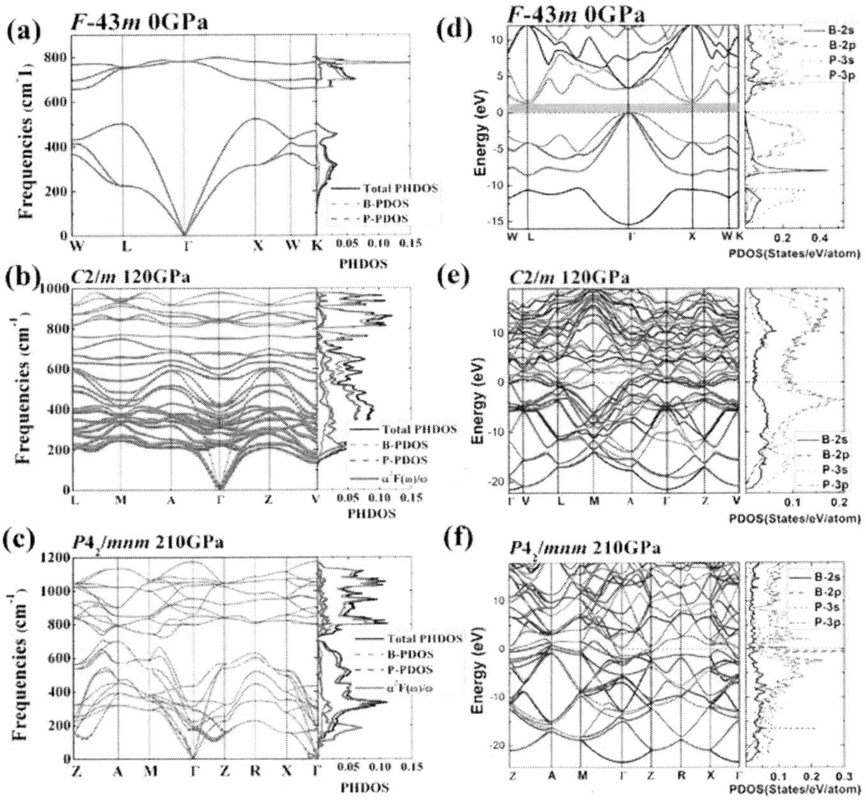

Figure 3: Phonon spectrum and electronic structures. (a) The phonon spectrum and partial atomic phonon density of states for phase *F*-43*m*. (b) The phonon spectrum, partial atomic phonon density of states and Eliashberg phonon spectral function $\alpha^2F(\omega)/\omega$ for phase *C*2/*m*. (c) The phonon spectrum, partial atomic phonon density of states and Eliashberg phonon spectral function $\alpha^2F(\omega)/\omega$ for phase*P*42/*mnm*. Red solid circles in (b) and (c) show the electron-phonon coupling with the radius proportional to their respective strength. Energy band and projected density of states (PDOS) for phase (d) *F*-43*m*, (e) *C*2/*m*, and (f)*P*42/*mnm*.

The high-pressure *C*2/*m* phase of BP exhibits interesting features in its electronic band structure. As shown in Figure 3(e), the electronic bands of the *C*2/*m* phase crossing the Fermi level along the Γ -V-L and A-Γ-Z-V are quite flat, and this section of band within a narrow energy window centered at the Fermi level gives rise to large electronic density of states near the Fermi energy. The corresponding confined conduction electrons near the band gap possess large effective mass with their group velocities approaching zero. In contrast, the bands along the L-M-A direction steeply

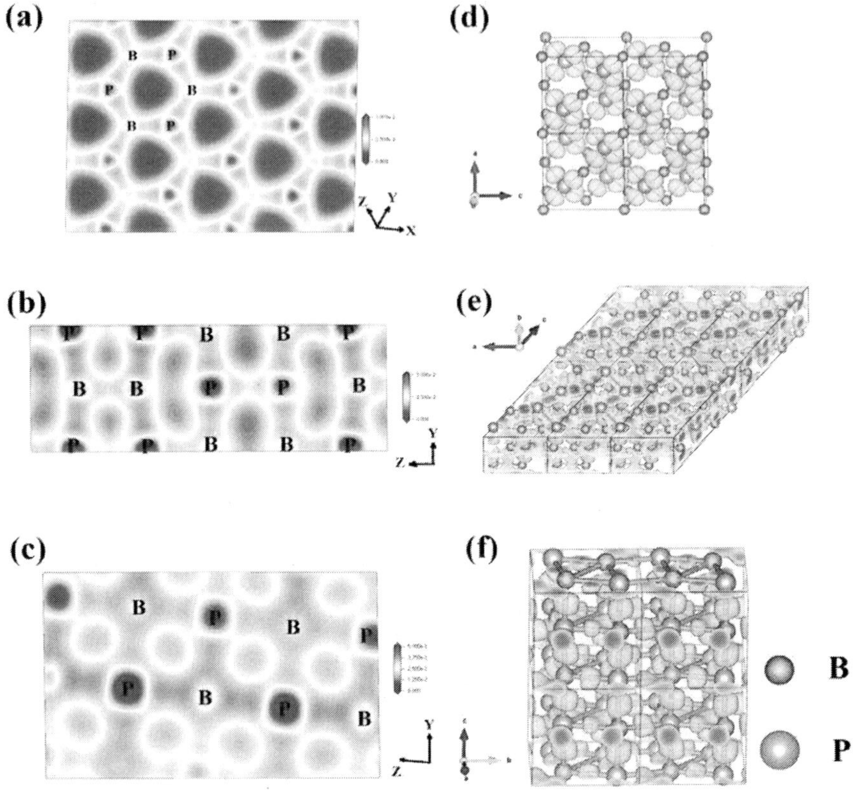

Figure 4: Electron density and Electron localization function(ELF). (a), (b), and (c) Electron density for phase *F-43m*, *C2/m*, and *P42/mnm*, respectively. (d) Calculated ELF with an isosurface ELF value of 0.8, 0.65, and 0.7 for phase *F-43m*, *C2/m*, and*P42/mnm*, respectively.

cross the Fermi level, providing itinerant electrons with high conduction electron velocity. Such flat bands with highly mobile and localized electrons have been shown to enhance superconductivity28′29′30. Similar electronic structures have previously been found in niobium that has the highest transition temperature T_C of 9.3 K31 in all element superconductors at ambient pressure, 9.8 K in $CaC_2$28 and 220–235 K in CaH_6 at high pressure of 150 GPa32. This motivates us to further investigate the superconductivity of BP at high pressures. The calculated spectral function $\alpha^2F(\omega)$ of BP phases *C2/m* and $P4_2/mnm$ at 120 GPa and 210 GPa were plotted inFigures 3(b) and 3(c). At 120 GPa (Figure 3(b)), the coupling parameter λ is 0.61 with the average phonon frequency ω_{ln} of 593 K. Using the strong coupling Allen-Dynes equation, an extension of the McMillan theory, with a nominal Coulomb pseudopotential parameter (μ*) of 0.12 the estimated

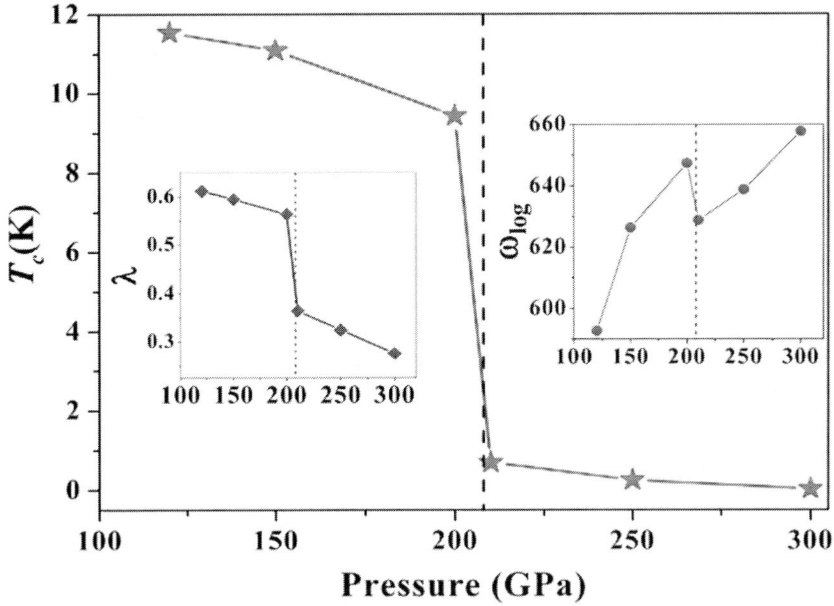

Figure 5: Superconducting temperature (TC) vs. pressure. The coupling parameters, the average phonon frequencies ω_{ln} and superconducting temperatures of BP as a function of pressure.

superconducting critical temperature T_C is 11.5 K. With increasing pressure, the calculated T_C become slightly lower as 11.1 K at 150 GPa and 9.4 K at 200 GPa due to slightly smaller λ of 0.59 at 150 GPa and 0.56 at 200 GPa, respectively. Here the T_C of two high-pressure phases for BP and T_C's dependence on pressure is shown in Figure 5. Calculated coupling parameter λ values and logarithmic phonon momentum ω_{log} vs. pressure curves are also presented in Figure 5. It can be seen that logarithmic phonon momentum ω_{log} increases with pressure, and coupling parameter λ decreases with pressure. Among these two high pressure phases, *C*2/*m* phase has the strongest electron-phonon coupling and so has the highest T_C. Apparently, the fairly high T_C is due to the large electron phonon coupling (λ). The origin can be traced from comparing the calculated Eliashberg spectral function ($\alpha^2F(\omega)/\omega$) with the projected phonon DOS. As shown in Figure 3(b), nearly 80% of the electron phonon coupling is contributed by the low-frequency vibrations in the frequency region from 100 to 600 cm^{-1}. The low-frequency vibrations in the frequency region from 100 to 600 cm^{-1}, which are mostly attributed to the P atom. At wide range of pressure, *C*2/*m* phase has comparative superconducting critical temperatures (from 11.5 K at 120 GPa to 9.3 K at 200 GPa) with 11.5 K value of $CaC_6$33′34. However, T_C of BP is close to zero above 210 GPa due to the smaller electron-phonon coupling parameter ($\lambda < 0.4$). We thus further analyze the coupling parameter in each q-point (Figures 3(b) and 3(c)). It is clearly seen that the coupling parameters in Γ-Z and X-Γ are very strong. This superconducting feature is typical anisotropy, leading a low superconductivity value.

To summarize, using the PSO technique on crystal structure prediction, we designed two new high-pressure phases of BP, *C*2/*m* and *P*4$_2$/*mnm*, which are stable in the pressure ranges of 113–208 GPa and 208–300 GPa, respectively. Elastic constants and phonon calculations have shown their mechanical and dynamical stability at the dominating pressure ranges. Our calculations reveal that the phosphorus atomic arrangement form from single atoms to 1D zigzag chains to discontinuous bonding. The high superconducting transition temperature of the *C*2/*m* phase benefits from the simultaneous presence of the steep bands (highly mobile electrons) and extremely flat bands (highly confined electrons), which is known to favor the electron paring and superconducting behavior.

COMPUTATIONAL METHODS

The search of high-pressure structures was performed with variable-cell PSO35 structure prediction simulations using CALYPSO approach22′23 in combined with Vienna *ab initio* simulation package (VASP)36. CALYPSO was designed to predict stable or metastable crystal structures requiring only chemical compositions of a given compound at specified external conditions. All structures were locally optimized using the DFT method. The 60% structures of each generation with lower enthalpies were selected to generate the structures for the next generation by PSO operation, and the other structures in new generation were randomly generated to increase the structural diversity. The underlying ab initio calculations were performed using density functional theory within the generalized gradient approximation (GGA)37, as implemented in the Vienna *ab initio*simulation package (VASP). The total energy calculations were carried out with the VASP code. The all-electron projector augmented wave (PAW) method38 was employed with a plane-wave cutoff energy of 600 eV for all phases. The k-point samplings in the Brillouin zone were performed using the Monkhorst-Pack scheme39. The total energy convergence tests showed that convergence to within 1 meV/atom was achieved with the above calculation parameters. Single crystal elastic constants were calculated via a strain-stress approach. i.e., by applying a small strain to the equilibrium lattice of the unit cell and fitting the dependence of the resulting stress on the strain. The bulk modulus, shear modulus, Young's modulus, and

Poisson's ratio were determined by using the Voigt-Reuss-Hill approximation40. The lattice-dynamical and superconducting properties are calculated using the density functional perturbation theory (DFPT)41 as implemented in the Quantum Espresso package42 using with the Troullier-Martins pseudopotentials with cutoff energies of 50 and 500 Ry for the wave functions and the charge density, respectively. Fine k-points mesh of MP was used and the estimated energy error in self-consistency was less than 10^{-4} a.u. The electron–phonon coupling was convergent with a finer grid and a Gaussian smearing of 0.02 Ry. In order to check the results, the phonon calculations were also carried out by using a supercell approach as implemented in the PHONOPY code43.

ACKNOWLEDGEMENTS

The authors acknowledge funding from NBRPC (grant 2013CB733000), NSFC (grants 51171160/51171163), J.Q. thanks the support from Ratchadaphisek sompoch Endowment Fund (2013), Chulalongkorn University (CU-56-805-FC), Ratchadapisek Sompoch Endowment Fund, Chulalongkorn University, granted to the Surface Coatings Technology for Metals and Materials Research Unit (GRU 57-005-62-001). This research was supported by Swedish Foundation for Strategic Research, Sweden and the Leading Foreign Research Institute Recruitment Program through theNational Research Foundation of Korea funded by Ministry of Education, Science and Technology Grant 2014-039452.

CONTRIBUTIONS

X.Z., J.Q., R.L. and R.A. designed and coordinated the research. X.Z., J.Q., H.L. and S.Z. did the calculations. X.Z., J.Q., H.L., S.Z., M.M., R.L., W.L. and R.A. analyzed all data. X.Z., J.Q., H.L., R.L. and R.A. wrote the manuscript.

COMPETING INTERESTS

The authors declare no competing financial interests.

REFERENCES

1. McMillan, P. F. Chemistry at high pressure. *Chem. Soc. Rev.* **35**, 855–857 (2006).
2. Haines, J., Léger, J. & Bocquillon, G. Synthesis and design of superhard materials. *Ann. Rev. Mater. Res.* **31**, 1–23 (2001).
3. Qin, J. *et al.* Is Rhenium Diboride a Superhard Material? *Adv. Mater.* **20**, 4780–4783 (2008).
4. Qin, J. *et al.* Polycrystalline γ-boron: As hard as polycrystalline cubic boron nitride. *Scripta Mater.* **67**, 257–260 (2012).
5. Qin, J. *et al.* Phase relations in boron at pressures up to 18 GPa and temperatures up to 2200°C. *Phys. Rev. B* **85**, 014107 (2012).
6. Raza, Z., Errea, I., Oganov, A. R. & Saitta, A. M. Novel superconducting skutterudite-type phosphorus nitride at high pressure from first-principles calculations. *Sci. Rep.* **4**, 5889; 10.1038/srep05889 (2014).
7. Oganov, A. R. *et al.* Ionic high-pressure form of elemental boron.*Nature* **457**, 863–867 (2009).
8. Ma, Y., Prewitt, C. T., Zou, G., Mao, H.-k. & Hemley, R. J. High-pressure high-temperature x-ray diffraction of β-boron to 30 GPa. *Phys. Rev. B* **67**, 174116 (2003).
9. Ma, Y., Tse, J. S., Klug, D. D. & Ahuja, R. Electron-phonon coupling of α-Ga boron. *Phys. Rev. B* **70**, 214107 (2004).
10. Kikegawa, T. & Iwasaki, H. An X-ray diffraction study of lattice compression and phase transition of crystalline phosphorus. *Acta Crystallogr. B* **39**, 158–164 (1983).
11. Akahama, Y., Kobayashi, M. & Kawamura, H. Simple-cubic–simple-hexagonal transition in phosphorus under pressure. *Phys. Rev. B* **59**, 8520–8525 (1999).
12. Akahama, Y., Kawamura, H., Carlson, S., Le Bihan, T. &Häusermann, D. Structural stability and equation of state of simple-hexagonal phosphorus to 280 GPa: Phase transition at 262 GPa. *Phys. Rev. B* **61**, 3139–3142 (2000).
13. Akahama, Y., Endo, S. & Narita, S. Electrical properties of single-crystal black phosphorus under pressure. *Physica B + C* **139–140**, 397–400 (1986).
14. Kawamura, H., Shirotani, I. & Tachikawa, K. Anomalous superconductivity in black phosphorus under high pressures.*Solid State Commun.* **49**, 879–881 (1984).
15. Brown, A. & Rundqvist, S. Refinement of the crystal structure of black phosphorus. *Acta Crystallographica* **19**, 684–685 (1965).
16. Peret, J. L. Preparation and Properties of the Boron Phosphides. *J. Am. Ceram. Soc.* **47**, 44–46 (1964).

17. Archer, R. J., Koyama, R. Y., Loebner, E. E. & Lucas, R. C. Optical Absorption, Electroluminescence, and the Band Gap of BP. *Phys. Rev. Lett.* **12**, 538–540 (1964).
18. Motojima, S., Ohtsuka, Y., Kawajiri, S., Takahashi, Y. & Sugiyama, K. Boron phosphide coatings on molybdenum by chemical vapour deposition, and their composition and microhardness. *J. Mater. Sci.* **14**, 496–498 (1979).
19. Schroten, E., Goossens, A. & Schoonman, J. Photo- and electroreflectance of cubic boron phosphide. *J. Appl. Phys.* **83**, 1660–1663 (1998).
20. Wentzcovitch, R. M., Cohen, M. L. & Lam, P. K. Theoretical study of BN, BP, and BAs at high pressures. *Phys. Rev. B* **36**, 6058–6068 (1987).
21. Wentzcovitch, R. M., Chang, K. J. & Cohen, M. L. Electronic and structural properties of BN and BP. *Phys. Rev. B* **34**, 1071–1079 (1986).
22. Wang, Y., Lv, J., Zhu, L. & Ma, Y. Crystal structure prediction via particle-swarm optimization. *Phys. Rev. B* **82**, 094116–094123 (2010).
23. Wang, Y., Lv, J., Zhu, L. & Ma, Y. CALYPSO: A method for crystal structure prediction. *Comput. Phys. Commun.* **183**, 2063–2070 (2012).
24. Godec, Y. *et al.* Equation of state of single-crystal cubic boron phosphide. *J. Superhard Mat.* **36**, 61–64 (2014).
25. Meradji, H. *et al.* First-principles elastic constants and electronic structure of BP, BAs, and BSb. *Phys. Status. Solidi. B* **241**, 2881–2885 (2004).
26. Lambrecht, W. R. L. & Segall, B. Electronic structure and bonding at SiC/AlN and SiC/BP interfaces. *Phys. Rev. B* **43**, 7070–7085 (1991).
27. Zaoui, A. & Hassan, F. E. H. Full potential linearized augmented plane wave calculations of structural and electronic properties of BN, BP, BAs and BSb. *J. Phys.: Condens. Matter* **13**, 253–262 (2001).
28. Li, Y.-L. *et al.* Pressure-induced superconductivity in CaC_2. *Proc. Natl. Acad. Sci. USA* **110**, 9289–9294 (2013).
29. Ranninger, J., Robin, J. M. & Eschrig, M. Superfluid Precursor Effects in a Model of Hybridized Bosons and Fermions. *Phys. Rev. Lett.* **74**, 4027–4030 (1995).
30. Simon, A. Superconductivity and chemistry. *Angew. Chem. Int. Ed.***36**, 1788–1806 (1997).
31. Broom, R. F., Raider, S. I., Oosenbrug, A., Drake, R. E. & Walter, W.Niobium oxide-barrier tunnel junction. *Ieee. T. Electron Dev.* **27**, 1998–2008 (1980).

32. Wang, H., Tse, J. S., Tanaka, K., Iitaka, T. & Ma, Y. Superconductive sodalite-like clathrate calcium hydride at high pressures. *Proc. Natl. Acad. Sci. USA* **109**, 6463–6466 (2012).
33. Emery, N. *et al.* Superconductivity of Bulk CaC_6. *Phys. Rev. Lett.***95**, 087003 (2005).
34. Weller, T. E., Ellerby, M., Saxena, S. S., Smith, R. P. & Skipper, N. T.Superconductivity in the intercalated graphite compounds C_6Yb and C_6Ca. *Nat. Phy.* **1**, 39–41 (2005).
35. Call, S. T., Zubarev, D. Y. & Boldyrev, A. I. Global minimum structure searches via particle swarm optimization. *J. Comput. Chem.* **28**, 1177–1186 (2007).
36. Kresse, G. & Furthmüller, J. Efficient iterative schemes for ab initio total-energy calculations using a plane-wave basis set. *Phys. Rev. B* **54**, 11169–11186 (1996).
37. Perdew, J. P., Burke, K. & Ernzerhof, M. Generalized gradient approximation made simple. *Phys. Rev. Lett.* **77**, 3865–3868 (1996).
38. Kresse, G. & Joubert, D. From ultrasoft pseudopotentials to the projector augmented-wave method. *Phys. Rev. B* **59**, 1758–1775 (1999).
39. Monkhorst, H. J. & Pack, J. D. Special points for Brillouin-zone integrations. *Phys. Rev. B* **13**, 5188–5192 (1976).
40. Hill, R. The Elastic Behaviour of a Crystalline Aggregate. *Proc. Phys. Soc. A* **65**, 349–355 (1952).
41. Baroni, S., de Gironcoli, S. & Dal Corso, A., Giannozzi P. Phonons and related crystal properties from density-functional perturbation theory. *Rev. Mod. Phys.* **73**, 515–562 (2001).
42. Giannozzi, P. *et al.* QUANTUM ESPRESSO: a modular and open-source software project for quantum simulations of materials. *J. Phys.: Condens. Matter* **21**, 395502–395530 (2009).
43. Togo, A., Oba, F. & Tanaka, I. First-principles calculations of the ferroelastic transition between rutile-type and $CaCl_2$-type SiO_2 at high pressures. *Phys. Rev. B* **78**, 134106–134114 (2008).

Chapter 15

Discovery of Fe_7O_9: a new Iron Oxide with a Complex Monoclinic Structure

Ryosuke Sinmyo, Elena Bykova, Sergey V. Ovsyannikov, Catherine McCammon, Ilya Kupenko,Leyla Ismailova & Leonid Dubrovinsky

Bayerisches Geoinstitut, Universität Bayreuth, D-95440 Bayreuth, Germany

ABSTRACT

Iron oxides are fundamentally important compounds for basic and applied sciences as well as in numerous industrial applications. In this work we report the synthesis and investigation of a new binary iron oxide with the hitherto unknown stoichiometry of Fe_7O_9. This new oxide was synthesized at high-pressure high-temperature (HP-HT) conditions, and its black single crystals were successfully recovered at ambient conditions. By means of single crystal X-ray diffraction we determined that Fe_7O_9 adopts a monoclinic *C2/m* lattice with the most distorted crystal structure among the binary iron oxides known to date. The synthesis of Fe_7O_9 opens a new portal to exotic iron-rich $(M,Fe)_7O_9$oxides with unusual stoichiometry and distorted crystal structures. Moreover, the crystal structure and phase relations of such new iron oxide groups may provide new insight into the cycling of volatiles in the Earth's interior.

INTRODUCTION

Iron oxides have great importance for all natural sciences as well as numerous industrial applications. Considering the high abundance of oxygen and iron in the Earth's crust and the mantle, binary iron oxides and their derivatives are important endmembers of phases that make a significant contribution to properties of the Earth. Many studies have been devoted to investigations of various properties of iron oxides at conditions relevant to the Earth's interior, i.e., at high pressure or at high pressure combined with high temperature (HP-HT). These studies reported a number of remarkable findings for the three basic iron oxides, $Fe_{1-x}O$ wüstite1′2, α-Fe_2O_3 hematite3′4′5′6′7′8, and Fe_3O_4magnetite9′10′11. Meanwhile, several experimental and theoretical studies indicated that the chemistry of iron oxides at extreme conditions of high pressure and high temperature may extend to intriguing behavior beyond these three well-known oxides12′13′14′15. For instance, recent experimental HP-HT investigations in the pressure range of 10–20 GPa reported two new orthorhombic *Cmcm* polymorphs with $Fe_4O_5$13 and $Fe_5O_6$14 stoichiometry, i.e., between Fe_3O_4 and $Fe_{1-x}O$, and hence, with mixed Fe^{2+} and Fe^{3+} oxidation states. More interestingly, using modern theoretical methods an opposite tendency in the stoichiometry changes in Fe-O systems was found, suggesting an extended stability of ferric hematite and a gradual shift to Fe^{4+}–bearing oxides at megabar pressures15. In particular, these studies predicted the existence of a new exotic FeO_2 oxide with highly charged Fe^{4+} ions that should have an extended range of stability against decomposition under megabar pressures15. Recently, a number of experimental studies have shown the important role of the iron oxides under HP-HT conditions in the Earth's deep interior16′17′18′19. Decomposition reactions of iron oxides and oxyhydroxides may induce the release of oxygen or hydrogen in the Earth's lower mantle17′18.

In addition to their obvious and primary importance for geosciences, iron oxides play crucial roles in many technological processes and applications, and remain among a handful of key materials with significant impact on the fundamental behavior of materials, including charge carrier transfer and interactions, spin dynamics of electrons, and other central topics. In other words, iron oxides are important prototype materials. For instance, the oldest-known natural magnet, magnetite, demonstrates a fascinating 'metal-insulator'-type transition near 120–125 K20, named later as the 'Verwey transition' after its discovery in 1939. This Verwey transition was believed to be related to charge ordering on octahedral sites in the spinel structure, and has been hotly debated for decades. Only recently it was revealed that this transition is linked to formation of hitherto unknown 'quasiparticles' consisting of three bonded Fe ions called 'trimerons'21. In hematite another landmark transition was discovered near 255 K with a related abrupt and drastic reorientation of spins of Fe^{3+} ions, named afterwards as a 'spin-flop' or Morin transition22. Also iron-deficient wüstite, $Fe_{1-x}O$, displays puzzling complexities with regard to stoichiometry, defect structure, and elastic and physical properties23′24′25′26′27′28, and serves as a prototype for systems with non-stoichiometry. Recently it was experimentally demonstrated that, similar to magnetite, the newly-discovered $Fe_4O_5$13 also undergoes an unprecedented 'metal-insulator'–type transition upon cooling below 150 K with competing dimeric and trimeric ordering in the Fe chains, leading to strong structural modulations29. Therefore, synthesis of new iron oxides with both mixed Fe^{2+} and Fe^{3+} valences and bearing highly charged Fe^{4+} ions is of significant interest for many scientific fields.

In the present work we investigated the phase stability of iron oxides at HP-HT conditions (see Methods) and discovered in samples recovered at ambient conditions the presence of crystals of a new iron oxide with unusual Fe_7O_9 stoichiometry. Fe_7O_9 with its ratio of Fe/O ~ 0.777 lies between Fe_3O_4 and newly-discovered $Fe_4O_5$13, but in contrast to both, it is non-magnetic at ambient conditions and adopts a monoclinic crystal structure with four sites for Fe cations. These observations combined with the difference in Fe^{3+}/Fe^{2+} ratios ($^4/_3$ in Fe_7O_9 *versus* 2 in $Fe_3O_4$and 1 in Fe_4O_5) suggest that the physical properties of Fe_7O_9 may be remarkably different. In addition, we also synthesized Fe_7O_9 containing a significant amount of Mg and discuss possible geological implications for this new polymorph.

RESULTS AND DISCUSSION

Single crystal XRD measurements

The single crystals of pure Fe_7O_9 and Mg-doped $(Mg,Fe^{2+})_3Fe^{3+}{}_4O_9$ had sizes less than 50 µm in their linear dimensions that restricted detailed investigations of their properties. The chemical composition of the samples was determined using conventional electron microprobe methods and from single crystal X-ray diffraction data. We also collected Mössbauer spectra from these samples to determine the oxidation states of the Fe ions.

By means of single crystal X-ray diffraction on the crystals (Fig. 1 andSupplementary Figure 1) we solved and refined the crystal structures of Fe_7O_9 and $(Mg,Fe^{2+})_3Fe^{3+}{}_4O_9$ at ambient conditions and confirmed their stoichiometry. Crystals of Fe_7O_9 and $(Mg, Fe^{2+})_3Fe^{3+}{}_4O_9$ had no pronounced asymmetry in their shape, but were rather small (0.03 × 0.02 × 0.01 and 0.03 × 0.02 × 0.02 mm^3), so it was not possible to perform an analytical absorption correction based on crystal shape. Crystals of Fe_7O_9 appeared to be twinned, and due to the high degree of peak overlap (>50%), we had to integrate both twin domains simultaneously to perform a refinement of the crystal structure against HKLF5 data (BASF value was about 49.8%). The twinning could also influence the quality of determination of the anisotropic parameters. For consistency, we also refined the structure in an isotropic approximation and demonstrated a negligible influence on the atomic positions (CIF-files in the Supplementary Materials). Technical details of the structure determinations are summarized in Tables S1 and S2 in theSupplementary Materials. We established that both compounds adopt the same monoclinic structure of the *C*2/*m* space group. The unit cell parameters in Fe_7O_9 are as follows: a = 9.696(2) Å, b = 2.8947(6) Å, c = 11.428(3) Å, β = 101.69(2)°, V = 314.10(12) $Å^3$, and Z = 2 (Fig. 2, and Tables S1 and S2 and CIF-files in the Supplementary Materials). The crystal structure of Fe_7O_9 has four different crystallographic sites for cations, three (Fe1, Fe2, Fe3) are octahedrally–coordinated and connected in a 3D network, while the fourth, Fe4, has a trigonal-prismatic arrangement (Fig. 2 and Table S2 in the Supplementary Materials). By analyzing the Fe-O bond distances in this polymorph using a bond valence sums (BVS) method30, we established the average oxidation state for Fe ions occupying the Fe1, Fe2, Fe3, and Fe4 sites to be +2.74, +2.72, +2.82, and +2.10, respectively. Thus, we conclude that the Fe4 sites are occupied almost exclusively by Fe^{2+} ions (Fig. 2). The other octahedral sites participate in electronic exchange between Fe^{2+} and Fe^{3+} ions through the polaron hopping mechanism, similar to the octahedral network of magnetite20 and other Fe-bearing oxides31. Thus, the BVS method shows the average charge for each of the Fe1-Fe3 sites. We note that various considerations lead to octahedral sites in the recently-discovered Fe_4O_5 phase having different charges29.

Mg-doped Fe_7O_9 crystals adopt the same crystal structure as $Fe_7O_9$with similar unit cell parameters: a = 9.6901(12) Å, b = 2.8943(5) Å, c = 11.4397(15) Å, β = 102.045(14)°, V = 313.77(8) $Å^3$, and Z = 2 (Fig. 2 and Table S1 in the Supplementary Materials). Electron microprobe analysis established their chemical composition to be $Mg_{1.06}Fe_{5.94}O_9$, i.e., nearly 15% of Fe ions are substituted by Mg. We used this chemical composition in the crystal structure refinement and found that Mg^{2+}ions occupy all four Fe sites, but with a noticeable preference for (*i*) the spacious Fe4 sites that are occupied by the larger Fe^{2+} ions in Fe_7O_9, and (*ii*) the Fe1 sites, located between the Fe4 sites (Fig. 2, Table S2 in the Supplementary Materials). In the case of Mg doping of Fe_4O_5, Mg ions were also found to occupy all Fe sites in the structure32. Repeating the same BVS analysis30 as above for Fe_7O_9 taking into account the Mg atom distribution determined by single crystal X-ray diffraction (Table S2 in the Supplementary Materials), we confirmed the ferrous nature of Fe4 ions and detected a minor increase in the BVS of all other octahedrally coordinated Fe ions as +2.81, +2.78, and +2.87 for the Fe1, Fe2, and Fe3 sites, respectively (Fig. 2). Thus, the results of BVS analysis suggest the persistence of charge transfer in the 3D octahedral network with the distributed Mg ions. However, these Mg impurities should dramatically lower the mobility of hopping polarons, and hence, the bulk electrical conductivity of Mg-doped Fe_7O_9 is expected to be much lower than that of Fe_7O_9.

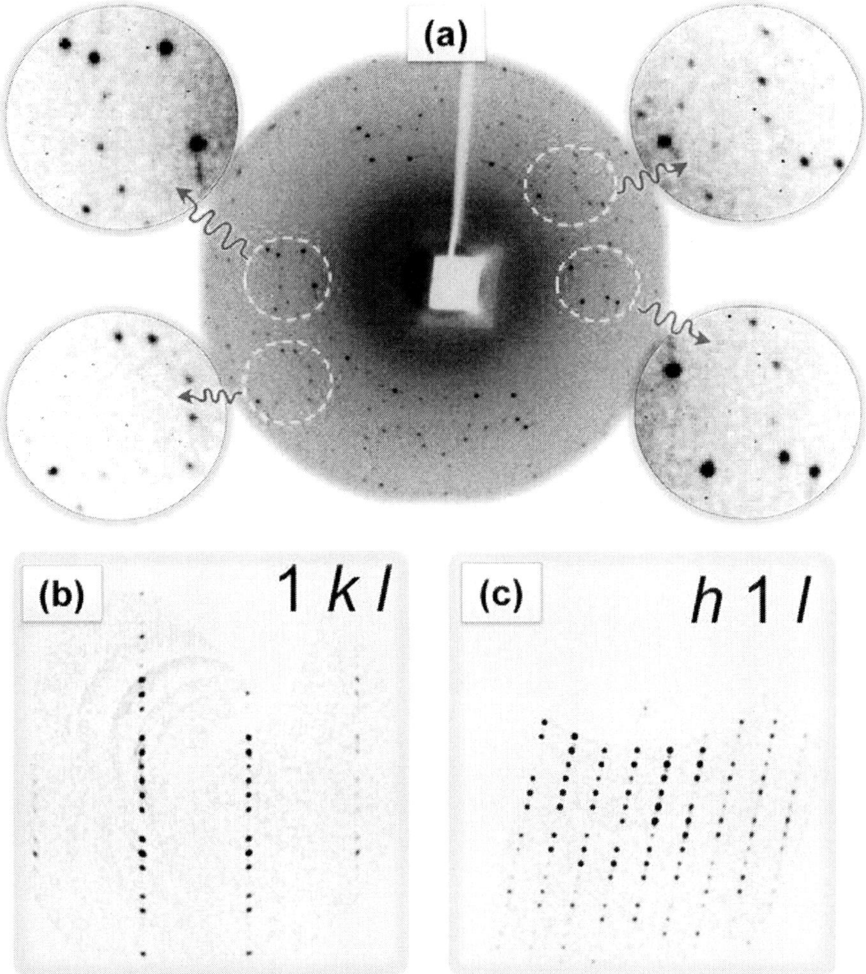

Figure 1: X-ray diffraction images. (**a**) Example of X-ray diffraction image of microscopic single crystal of $(Mg,Fe)_3Fe_4O_9$collected at ambient conditions under rotation of a sample over 360 degrees in a beam. This image contains several hundreds of small well-resolved reflections. Insets show selected magnified areas of this image that better show indistinct reflections. (**b**,**c**) Projections of these X-ray diffraction data in reciprocal space, for two selected planes, *1kl* (**b**) and *h1l* (**c**).

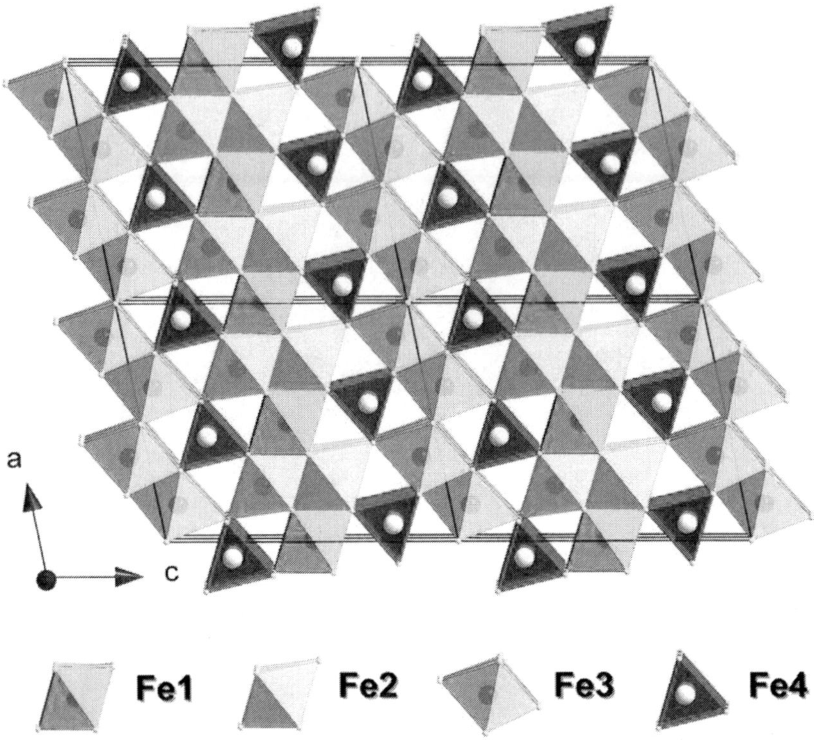

Figure 2: The crystal structure of Fe_7O_9. This structure corresponds to ambient conditions and is shown in a projection along the *b*axis.

Mössbauer spectroscopy

Both Fe_7O_9 and $(Mg,Fe)_7O_9$ samples were analyzed by Mössbauer spectroscopy using a synchrotron Mössbauer source that gave excellent signal to noise ratios (Fig. 3 and Table S3 in the Supplementary Materials). We did not observe any magnetic component in these spectra, and hence conclude that these materials are non-magnetic at ambient conditions. The spectrum of Fe_7O_9 (Fig. 3a) could be fitted by a superposition of two basic components, including (*i*) Fe^{2+} ions at the prismatically-coordinated Fe4 sites in the crystal structure (Fig. 2) and (*ii*) a merged component related to octahedrally-coordinated Fe1, Fe2, and Fe3 ions with an average oxidation state of $Fe^{2.8+}$ (Fig. 3a and Table S3 in the Supplementary Materials). We note that this $Fe^{2.8+}$ component is an average because of the above-mentioned rapid charge exchange between Fe^{2+} and Fe^{3+} ions at the octahedral sites, similarly to $Fe_3O_4$33. This finding is in excellent agreement with the above BVS oxidation states of the Fe ions obtained from the single crystal XRD data. It should be noted here that the three slightly structurally-

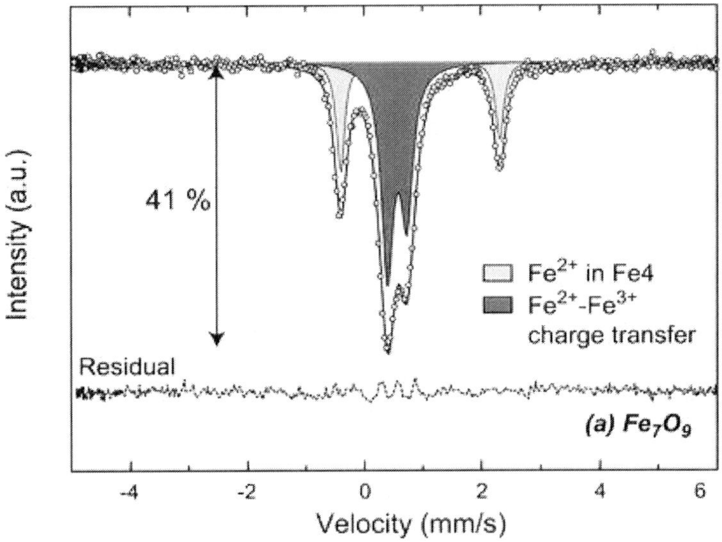

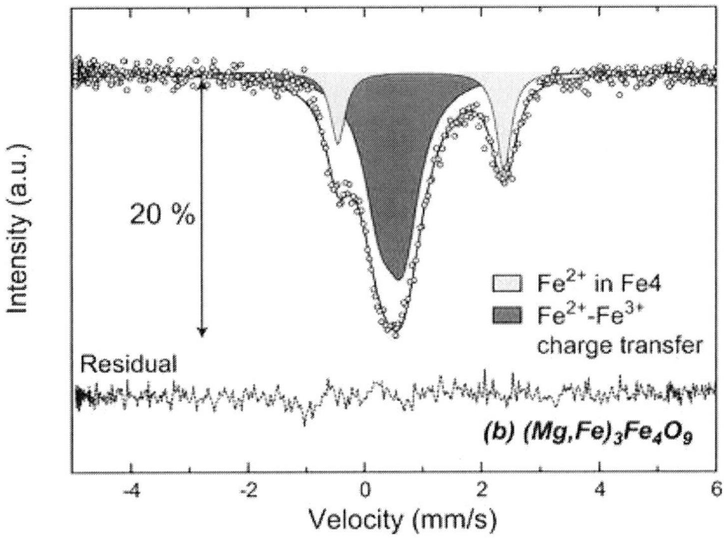

Figure 3: Mössbauer spectra of Fe_7O_9 samples. (**a**) Single crystal of Fe_7O_9. (**b**) Single crystal of $(Mg,Fe)_3Fe_4O_9$. Both spectra were collected at ambient conditions. Black open circles, experimental spectrum; lines, fitted spectra; broken line, residual.

inequivalent Fe1, Fe2 and Fe3 sites (Fig. 2) give similar contributions to the Mössbauer spectrum because of the very similar environment (edge-sharing FeO_6polyhedron) of the Fe ions, so it is not possible to separate their individual components in the spectrum (Fig. 3a). Spectra obtained from $(Mg,Fe)_7O_9$ were quite similar to Fe_7O_9 (Fig. 3b), while the merged component in the Mg-bearing sample spectrum shows a slightly higher average oxidation state of $Fe^{2.9+}$ due to incorporation of Mg.

DISCUSSION

In our work we synthesized Fe_7O_9 and $(Mg,Fe^{2+})_3Fe^{3+}{}_4O_9$ crystals at high pressures around 24–26 GPa. In previous studies the orthorhombic polymorphs of $Fe_4O_5$13 and $Fe_5O_6$14 were prepared at substantially lower pressures, between 10 and 20 GPa. It is interesting to note that the first pressure-driven structural phase transitions in the known iron oxides were detected at similar pressures around 20–25 GPa. For instance, at room temperature cubic $Fe_{1-x}O$ wüstite with the rocksalt structure transforms to a rhombohedral lattice above 20 GPa1, while at high temperatures the rocksalt structure of $Fe_{1-x}O$ is stable to at least 60 GPa1′2. High temperature-assisted phase transitions in Fe_3O_4 from cubic spinel to an orthorhombic phase34′35, and from corundum-type α-$Fe_2O_3$to an orthorhombic Rh_2O_3(II)-type or to other phases4′5′6′7′8 were also observed in some studies already above 20–25 GPa, although these phase transitions are still hotly debated. These observations suggest that all conventional iron oxides (α-Fe_2O_3, Fe_3O_4, and $Fe_{1-x}O$) become unstable with respect to structural transformations in approximately similar pressure ranges that might be related to similar shortening of Fe-O bond lengths in their structures. Hence, the resultant high-pressure polymorph of a compressed and heated iron oxide could depend on its stoichiometry, thereby suggesting chemical tuning as a route to new structural phases. In general, one could expect that a minor tuning in stoichiometry could lead either to structures with vacancies (ordered or disordered) or to modified, Fe_3O_4-like or Fe_2O_3-like high-pressure phases in new oxides. Likewise, significant shifts from known stoichiometry could potentially lead to hitherto unknown structures. For instance, the newly-discovered orthorhombic *Cmcm*polymorphs of $Fe_4O_5$13 and $Fe_5O_6$14 crystalize in structures that are linked to the high-pressure orthorhombic polymorph of $Fe_3O_4$34′35 (Fig. 4). By analogy with the existing family of calcium ferrites, $CaFe^{2+}_nFe^{3+}_2O_{4+n}$36, it was proposed that iron oxides with this *Cmcm* structure could also form such a family as $Fe^{2+}_{1+n}Fe^{3+}_2O_{4+n}$37, which includes Fe_3O_4, $Fe_4O_5$13 and $Fe_5O_6$14. However, the present discovery of Fe_7O_9 that does not belong to this family on the one hand, but having a certain structural similarity with the above oxides on the other hand (Fig. 4), suggests that the family of iron oxides that are structurally linked to the high-pressure polymorph of Fe_3O_4 may be more broad, e.g., like $Fe^{2+}_{3\pm n}Fe^{3+}_{4}O_{9\pm n}$, thereby suggesting a potentially greater diversity than in the calcium ferrite oxides. For instance, Fe_9O_{11} (n = 2) or $Fe_{11}O_{13}$ (n = 3) might be hypothetically stable under certain HP-HT conditions.

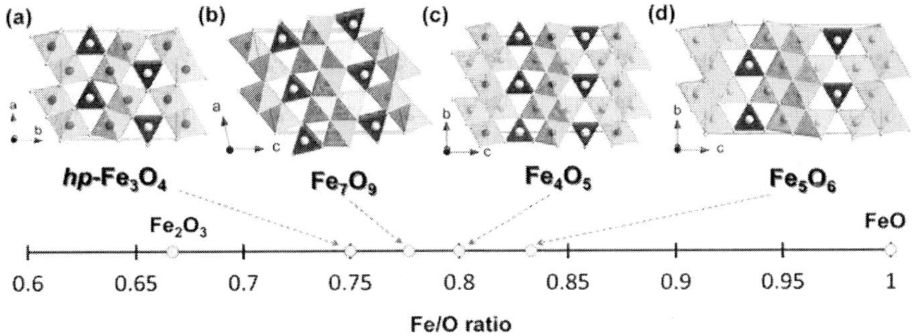

Figure 4: Comparison of unit cells of crystal structures of iron oxides. (**a**) High-pressure orthorhombic polymorph of Fe_3O_4[34], (**b**) Monoclinic Fe_7O_9 polymorph discovered in the present work. (**c**) Orthorhombic *Cmcm* polymorph of Fe_4O_5 discovered in ref. 13. (**d**) Orthorhombic *Cmcm* polymorph of Fe_5O_6 discovered in ref. 14. Different colors of the octahedra denote different crystallographic sites for Fe ions.

At the moment the stability field of this new Fe_7O_9 polymorph is not well defined, although it appears to lie at pressures higher than those of Fe_4O_5[13] and Fe_5O_6[14]. The chemical compositions and the structural phases of iron-magnesium oxides in the Earth's mantle remain a disputed issue that requires *in situ* investigations at HP-HT conditions under different oxygen fugacities. In this regard, the unexpected discovery of a Fe_7O_9 polymorph provides new insight into possible compositions of mantle phases, and provides a new type compound that may play a key role in determining physical and chemical properties.

The new oxide, Fe_7O_9, is a compound with a ratio of Fe^{3+}/Fe^{2+} intermediate between those of Fe_3O_4 and Fe_4O_5 (Fig. 4). Both Fe_3O_4 and Fe_4O_5 are model systems for investigations of Fe^{2+}–Fe^{3+} interactions in solids, demonstrating enigmatic low-temperature phase transitions of 'metal-insulator'-type that lead to the formation of exotic 'trimeron quasiparticles' in Fe_3O_4[21] or to even more intricate ordering patterns in Fe_4O_5[29]. We note that the low-temperature Verwey transition in Fe_3O_4 has had a strong impact on solid state physics and chemistry for decades. Thus, Fe_7O_9 presents an exciting compound that promises important implications for geosciences, solid state physics and chemistry with potential for industrial applications.

The recently discovered iron oxides may play important roles in the cycling of volatiles in the Earth's deep interior[16,17,18,19]. For instance, oxygen could be released in the deeper part of the lower mantle via decomposition reactions of Fe_2O_3 and Fe_3O_4 into Fe_5O_7 and $Fe_{25}O_{32}$above 60 GPa[17]. Further, it was recently reported that FeO_2 could be formed in the lower mantle as a product of FeOOH goethite decomposition[18]. Moreover, this reaction could supply hydrogen to the surrounding mantle[18]. Physical and chemical properties of the newly-discovered iron oxides can therefore provide novel insights into the chemical evolution of the Earth's interior.

ADDITIONAL INFORMATION

How to cite this article: Sinmyo, R. *et al.* Discovery of Fe_7O_9: a new iron oxide with a complex monoclinic structure. *Sci. Rep.* **6**, 32852; doi: 10.1038/srep32852 (2016).

ACKNOWLEDGEMENTS

R.S. was supported by the Alexander von Humboldt Foundation. S.V.O. acknowledges the financial support of the Deutsche Forschungsgemeinschaft (DFG) under project OV-110/1–3. We acknowledge the European Synchrotron Radiation Facility for provision of synchrotron radiation facilities and we would like to thank V. Cerantola for assistance in using beamline ID18.

CONTRIBUTIONS

The samples were synthesized by R.S. and S.V.O. E.B. and L.I. conducted the single crystal XRD measurements. Mössbauer spectra were obtained and analyzed by S.V.O., C.M., I.K. and L.D. Microprobe analysis were carried out by R.S. R.S., S.V.O. and L.D. designed the research project. The manuscript was written by R.S., S.V.O. and C.M. with contributions from all co-authors.

COMPETING INTERESTS

The authors declare no competing financial interests.

REFERENCES

1. Y. W. Fei & H. K. Mao. *In-situ* determination of the NiAs phase of FeO at high-pressure and temperature. *Science* **266**, 1678–1680 (1994).
2. K. Ohta, K. Hirose, K. Shimizu & Y. Ohishi. High-pressure experimental evidence for metal FeO with normal NiAs-type structure. *Phys. Rev. B* **82**, doi: 10.1103/PhysRevB.82.174120 (2010).
3. M. P. Pasternak *et al.* Breakdown of the Mott-Hubbard state in Fe_2O_3: A first-order insulator-metal transition with collapse of magnetism at 50 GPa. *Phys. Rev. Lett.* **82**, 4663–4666 (1999).
4. J. Badro *et al.* Nature of the high-pressure transition in Fe_2O_3hematite. *Phys. Rev. Lett.* **89**, doi: 10.1103/PhysRevLett.89.205504 (2002).
5. G. K. Rozenberg *et al.* High-pressure structural studies of hematite Fe_2O_3. *Phys. Rev. B* **65**, doi: 10.1103/PhysRevB.65.064112 (2002).
6. S. Ono, T. Kikegawa & Y. Ohishi. High-pressure phase transition of hematite, Fe_2O_3. *J. Phys. Chem. Solids* **65**, 1527–1530 (2004).
7. E. Ito *et al.* Determination of high-pressure phase equilibria of Fe_2O_3 using the Kawai-type apparatus equipped with sintered diamond anvils. *Am. Mineral.* **94**, 205–209 (2009).
8. E. Bykova *et al.* Novel high pressure monoclinic Fe_2O_3 polymorph revealed by single-crystal synchrotron X-ray diffraction studies.*High Pressure Res.* **33**, 534–545 (2013).
9. Y. W. Fei, D. J. Frost, H. K. Mao, C. T. Prewitt & D. Hausermann. *In situ* structure determination of the high-pressure phase of Fe_3O_4.*Am. Mineral.* **84**, 203–206 (1999).
10. C. Haavik, S. Stolen, H. Fjellvag, M. Hanfland & D. Hausermann.Equation of state of magnetite and its high-pressure modification: Thermodynamics of the Fe-O system at high pressure. *Am. Mineral.* **85**, 514–523 (2000).
11. M. P. Pasternak, S. Nasu, K. Wada & S. Endo. High-pressure phase of magnetite. *Phy. Rev. B* **50**, 6446–6449 (1994).
12. M. Koch, A. B. Woodland & R. J. Angel. Stability of spinelloid phases in the system Mg_2SiO_4–Fe_2SiO_4–Fe_3O_4 at 1100 °C and up to 10.5 GPa. *Phys. Earth Planet. Inter.* **143–144**, 171–183 (2004).
13. B. Lavina *et al.* Discovery of the recoverable high-pressure iron oxide Fe_4O_5. *Proc. Natl. Acad. Sci. USA* **108**, 17281–17285 (2011).
14. B. Lavina & Y. Meng. Unraveling the complexity of iron oxides at high pressure and temperature: synthesis of Fe_5O_6. *Sci. Adv.* **1**, e1400260, doi: 10.1126/sciadv.1400260 (2015).

15. G. L. Weerasinghe, C. J. Pickard & R. J. Needs. Computational searches for iron oxides at high pressures. *J. Phys.: Condens. Matter* **27**, 455501 (2015).
16. B. Woodland *et al.* Fe_4O_5 and its solid solutions in several simple systems. *Contrib. Mineral. Petrol.* **166**, 1677–1686 (2013).
17. E. Bykova *et al.* Structural complexity of simple Fe_2O_3 at high pressures and temperatures. *Nat. Commun.* **7**, 10661, doi: 10.1038/comms10661 (2016).
18. Q. Hu *et al.* FeO_2 and FeOOH under deep lower-mantle conditions and Earth's oxygen-hydrogen cycles. *Nature* **534**, 241–244 (2016).
19. R. Myhill *et al.* On the *P-T-*fO_2 stability of Fe_4O_5, Fe_5O_6 and Fe_4O_5-rich solid solutions. *Contrib. Mineral. Petrol.* **171**, 51 (2016).
20. E. J. W. Verwey. Electronic conduction of magnetite (Fe_3O_4) and its transition point at low temperatures. *Nature* **144**, 327–328 (1939).
21. M. S. Senn, J. P. Wright & J. P. Attfield. Charge order and three-site distortions in the Verwey structure of magnetite. *Nature* **481**, 173–176 (2012).
22. F. J. Morin. Magnetic Susceptibility of α-Fe_2O_3 and α-Fe_2O_3 with Added Titanium. *Phys. Rev.* **78**, 819–820 (1950).
23. R. Jeanloz & R. M. Hazen. Compression, nonstoichiometry and bulk viscosity of wüstite. *Nature* **304**, 620–622 (1983).
24. C. A. McCammon & L. Liu. The Effects of Pressure and Temperature on Nonstoichiometric Wustite, Fe_xO: The Iron-rich Phase Boundary. *Phys. Chem. Minerals* **10**, 106–113 (1984).
25. C. McCammon. Effect of Pressure on the Composition of the Lower Mantle End Member FexO. *Science* **259**, 66–68 (1993).
26. J. Zhang. Effect of Defects on the Elastic Properties of Wüstite.*Phys. Rev. Lett.* **84**, 507–510 (2000).
27. Kantor *et al.* Anelasticity of Fe*x*O at high pressure. *Appl. Phys. Lett.* **93**, 034106-1-3 (2008).
28. S. V. Ovsyannikov, V. V. Shchennikov, M. A. Shvetsova, L. S. Dubrovinsky & A. Polian. Tuning of the stoichiometry of $Fe_{1-x}O$ wüstite by compression. *Phys. Rev. B* **81**, 060101 (2010).
29. S. V. Ovsyannikov *et al.* Charge-ordering transition in iron oxide Fe_4O_5 involving competing dimer and trimer formation. *Nature Chemistry* **8**, 501–508 (2016).
30. N. E. Brese & M. O'Keffee. Bond-valence parameters for solids.*Acta Cryst.* **B47**, 192–197 (1991).
31. N. F. Mott. *Metal–insulator transitions* (London, Taylor and Francis Ltd, 1974).
32. T. Boffa Ballaran, L. Uenver-Thiele & A. B. Woodland. Complete substitution of Fe^{2+} by Mg in Fe_4O_5: The crystal structure of the $Mg_2Fe_2O_5$ end-member. *Am. Mineral.* **100**, 628–631 (2015).

33. R. G. Burns, D. A. Nolet, K. M. Parkin, C. A. McCammon & K. B. Schwartz. In *Mixed-Valence Compounds* D. B. Brown Ed., pp. 295–336 (D. Reidel Publishing Company, Dordrecht, 1980).
34. Y. W. Fei, D. J. Frost, H. K. Mao, C. T. Prewitt & D. Hausermann. *In situ* structure determination of the high-pressure phase of Fe_3O_4.*Am. Mineral.* **84**, 203–206 (1999).
35. C. Haavik, S. Stolen, H. Fjellvag, M. Hanfland & D. Hausermann.Equation of state of magnetite and its high-pressure modification: Thermodynamics of the Fe-O system at high pressure. *Am. Mineral.* **85**, 514–523 (2000).
36. O. Evrard *et al.* Mise en évidence de $CaFe_4O_6$ et détermination des structures cristallines des ferrites de calcium $CaFe_{2+n}O_{4+n}$(n = 1, 2, 3): Un nouvel exemple d'intercroissance. *J. Solid State Chem.*.**35**, 112–119 (1980).
37. J. Guignard & W. A. Crichton. Synthesis and recovery of bulk Fe_4O_5 from magnetite, Fe_3O_4. A member of a self-similar series of structures for the lower mantle and transition zone. *Mineral. Mag.* **78**, 361–371 (2014).

Chapter 16

Delocalization of Electrons in Strong Insulators at High Dynamic Pressures

William J. Nellis

Department of Physics, Harvard University, Cambridge, MA 02138, USA;

ABSTRACT

Systematics of material responses to shock flows at high dynamic pressures are discussed. Dissipation in shock flows drives structural and electronic transitions or crossovers, such as used to synthesize metallic liquid hydrogen and most probably Al_2O_3 metallic glass. The term "metal" here means electrical conduction in a degenerate system, which occurs by band overlap in degenerate condensed matter, rather than by thermal ionization in a non-degenerate plasma. Since H_2 and probably disordered Al_2O_3 become poor metals with minimum metallic conductivity (MMC) virtually all insulators with intermediate strengths do so as well under dynamic compression. That is, the magnitude of strength determines the split between thermal energy and disorder, which determines material response. These crossovers occur via a transition from insulators with electrons localized in chemical bonds to poor metals with electron energy bands. For example, radial extents of outermost electrons of Al and O atoms are 7 a_0 and 4 a_0, respectively, much greater than 1.7 $a_0$needed for onset of hybridization at 300 GPa. All such insulators are Mott insulators, provided the term "correlated electrons" includes chemical bonds.

Keywords: dynamic pressure; metallic liquid H; oxide metallic glasses

1. INTRODUCTION

Electrons in strong insulators are delocalized at high pressures. A strong insulator is defined to be one that has a high bulk modulus and a wide band-gap at ambient. A weak insulator is defined here to be one that has a small bulk modulus and a wide band-gap at ambient. Examples are sapphire (single-crystal Al_2O_3) and liquid H_2, respectively. Their band-gaps at ambient are 10 eV and 15 eV, respectively. Sapphire is one of the most incompressible materials known and condensed H_2 is one of the most compressible materials. Despite the great disparity in strength and compressibility, both of these wide band-gap insulators undergo a continuous transition or crossover to poor metals at ~100 GPa shock pressures and probably by the same general mechanism. The purpose of this paper is to discuss the likely reason why such different insulators would behave so similarly with respect to the onset of electrical conduction with essentially the same value of conductivity—minimum metallic conductivity.

1.1. Basic Idea under Dynamic Pressure

An electronic transition to a metallic state is called an insulator-metal transition (IMT). Many IMTs are associated with structural phase transitions or crossovers, as likely occurs in sapphire to an amorphous state at a shock pressure of ~300 GPa (3 million bar = 3 Mbar) [1] and does occur in liquid H_2 at 140 GPa [2]. While one would normally not expect materials with such an extreme difference in strength to have much in common in terms of an IMT, the metallization processes in these two at high shock pressures are probably extreme limits of the same general process.

One key to recognizing this situation is the fact that crossovers to poor metals in both fluid hydrogen and amorphous Al_2O_3 are accomplished by dynamic or shock compression. Shock compression generally means compression by a single shock wave; dynamic compression is a more general term that often means compression by multiple shock waves or by ramp waves in time. In this materials context, the difference is a detail. The important point with respect to materials is that shock and dynamic compression are dissipative and dissipation in shock flows drives atomic disorder and electron delocalization in amorphous solids and in fluids.

Shock compression is dissipative, in the sense that not all the energy deposited by shock compression goes into compression; *i.e.*, into PdV work, where P is pressure and V is volume. Dissipation energy goes into temperature T, or non-equilibrated thermal energy, and entropy S, thermally equilibrated or not. Entropy is a word used here to denote the concept of damage or disorder. Shock-induced IMTs have been observed at elevated shock temperatures (T ≥ 1,000 K) and are thus often called semiconductor-metal transitions (SCMT).

The basic idea is that at sufficiently high dynamic pressure there is sufficient dissipation in the form of thermal energy and entropy to destroy chemical bonds, which localize electrons at ambient. Once chemical bonds are destroyed with high dynamic pressures, wave functions of disordered atoms overlap at sufficiently high densities for itinerant energy bands to form. In Al_2O_3 and H_2 the bonds are Al-O and H-H and both bonds have the same bond strength, 4.5 eV [3]. On the other hand, their material structures are totally different. At low pressures Al_2O_3 has a rigid, three-dimensional (3D), ordered crystal structure with a strong bond between all Al-O pairs in the 3D lattice. In contrast liquid H_2 is a disordered structure in which H_2 molecules interact via a weak Vander Waals pair potential with a well-depth of 0.004 eV [4].

The different initial structures significantly affect the way these two materials respond on a microstructural scale to dynamic pressure. However, in both cases chemical bonds eventually dissociate under sufficiently high dynamic pressure. Once the bonds are destroyed, electrons that were correlated in local chemical bonds have spherically symmetric or spherically-averaged atomic wave functions that hybridize into itinerant energy bands of a metal at sufficiently high densities. Shock pressures in the 100 GPa range produce sufficiently high densities. This process is termed Bonds to Bands (BtB). Material strength determines the split between thermal energy and entropy, but not whether or not a BtB can occur. Because of the highly disordered metallic state, conduction electron scattering is strong and so "metallic conductivity" means minimum metallic conductivity (MMC ≈ 2,000/(Ω-cm) = 500 μΩ-cm.), as observed in fluid H_2 [2] and as predicted based on modest extrapolation of measured electrical conductivities of Al_2O_3 shock-compressed up 220 GPa [1].

1.2. Possibilities under Static Pressure

Several strong insulators also probably reach MMC in a diamond-anvil cell (DAC) at 100 GPa static pressures at 300 K. These are ones with frustrated phase transitions that produce amorphous samples at sufficiently high static pressures. An example is Al_2O_3, which has been compressed in a DAC up to 200 GPa [5]. In the as-compressed state, only a disordered α-corundum phase was observed up to 180 GPa in a DAC. Laser heating up to ~2,000 K was required at high pressures in order to drive and then thermally quench high-pressure phases to ambient temperature at high pressure. For the Al_2O_3 sample compressed to 200 GPa, no crystalline symmetry was reported, which presumably means the Al_2O_3sample compressed statically to 200 GPa was amorphous. It is possible that amorphous Al_2O_3 reaches MMC at ~300

GPa in a DAC, as suggested previously [1]. A similar situation occurs in $Gd_3Ga_5O_{12}$(GGG) [6,7,8], as discussed below.

2. MOTT-LIKE INSULATORS

The notion that chemical bonds in electrical insulators localize electrons and that electron delocalization is achieved by breaking those bonds to form metallic energy bands is similar to the idea of a Mott insulator, of which NiO is a paradigm [9]. Based on its band structure, NiO is expected to be a metal. However, it is actually a transparent, anti-ferromagnetic insulator. Mott proposed that many-body electron correlations localize electrons at ambient, and that an IMT probably occurs at high pressures. A chemical bond consists of correlated-electron wave functions and their associated correlated electron spins. The cases considered here involve nonmagnetic insulators, which traditionally have not been considered to be Mott insulators. To distinguish classical magnetic Mott insulators from ones with shock-induced SCMTs in nonmagnetic crystals, I call the latter ones Mott-like. However, Mott and Mott-like insulators are essentially the same, provided the definition of Mott insulator is generalized to include all insulators with electron correlations that localize electrons at ambient, magnetic or not, and become metallic at high pressures.

3. DYNAMIC COMPRESSION

Extreme states of condensed matter are generated via the coupling between shock hydrodynamics and either equilibrium thermodynamics or non-equilibrium microstructures. Pressures used in experiments discussed herein were single-shock pressures in the case of Al_2O_3 and multiple-shock compression in the case of H_2. Shock pressures were generated by high-velocity (1 to 8 km/s) impacts of a planar solid disc onto a sample target at rest. Experimental lifetimes were typically ~100 ns. Rise times of shock pressures ranged from a few tenths to a few tens of ns. Because of the fast time scale, the method is called dynamic compression. Because of the 100 ns lifetime, dynamic compression is adiabatic (too brief for heat to diffuse out of a sample), too brief for highly-mobile shock-compressed and heated hydrogen to diffuse out of its sample holder, sufficiently long for fluids and metals to equilibrate thermally, and too brief for strongly covalently-bonded materials to equilibrate thermally. A general discussion of dynamic compression has been published [10].

In this paper we present arguments for the fact that metallization of all insulators at high dynamic pressures probably occurs via a single general mechanism. Strong and weak insulators, such as sapphire and liquid H_2, are two extreme cases of this general mechanism and these extreme cases facilitated its recognition. In Section 4 the classical IMT in H_2 is discussed. In Section 5 the SCMT in fluid hydrogen is discussed because it is the first experimental observation of a metallic phase of highly condensed hydrogen, is a reference point for discussion of metallization of strong insulators, and as an illustration of how shock dissipation causes the Bonds-to-Bands Transition (BtBT) in H_2. It is shock dissipation that enabled metallic fluid H to exist at pressures sufficiently low to be observed in a laboratory. The IMT from solid H_2 to metallic hydrogen at low temperatures ($T < 300$ K) is yet to be observed experimentally. In Section 6 general shock synthesis of amorphous metallic oxides is discussed. Experimental results for Al_2O_3 and $Gd_3Ga_5O_{12}$, are reviewed, which includes comparing radial extents of electron charge-densities of the various atoms to the size of the average volumes into which they must fit at high pressures. The latter provides a basis for determining if a material is likely to be metallic at a given compression. Basic conclusions are given in Section 7.

4. CLASSICAL IMT IN H_2

The IMT from H_2 to H at $T = 0$ K was proposed by Wigner and Huntington (WH) in 1935 [11] and remains today the paradigm of a pressure-driven first-order phase transition. In particular, WH demonstrates the intrinsic importance and simplicity of the dissociative phase transition for achieving a metallic state. The classical IMT of WH was a paradigm for understanding the crossover under dynamic compression from insulating liquid H_2 to metallic fluid H [2], which in turn provided a basis for understanding the likely crossover from strong insulators to disordered poor metals [1]. So as background for discussions below, we begin with a discussion of the WH IMT in H_2.

WH's classical view of an IMT under pressure is one of a crystalline insulator compressed hydrostatically at temperature $T = 0$ K and at some high pressure a first-order dissociative phase transition to a metallic state occurs [11]. H_2 is an insulator because two electrons are localized in each intramolecular H-H bond. However, on dissociation to H one electron per atom at sufficiently high density means one electron per energy band, which is a band structure of a metal. The dissociative phase transition from H_2 to metallic H occurs simply because of dissociation and the resulting half-filled energy band, independent of band shape.

In 1935 WH predicted theoretically the classical IMT in H_2 would occur at an estimated pressure of ~25 GPa. The IMT in solid H_2 has yet to be observed experimentally under static pressures up to ~300 GPa. Optical studies on solid H_2 in a DAC suggest metallic H_2 should be observed at about 450 GPa near 100 K [12]. The current theoretical estimate of pressure for the dissociative transition from solid $H_2$to solid metallic H at T = 0 K is 500 GPa [13]. Both of these predicted pressures are well beyond current capabilities of a diamond anvil cell (DAC).

It is theoretically possible, in principal, for H_2 to metalize at T = 0 K if its energy gap closes to $k_BT = 0$, where k_B is Boltzmann's constant. However, H_2 dissociation energy and band gap near ambient are 4.5 eV [3] and 15 eV [14], respectively. While the density-dependences of these parameters are not yet known exactly, the fact that the band gap at ambient is a factor of 3.3 times larger than dissociation energy, implies that dissociation at T = 0 K, and thus metallization, is more likely to occur at a lower density/pressure than band-gap closure of H_2. Thus, the IMT in H_2 probably occurs by dissociation, though this point is yet to be answered by experiment. One likely reason the IMT of WH is yet to be observed is the large H-H bond strength of 4.5 eV, which is also the strength of the Al-O bond. To date it has not been possible to deposit 4.5 eV into solid H_2 to dissociate it by compression alone at T = 0 K.

5. SHOCK DISSIPATION IN LIQUID H_2: METALLIC FLUID H

Since predicted pressures required to metalize hydrogen in a DAC at low temperatures are beyond current DAC capability, it was appropriate to look for the IMT in H_2 by trying something in addition to pure compression, something that might induce the IMT in H_2 at pressures that can be achieved in a laboratory. Heating compressed H_2 in a DAC is the obvious choice because heating might drive dissociation, which produces metal. However, if H_2 at ~100 GPa in a DAC is heated above ~300 K, H_2diffuses out of a DAC in a few minutes, which thus far has been a major impediment to making metallic hydrogen in a DAC.

So the issue then becomes the more complex one of finding a method at finite T. The free energy F of a system is F = U–TS, where U is internal energy and dissipation energy is E_d = TS. The method used would need to compress condensed hydrogen adiabatically to a temperature T with a ~100 GPa pressure pulse to increase its density by a factor of ~10 in an experimental lifetime such that hydrogen has insufficient time to diffuse out of the sample holder but sufficient time to thermally equilibrate to T and S. In H_2, dissociation is the dominant contribution to ΔS. Because the method must be adiabatic, pressurization, compression, and heating must be simultaneous. Irreversible shock energy deposited in the multiple-shock compression process is divided between T and S such that their product is maximized in order to minimize the free energy. Thus, at sufficiently large T > 0 and entropy increase ΔS > 0, the SCMT in H_2 might be achieved to a metallic H phase at lower pressures than can be achieved at T = 0 K at the same density.

Dynamic compression achieves ~100 GPa pressures, up to 10-fold compression of initial density of liquid H_2, at temperatures up to ~3,000 K simultaneously for experimental lifetimes of ~100 ns. Because of the fast rise time of pressure (~10 ns) and the short experimental lifetime, dynamic compression is dissipative and adiabatic, respectively. That is, experimental lifetime is too short for heat and hydrogen to diffuse out of the compressed hydrogen sample. Shock dissipation energy E_d goes into T and S. This SCMT has in fact been observed experimentally. By simultaneously pressurizing, compressing, and heating liquid H_2 initially at 20 K with multiple-shock compression, MMC of liquid H is achieved at 140 GPa, 9-fold compressed initial liquid-H_2 density, and ~3,000 K [2,15].

Ross has shown this SCMT is facilitated by the fact that H_2 dissociation energy decreases with compression [16]. At the same time temperature increases with dynamic compression. Thus, at a sufficiently high dynamic pressure, H_2 molecules dissociate to metallic fluid H in a crossover region. The H temperature of 3,000 K is much higher than melting temperatures of H_2, measured experimentally and calculated theoretically at 140 GPa [17,18,19,20,21]. Thus, dense metallic H is a fluid. Because of the large compression, the Fermi temperature T_F exceeds 200,000 K, $T/T_F \sim 0.01$, and metallic fluid H is highly degenerate. The highest measured value of electrical conductivity in the fluid is 2,000/(Ω-cm) at pressures from 140 to 180 GPa, which is MMC and consistent with theory of dense hydrogen [22,23].

Dissociation of the H-H bond achieves spherically symmetric H atoms. Sufficient pressure was available in those experiments to achieve sufficient overlap of electronic wave functions on adjacent H atoms to form an itinerant energy band of a metal. N_2 and O_2 undergo a similar crossover from a diatomic insulator to monatomic metallic fluid. Fluid H, N, and O have similar measured values of MMC (2,000 $(\Omega\text{-cm})^{-1}$) at highest pressures (100 to 140 GPa) [2,24,25,26]. Because T is finite, at dynamic pressures lower than required to achieve MMC, hydrogen, nitrogen, and oxygen are semiconductors. Expanded-fluid Cs and Rb near their liquid-vapor coexistence curves at ~2,000 K reach MMC at ~10 MPa (100 bar) static pressures [27]. Thus, five elemental monatomic fluids, H, N, O, Rb, and Cs, undergo a Mott-like SCMT under pressure.

Figure 1 is a plot of electrical conductivities of H, N, O, Rb, and Cs *versus* $a^*/D^{1/3}$, the ratio of atom size, assumed to be the effective Bohr radius a*, to the average distance between adjacent atoms in the fluid, which is determined by density and thus by pressure. D is the volume of the average cube around each atom. $D^{1/3}$ is average distance between adjacent nuclei. Plotting data this way for systems near an IMT was suggested by Mott [28], and applied by Hensel to his Rb and Cs conductivity data [27]. Figure 1implies the radial extents of wave functions of Rb and Cs are relatively large because relatively little compression of Rb and Cs is required for them to conduct. Similarly, H requires substantial compression in order to conduct. N and O are intermediate but closer to H. Figure 1 also shows that as atoms are pushed together by pressure, conductivity increases until at highest pressures all five elements have MMC (2,000 $(\Omega\text{-cm})^{-1}$). Since $a^*/D^{1/3}$~0.35–0.38 for 5 elements at MMC, overlap of wave functions on adjacent atoms is substantial for all of them. The maximum value $a^*/D^{1/3}$ can have is 0.5, which corresponds to coincidence of maxima in electron densities on adjacent atoms. The idea that large overlap is also probably required for oxides to become poor metals at high pressures arises below in the discussion of Al_2O_3 and $Gd_3Ga_5O_{12}$.

The trend above is as expected. For example, attraction by the Cs nucleus of the outer $6s^1$ conduction electron is screened by a Xe core. Substantial screening means the radial extent of the $6s^1$ electron is relatively large and so relatively little compression is needed to obtain sufficient overlap of wave functions on adjacent sites to cause the onset of electrical conduction. For H, the opposite is true. The attraction between electron and proton is unscreened and so substantial compression is expected before the onset of conduction in fluid H. Screening in N and O is relatively small compared to that in Cs. That is, screening in N and O is caused by filled $1s^2$ and $2s^2$ electron shells, which are relatively small compared to the size of a Xe core in Cs. Thus, onset density of conductivity in N and O is closer to that in H.

Statements about expected radial extents of atomic wave functions derived from Figure 1 are readily checked by comparing radial electron-density distributions calculated in the Hartree-Fock-Slater approximation for the five elements [24,29]. These results are shown in Figure 2. Inspection of (b) shows that the radial extent of the electron density distribution for H is least, greatest for Rb and Cs, and intermediate for N and O as well as being closer to that of H than to that of Rb and Cs, as expected.

Fluid H, N, O, Rb, and Cs have been discussed in some detail to demonstrate that Mott plots of measured electrical conductivities *versus* Mott parameter $a^*/D^{1/3}$ provide qualitative estimates of the relative radial extents of electron density distributions, calculated in the Hartree-Fock-Slater approximation. This extensive self-consistency between experimental results and atomic structure calculations has two significant implications. First, it is reasonable to use radial electron-density distributions of atoms calculated in the Hartree-Fock-Slater approximation to make estimates about densities at

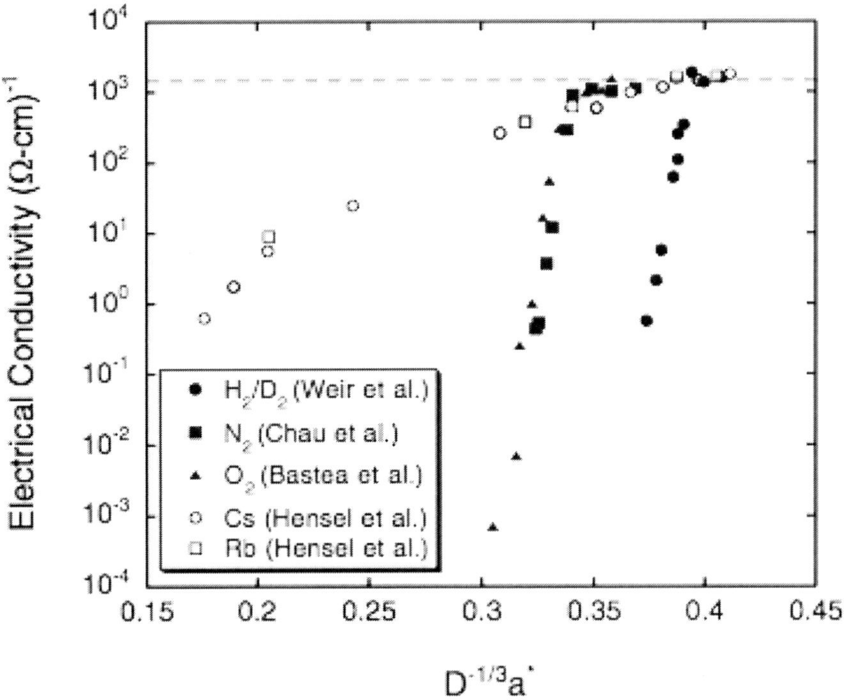

Figure 1. Electrical conductivities of H, N, O, Rb, and Cs plotted*versus* a*/D$^{1/3}$, where a* is size of atom (effective Bohr radius) and D is volume of average cube around each atom. D$^{1/3}$ is average distance between adjacent nuclei. Each point is measured electrical conductivity at specific pressure [24].

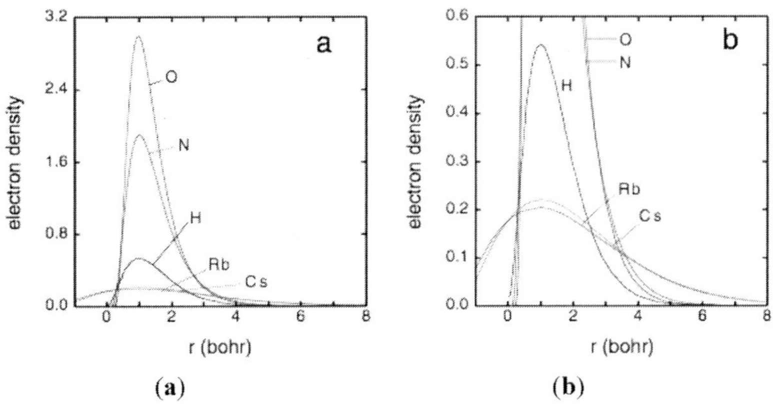

Figure 2. Electron densities ($4\pi r^2\psi^*\psi$) plotted *versus* radii [24,29]. To look at radial extents of outer electrons, for comparison purposes peak of each distribution was shifted to r = 1 bohr in plots. Curves in (**b**) are curves in (**a**) plotted on expanded scale.

which materials more complex than simple fluids might be metallic at extreme conditions. In this regard, likely metallization pressures and densities of Al_2O_3 and $Gd_3Ga_5O_{12}$ are discussed in the next section. Second, electrical conductivities of elemental fluids measured with a 20-m long two-stage light-gas gun enable resolution of radial extents of quantum mechanical electron density distributions on the spatial scale of ~Bohr.

If metastable solid metallic hydrogen (MSMH) could be quenched to ambient pressure and temperature, this material might have several scientific and technological uses including: a quantum solid with novel physical properties, including room-temperature superconductivity; a very light-weight structural material; a chemical fuel, propellant, or explosive, depending on the rate of release of stored energy; a dense nuclear-fusion fuel made with isotopes deuterium and tritium, rather than hydrogen, to obtain higher energy yields in inertial confinement fusion [30].

In summary, under dynamic compression, “soft” fluids, such as H_2 rapidly undergo large compressions, which induce high dynamic temperatures. In such systems thermal equilibrium is generally achieved in a sub-ns time scale. As dynamic pressure increases H_2 dissociates which causes entropy to increase and thermally equilibrate also on a sub-ns time scale. Thus, dynamically compressed fluids can be treated with equilibrium thermodynamics. In this case relatively simple assumptions and approximations give theoretical results in generally good agreement with experiment. Representative examples are Ross’ work on dissociation of H_2 [16] and Mott’s MMC [28].

6. SHOCK DISSIPATION IN STRONG INSULATORS: LIKELY SYNTHESIS OF METALLIC OXIDE GLASSES

Strong insulators differ substantially in most respects from weak fluid insulators discussed in Section 5. Interaction potentials differ substantially, breaking strong bonds in a rigid 3D lattice requires several eV, and atom densities in strong insulators near ambient are quite high compared to fluids. As a result, up to 10–100 GPa shock pressures, shock dissipation in dense strong insulators is absorbed substantially by mechanically breaking and bending inter-atomic bonds. In fact in this range of shock pressures, shock-induced damage and heating are often heterogeneous and T and S do not equilibrate thermally in bulk during experimental lifetimes until shock pressures exceed ~100 GPa. Nevertheless, with modest extrapolations of existing conductivity data of Al_2O_3 and $Gd_3Ga_5O_{12}$ up to ~250 GPa, both likely reach MMC at 300–400 GPa, as do compressible fluid H, N, and O at ~100 GPa.

6.1. Al_2O_3 (Sapphire)

Sapphire with a density of 3.98 g/cm^3 disorders substantially under shock pressures up to ~100 GPa ([31] and references therein). It is likely that shock heating of Al_2O_3 is not uniform in bulk until shock pressures exceed ~200 GPa. Moreover, the Hugoniot and 300-K isotherm of sapphire are nearly coincident up to 400 GPa. Above ~400 GPa, Hugoniot pressure increases dramatically and diverges from the 300-K isotherm. These observations suggest a picture in which entropy dominates dissipation below ~400 GPa and once entropy is maximized total pressure and thus shock temperature and thermal pressure increase rapidly with additional shock compression.

In a DAC, the corundum (α-Al_2O_3) to Rh_2O_3(II)-type phase transition occurs at 103 GPa and the Rh_2O_3(II)-type to $CaIrO_3$-type transition occurs at 130 GPa and persists to at least 180 GPa. However, these transitions in a DAC are sluggish. They occur only with laser-heating and thermal quenching at high pressures. X-ray spectra of laser-heated and quenched samples consist of broad individual diffraction peaks superimposed on a significant broad background, indicative of disordered structures with short-range order [5]. Those samples in a DAC are substantially disordered, which means a substantial amount of entropy.

The effect of shock-induced disorder on measured electrical resistivities of sapphire at shock pressure from 91–220 GPa [32] is illustrated in Figure 3. Up to ~130 GPa electrical resistivity is large and essentially constant. From 130–220 GPa, resistivity decreases by a factor of 10^3 and extrapolates to 500 μΩ-cm around 280 GPa. 500 μΩ-cm corresponds to electrical conductivity of 2,000 $(\Omega\text{-cm})^{-1}$, which is MMC, the same MMC as metallic fluid H reaches at 140 GPa, 9-fold compressed liquid-H_2 density, and ~3,000 K (Section 5). In sapphire a static pressure of 130 GPa is the pressure of the Rh_2O_3(II)-type to $CaIrO_3$-type transition in a DAC, and $CaIrO_3$-type disorders increasingly at higher pressures in a DAC. The picture that emerges from all these experiments is that shocked sapphire damages in the corundum (α-Al_2O_3) and Rh_2O_3(II)-type phases; from 130–280 GPa shocked sapphire undergoes a crossover with increasing disordering. At ~280 GPa shocked sapphire has MMC, which suggests it is an amorphous atomic metal or metallic glass, basically a frozen fluid. Of course, the possibility exists that sapphire is a fluid at ~280 GPa because the melting curve of sapphire has yet to be measured at these pressures. Alternatively, Al and O might phase separate at these extreme conditions, which might mean that current is conducted through Al filaments surrounded by non-conductive oxygen. This is conceivable because Al composition is 40 at.% which exceeds the percolation limit for filamentary conduction.

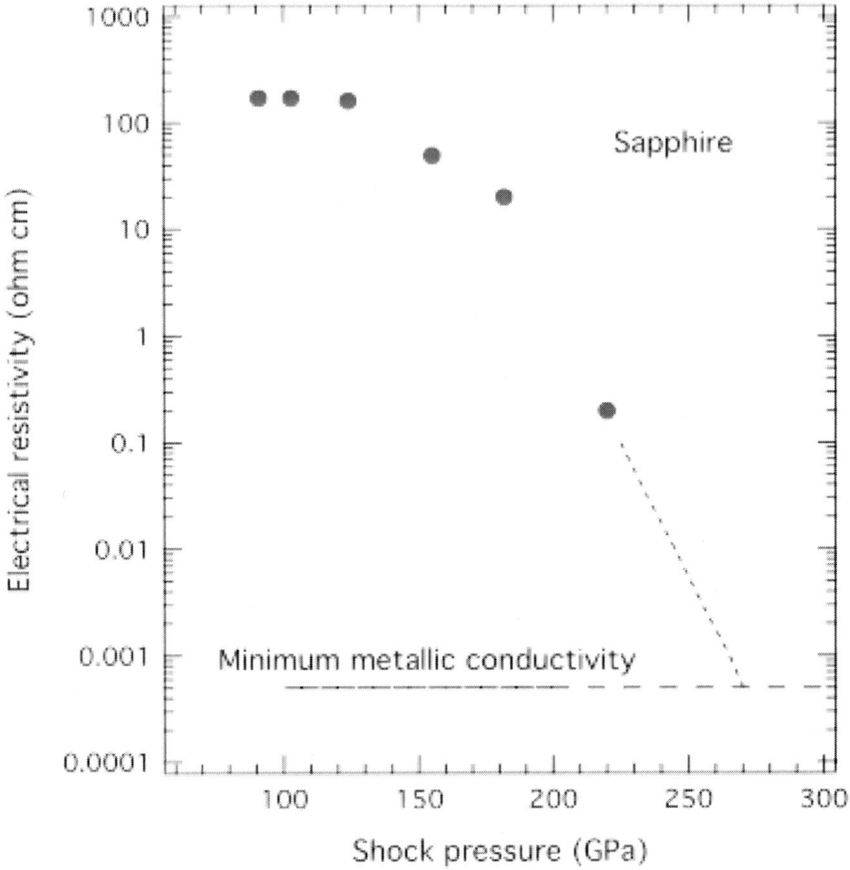

Figure 3. Measured electrical resistivity of sapphire plotted *versus*shock pressure up to 220 GPa [32].

A key issue to consider is whether or not it is possible for Al and O to form hybridized energy bands and, thus, have a band structure of a disordered Al-O metallic alloy. Chemical bonds have a characteristic length and energy. Shock compression is about a factor of 1.6 in density, which would shorten and distort bond lengths. Shock dissipation supplies energy to substantially disorder strong insulators. So it is reasonable to assume that all bonds are broken at sufficiently high shock pressures, which produces a disordered collection of Al and O atoms whose wave functions might hybridize. At shock compression of 1.6 in density and under the assumption that both Al and O atoms must fit into an average-size cube with the same volume, then each atom must fit into a cube with an edge length of 3.3 a_0, where a_0 is the Bohr radius. Thus, if the radial extents of the electron density distributions of Al and O atoms exceed 1.7 a_0, then wave functions on adjacent atoms will overlap and it will be possible for them to hybridize in the band structure of a disordered solid solution.

Radial extents of the outermost electrons of Al ($3s^2 3p^1$) and O ($2p^4$) atoms were calculated in the Hartree-Fock-Slater approximation [1,29] and are shown in Figure 4. The radial extents of outermost electrons of Al and O atoms are 7 a_0 and 4 a_0, respectively, both of which are much greater than 1.7 $a_0$needed for onset of hybridization. Overlap would be substantial, as in the case of compressible fluids discussed in Section 5. Thus, it is spatially possible for Al and O to form hybridized energy bands of a metal in such a dense atomic glass. Thus it is possible that MMC at ~280 GPa in Figure 3 is caused by hybridization of Al and O wave functions and strong electron scattering in an amorphous metallic alloy.

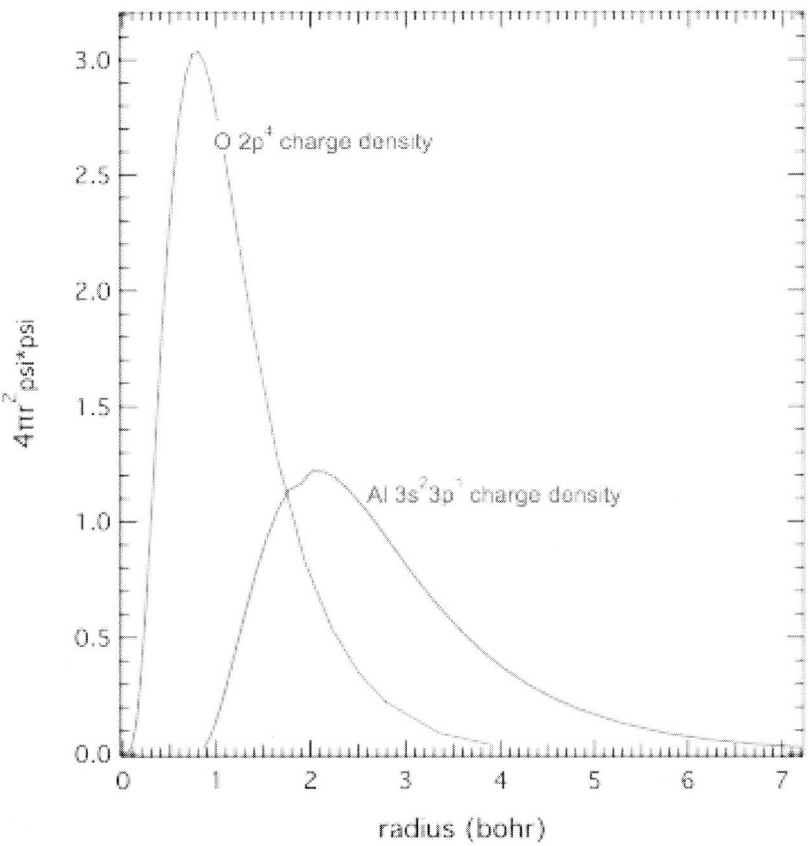

Figure 4. Spherically averaged atomic electron densities *versus*radius for O *2p*^{4} and Al *3s*^{2}*3p*^{1} electrons calculated with Hartree-Fock-Slater method [1,29].

Figure 5 compares the Hugoniot [33], DAC data on 300-K isotherm [5], and theory [34] for Al_2O_3 up to 340 GPa. Static compression to all data points shown produced only disordered corundum [5]. Laser-heating and thermal quenching at high pressures to 300 K were required to obtain those data points inFigure 5. Static compression of Al_2O_3 without a pressure medium and without laser-heating might produce a sample in a DAC with similar disorder as states on the Hugoniot at comparable compression. In this way it might be possible to synthesize an amorphous Al_2O_3 sample in a DAC for characterization. An amorphous Al_2O_3 sample in a DAC might have MMC at ~300 GPa, as well as under shock compression.

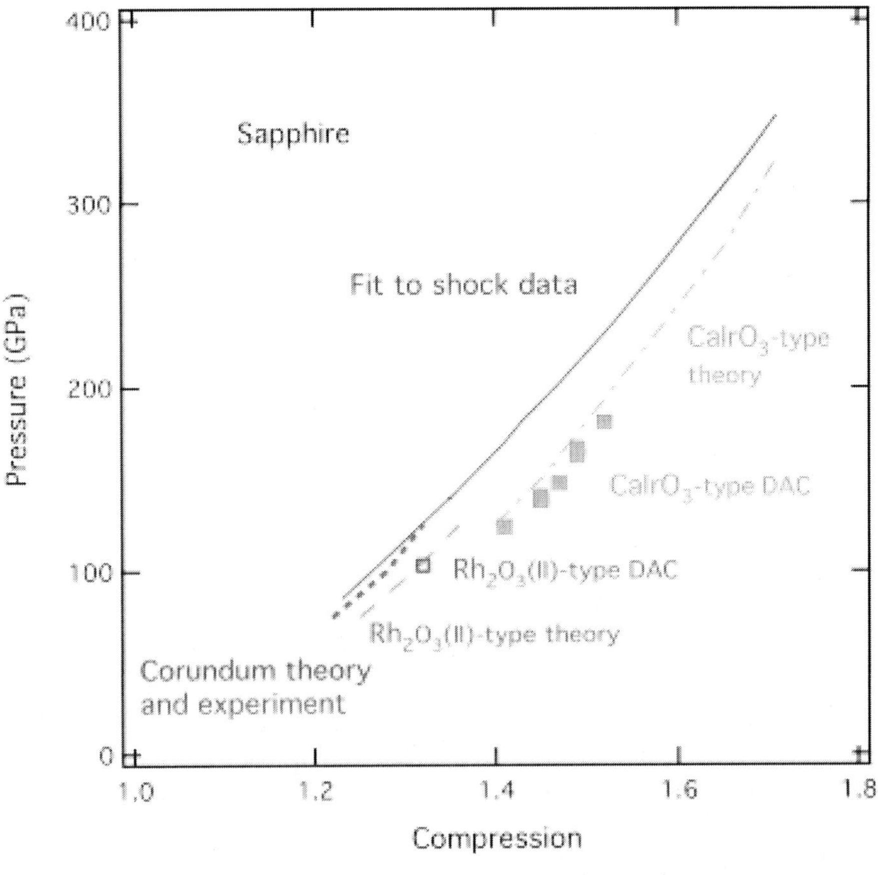

Figure 5. Pressure *versus* compression on shock Hugoniot [33], DAC data on 300-K isotherm [5], and theory [34] for Al_2O_3 up to 340 GPa.

6.2. $Gd_3Ga_5O_{12}$ (GGG)

GGG, with a density of 7.10 g/cm^3 and no long-range magnetic order, disorders substantially under static and shock compression. In a DAC, GGG is crystalline up to 74 GPa, above which x-ray diffraction peaks broaden up to 84 GPa, above which GGG is amorphous [6]. Hugoniot data of GGG single crystals have been measured from 30 to 260 GPa. The Hugoniot elastic limit (HEL) of GGG was found to be 30 GPa, a transition to an intermediate (IM) phase occurs at 65 GPa and extends up to 120 GPa, followed by the onset of a virtually incompressible high-pressure (HP) phase that extends up to 260 GPa. Calculated shock temperatures of GGG reach 6500 K at 260 GPa. Up to ~70 GPa the pressure-volume data measured in a DAC at 300 K and the Hugoniot are virtually identical. From shock pressures of 120 up to 260 GPa (HP phase), electrical conductivity measurements show that GGG is semiconducting [7].

The sharp increase in slope of shock pressure *versus* compression of GGG at ~120 GPa and ~1,000 K [7] could be caused by a transition to a virtually incompressible phase (a density effect) or by onset of substantial shock temperature and thus thermal and total pressure. Mao *et al.* found that GGG becomes amorphous in a DAC at 88 GPa and transforms to a new high-pressure phase at 88 GPa on laser heating to 1500 K [8]. The high-pressure phase in a DAC is cubic, consistent with a perovskite structure and stoichiometry of $(Gd_{0.75}Ga_{0.25})GaO_3$, and persists up to 180 GPa at 1,500 K. No rapid increase in pressure with compression is observed up to 180 GPa in the DAC data. The DAC data imply the rapid increase in shock pressure with compression at 120 GPa is caused by a combination of fast incomplete phase transitions and increasing shock heating, which is consistent with observed semiconductivity. It is interesting to note that Al_2O_3, as well as GGG, must also be laser heated at 100 GPa pressures in a DAC to induce high-pressure crystalline phases [5].

Time-resolved shock-wave profiles and Hugoniot data of single-crystal GGG have also been measured from 8.5 to 113 GPa [35]. The HEL increases from 8 to 24 GPa as final shock pressure increases from 8.5 to 89 GPa. Such a strong pressure dependence of the HEL has not been reported previously and suggests metastability and disorder during experimental lifetimes at those pressures. The phase transition observed at 76 GPa is probably the one observed by Mashimo *et al.* at somewhat lower pressure. It is interesting to note that single-crystal Al_2O_3 also has very unusual shock-wave profiles as GGG at comparable shock pressures, which are also indicative of disorder [36].

Extrapolation of electrical conductivities of GGG measured up to 260 GPa suggest GGG reaches MMC at ~400 GPa (0.4 TPa). In order to determine if this metallization is reasonable, once again the radial extents of atomic charge densities were compared to the size of the average cube into which each must fit at pressure. The highest pressure phase of GGG in a DAC is cubic with an effective stoichiometry of $(Gd_{0.75}Ga_{0.25})GaO_3$. As for Al_2O_3, radial extents of atomic charge densities of outermost electrons of Gd ($5d^1 6s^2$) and Ga ($4s^2 4p^1$) were calculated with the Hartree-Fock-Slater method. Atomic Gd has a radial charge density distribution that is more extended with a lower broad maximum than that of Ga, which is very similar to that of Al. We assume that at 0.4 TPa and compression of 1.9 over initial crystal density, bonds in crystalline $Gd_3Ga_5O_{12}$ are broken in a glass and Gd, Ga and O atoms can be treated as spherically averaged and symmetric. Calculated radial extents of $5d^1 6s^2$, $4s^1 4p^2$, and $2p^4$ electrons of Gd, Ga and O atoms are 8 a_0, 7 a_0, and 4 a_0, respectively. These wave functions must fit into an average cube with an edge length of 3.5 a_0, which means the distance from the center of the cube to a face is 1.8 a_0, which is substantially less than the radial extents of outer electrons Gd, Ga, and O. Thus, substantial overlap of atomic wave functions occurs at 0.4 TPa in GGG and it too is likely a metal by band overlap at these extreme conditions.

7. CONCLUSIONS

Dissipation in shock flows at high dynamic pressures drives structural and electronic transitions or crossovers, such as to metallic liquid hydrogen and most probably Al_2O_3 and $Gd_3Ga_5O_{12}$ metallic glasses.

These crossovers occur via Mott-like transitions from chemical bonds in which electrons are localized to disordered metallic systems in which electrons hybridize into itinerant energy bands and have MMC by virtue of strong electron scattering.

Dissipation is energy that does not go into compression at pressure-thermal energy and disorder or temperature T and entropy S if in thermal equilibrium. Contrary to popular opinion, dissipation is more than shock heating and significant shock heating is not required to make a poor metal.

The term "metal" here means electrical conduction in a degenerate system, which occurs by band overlap, rather than thermal ionization.

Since H_2 and probably disordered Al_2O_3 become poor metals with MMC, virtually all insulators with intermediate strengths probably do so as well.

REFERENCES AND NOTES

1. Nellis, W.J. Al_2O_3 as a metallic glass at 300 GPa. *Phys. Rev. B* **2010**, *82*, 092101.

2. Nellis, W.J.; Weir, S.T.; Mitchell, A.C. Minimum metallic conductivity of fluid hydrogen at 140 GPa (1.4 Mbar). *Phys. Rev. B* **1999**, *59*, 3434–3449.

3. Weast, R.C. *Handbook of Chemistry and Physics*, 52nd ed.; Chemical Rubber Co.: Cleveland, OH, USA, 1971; pp. F177–F178.

4. Ross, M.; Ree, F.H.; Young, D.A. The equation of state of molecular hydrogen at very high density.*J. Chem. Phys.* **1983**, *79*, 1487–1494.

5. Ono, S.; Oganov, A.R.; Koyama, T.; Shimizu, H. Stability and compressibility of the high-pressure phases of Al_2O_3 up to 200 GPa: Implications for the electrical conductivity of the base of the lower mantle. *Earth Planet. Sci. Lett.* **2006**, *246*, 326–335.

6. Hua, H.; Mirov, S.; Vohra, Y.K. High-pressure and high-temperature studies on oxide garnets.*Phys. Rev. B* **1996**, *54*, 6200–6209.

7. Mashimo, T.; Chau, R.; Zhang, Y.; Kobayoshi, T.; Sekine, T.; Fukuoka, K.; Syono, Y.; Kodama, M.; Nellis, W.J. Transition to a virtually incompressible oxide phase at a shock pressure of 120 GPa (1.2 Mbar): $Gd_3Ga_5O_{12}$. *Phys. Rev. Lett.* **2006**, *96*, 105504.

8. Mao, Z.; Dorfman, S.M.; Shieh, S.R.; Lin, J.F.; Prakapenka, V.B.; Meng, Y.; Duffy, T.S. Equation of state of a high-pressure phase of $Gd_3Ga_5O_{12}$. *Phys. Rev. B* **2011**, *83*, 054114.

9. Mott, N.F. The basis of the electron theory of metals, with special reference to the transition metals. *Proc. Phys. Soc., London, Sect. A* **1949**, *62*, 416–422.

10. Nellis, W.J. Dynamic compression of materials: Metallization of fluid hydrogen at high pressures.*Rep. Prog. Phys.* **2006**, *69*, 1479–1580.

11. Wigner, E.; Huntington, H.B. On the possibility of a metallic modification of hydrogen. *J. Chem. Phys.* **1935**, *3*, 764–770.

12. Loubeyre, P.; Occelli, F.; LeToullec, R. Optical studies of solid hydrogen to 320 GPa and evidence for black hydrogen. *Nature* **2002**, *416*, 613–616.

13. McMahon, J.; Ceperley, D.M. Ground-state structures of atomic metallic hydrogen. *Phys. Rev. Lett.***2011**, *106*, 165302.

14. Inoue, K.; Kanzaki, H.; Suga, S. Fundamental absorption spectra of solid hydrogen. *Solid State Commun.* **1979**, *30*, 627–629.

15. Weir, S.T.; Mitchell, A.C.; Nellis, W.J. Metallization of fluid molecular hydrogen at 140 GPa (1.4 Mbar). *Phys. Rev. Lett.* **1996**, *76*, 1860–1863.

16. Holmes, N.C.; Ross, M.; Nellis, W.J. Temperature measurements and dissociation of shock-compressed liquid deuterium and hydrogen. *Phys. Rev. B* **1995**, *52*, 15835–15845.

17. Gregoryanz, E.; Goncharov, A.; Matsuishi, K.; Mao, H.-K.; Hemley, R.J. Raman spectroscopy of hot dense hydrogen. *Phys. Rev. Lett.* **2003**, *90*, 175701.

18. Bonev, S.A.; Schwegler, E.; Ogitsu, T.; Galli, G. A quantum fluid of metallic hydrogen suggested by first-principles calculations. *Nature* **2004**, *431*, 669–672.

19. Deemyad, S.; Silvera, I.F. Melting line of hydrogen at high pressures. *Phys. Rev. Lett.* **2008**, *100*, 155701:1–155701:4.

20. Eremets, M.I.; Trojan, I.A. Evidence of maximum in the melting curve of hydrogen at megabar pressures. *JETP Lett.* **2009**, *89*, 174–179.

21. Morales, M.A.; Pierleoni, C.; Schwegler, E.; Ceperley, D.M. Evidence for a first-order liquid-liquid transition in high-pressure hydrogen from ab initio simulations. *Proc. Natl. Acad. Sci. USA* **2010**,*107*, 12799–12803.

22. Lin, F.; Morales, M.A.; Delaney, K.T.; Pierleoni, C.; Martin, R.M.; Ceperley, D.M. Electrical conductivity of high-pressure liquid hydrogen by Quantum Monte Carlo methods. *Phys. Rev. Lett.***2009**, *103*, 256401.

23. Tamblyn, I.; Bonev, S.A. A note on the metallization of compressed liquid hydrogen. *J. Chem. Phys.***2010**, *132*, 134503.

24. Chau, R.; Mitchell, A.C.; Minich, R.W.; Nellis, W.J. Metallization of fluid nitrogen and the Mott transition in highly compressed low-Z fluids. *Phys. Rev. Lett.* **2003**, *90*, 245501.

25. Bastea, M.; Mitchell, A.C.; Nellis, W.J. High-pressure insulator-metal transition in molecular fluid oxygen. *Phys. Rev. Lett.* **2001**, *86*, 3108–3111.

26. Fortov, V.E.; Ternovoi, V.Ya.; Zhernokletov, M.V.; Mochalov, M.A.; Mikhailov, A.L.; Filimonov, A.S.; Pyallinga, A.A.; Mintsev, V.B.; Gryaznov, V.K.; Iosilevskii, I.L. Pressure-produced ionization on nonideal plasma in a megabar range of dynamic pressures. *J. Exp. Theor. Phys.* **2003**, *97*, 259–278.

27. Hensel, F.; Edwards, P. The changing phase of liquid metals. *Phys. World* **1996**, *4*, 43–46.

28. Mott, N.F.; Davis, E.A. *Electronic Processes in Non-Crystalline Materials*; Oxford University Press: Oxford, UK, 1971.

29. Herman, F.; Skillman, S. *Atomic Structure Calculations*; Prentice-Hall: Englewood Cliffs, NJ, USA, 1963.

30. Nellis, W.J. Metastable solid metallic hydrogen. *Philos. Mag. B* **1999**, *79*, 655–661.

31. Nellis, W.J.; Kanel, G.I.; Razorenov, S.V.; Savinykh, A.S.; Rajendran, A.M. Entropy-dominated dissipation in sapphire shock-compressed up to 400 GPa (4 Mbar). *J. Phys.: Conf. Ser.* **2010**, *215*, 012148.

32. Weir, S.T.; Mitchell, A.C.; Nellis, W.J. Electrical resistivity of single-crystal Al_2O_3 shock-compressed in the pressure range 91–220 GPa (0.91–2.20 Mbar). *J. Appl. Phys.* **1996**, *80*, 1522–1525.

33. Erskine, D. *High-Pressure Science and Technology-1993*; Schmidt, S.C., Shaner, J.W., Samara, G.A., Ross, M., Eds.; AIP Press: New York, NY, USA, 1994; pp. 141–143.

34. Umemoto, K.; Wentzcovitch, R.M. Prediction of an U_2S_3-type polymorph of Al_2O_3 at 3.7 Mbar.*Proc. Natl. Acad. Sci. USA* **2008**, *105*, 6526–6530.

35. Zhou, X.; Li, J.; Nellis, W.J.; Wang, X.; Li, J.; He, H.; Wu, Q. Pressure-dependent Hugoniot elastic limit of $Gd_3Ga_5O_{12}$ single crystals. *J. Appl. Phys.* **2011**, *109*, 083536.

36. Kanel, G.I.; Nellis, W.J.; Savinykh, A.S.; Razorenov, S.V.; Rajendran, A.M. Response of seven crystallographic orientations of sapphire crystals to shock stresses of 16–86 GPa. *J. Appl. Phys.***2009**, *106*, 043524.

Chapter 17

Identification of new Pillared-Layered Carbon Nitride Materials at High Pressure

Ashkan Salamat, Malek Deifallah, Raul Quesada Cabrera, Furio Corà & Paul F. McMillan

Lyman Laboratory of Physics, Harvard University, Cambridge, MA 02138, USA

ABSTRACT

The compression of the layered carbon nitride $C_6N_9H_3{\cdot}HCl$ was studied experimentally and with density functional theory (DFT) methods. This material has a polytriazine imide structure with Cl^- ions contained within $C_{12}N_{12}$ voids in the layers. The data indicate the onset of layer buckling accompanied by movement of the Cl^- ions out of the planes beginning above 10–20 GPa followed by an abrupt change in the diffraction pattern and c axis spacing associated with formation of a new interlayer bonded phase. The transition pressure is calculated to be 47 GPa for the ideal structures. The new material has mixed sp^2–sp^3hybridization among the C and N atoms and it provides the first example of a pillared-layered carbon nitride material that combines the functional properties of the graphitic-like form with improved mechanical strength. Similar behavior is predicted to occur for Cl-free structures at lower pressures.

INTRODUCTION

High pressure investigations of carbon nitride materials have been stimulated by theoretical predictions that dense sp^3-bonded C_3N_4phases would display low compressibility and high hardness values1′2. Despite many synthesis attempts including use of high pressure the formation of crystalline C_3N_4 polymorphs remains an elusive goal. However an sp^3-bonded carbon nitride imide C_2N_3H with a defective wurtzite structure has been produced in laser-heated diamond anvil cell (DAC) experiments and this material was recoverable to ambient conditions3′4. Low density sp^2 bonded polymeric or layered graphitic carbon nitride materials (g-CNMs) are also well known5 and these have attracted attention as metal-free redox catalysts, photocatalysts and electroceramic materials6′7. They also provide precursors for investigation of pressure-induced transformations into dense sp^3bonded phases. We investigated the room temperature compression behavior of one such well-characterized g-CNM of composition $C_6N_9H_3{\cdot}HCl$8 to 70 GPa using synchrotron X-ray diffraction experiments in a diamond anvil cell (DAC) combined with density functional theory (DFT) calculations carried out to 100 GPa. The combined experimental and theoretical results indicate the initial onset of layer buckling and movement of the Cl^- ions out of their sites within the planes followed by a phase transformation into a structure containing interlayer C-N bonds between sp^3 hybridized atoms. The new material constitutes a new example of a pillared-layered gCNM with mixed sp^2–sp^3 bonding. Related bonding changes have been recorded in graphitic C and BN materials at high pressure9′10′11′12′13′14.

A large class of oligomeric, polymeric and graphitic carbon nitride materials are determined to have structures based on heptazine (tri-*s*-triazine, C_6N_7) rings linked by -N = or -NH- groups15. A second structural motif based on polymerized triazine (C_3N_3) units is found among graphitic nanocrystalline C_3N_4 prepared by chemical vapor deposition16. Triazine units are also the building blocks in bulk crystalline compounds synthesized from precursors including melamine ($C_3N_6H_6$) and cyanuric chloride ($C_3N_3Cl_3$) under high-P,T conditions (0.5–3 GPa, 300–500°C). Further studies have obtained crystalline graphitic polytriazine imides (PTI) with related structures by reactions in molten salt mixtures (e.g. KCl-LiCl eutectics)5′17. The *s*-triazine rings linked by bridging -N = or -NH- units lead to large voids within the graphitic layers. In the case of $C_6N_9H_3{\cdot}HCl$ studied here the Cl^- ions occupy the centers of the voids at ambient pressure. The layers are stacked to provide an alternating pattern of triazine and $C_{12}N_{12}$ rings along the *c* axis (Fig. 1).

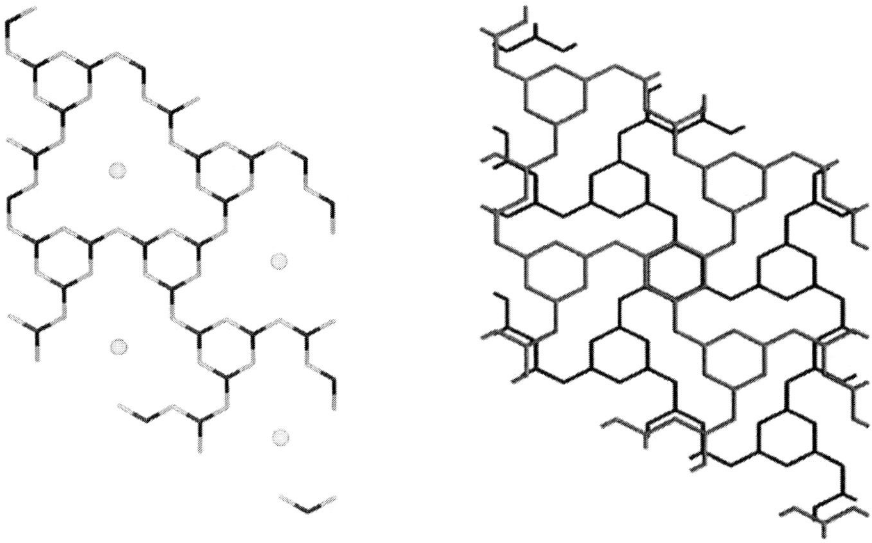

Figure 1: Structural features of graphitic $C_6N_9H_3{\cdot}HCl$ at ambient pressure. (left) Top view of one plane of the structure showing the triazine (C_3N_3) rings connected by -NH- groups to form large $C_{12}N_{12}$ voids that are occupied by Cl^- ions. The accompanying extra H^+ ion is attached to one of six possible N positions from the triazine units surrounding the large ring6′16. The H atoms have been omitted for clarity. (right) View down *c* of two adjacent layers of the structure showing the ABAB stacking sequence that places one triazine ring above the $C_{12}N_{12}$ void in successive planes. The H and Cl species are not shown for clarity.

RESULTS

Experimental findings

The starting compound g-$C_6N_9H_3{\cdot}HCl$ is descibed by a hexagonal unit cell with $P6_3/m$ symmetry (Fig. 1)8′18. The X-ray patterns at low pressure (Fig. 2) are dominated by the (002) interlayer reflection at 2θ ~ 8° that is indicative of a layered graphite-like structure with an ABAB stacking along the c axis. The polytriazine imide layers contain $C_{12}N_{12}H_3$ voids hosting Cl^- ions derived from the synthesis reaction between melamine and cyanuric chloride, and additional H^+ ions are bound to the N atoms of the triazine rings8′18. Only one of six available N sites is protonated in this way so that the $P6_3/m$ space group used to analyze the data represents a spatially averaged solution. The ABAB stacking of the graphitic layers creates a structure in which one half of the triazine rings within each layer is positioned above or below $C_{12}N_{12}$ voids in adjacent layers, while the others overlap triazine rings in the layers above and below (Fig. 1).

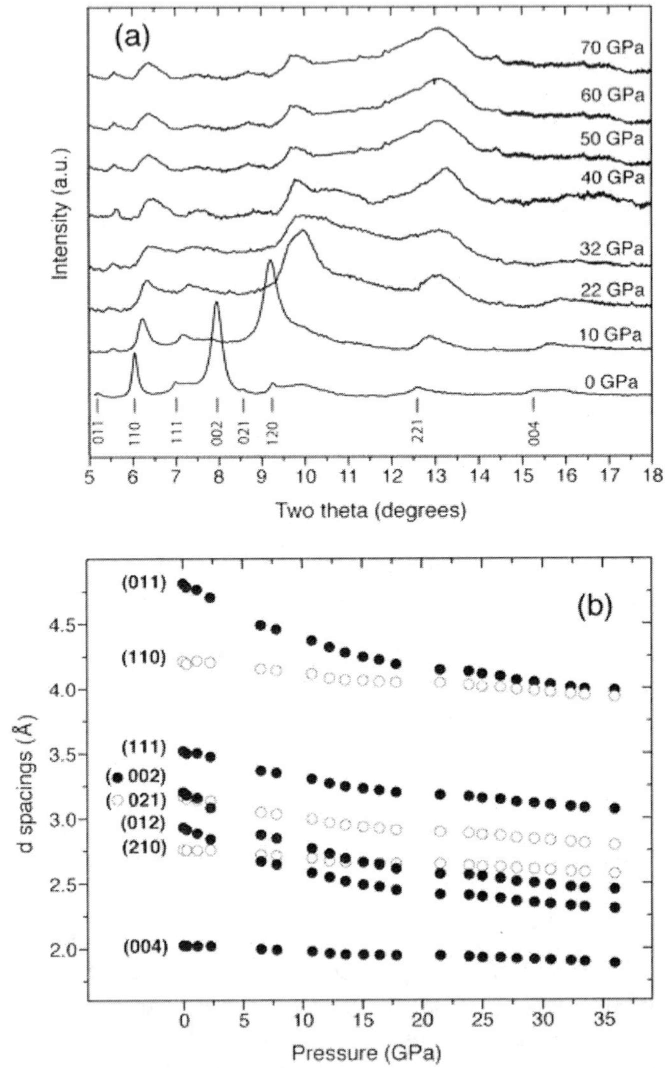

Figure 2: Angle dispersive synchrotron X-ray diffraction data for $C_6N_9H_3{\cdot}HCl$ obtained up to P = 70 GPa (λ_o = 0.4441 Å). Peaks of interest are labelled with *hkl* designations according to the P6_3/m space group. (a) Selected patterns showing the onset of peak broadening for the layered phase above 10 GPa followed by the rapid emergence of the new pattern above 40 GPa. (b) Variation of *d*spacings measured for the principal peak maxima assigned to the low pressure graphitic phase as a function of pressure.

The prominent (002) interlayer reflection shifts rapidly to smaller d spacing (2θ ~ 10°) with increasing pressure (Fig. 2). Above 10 GPa the diffraction peaks become significantly broadened but the general features of the graphitic material are still recognizable up to P ~ 36 GPa. However at P = 40 GPa the nature of the pattern has changed to become dominated by a main broad asymmetric peak near 13° 2θ. This result signals a structural change into a new high-density form (Fig. 2). After the transformation has occurred the peak positions vary little with continued compression up to 70 GPa indicating that the high pressure structure is significantly less compressible than the graphitic layered phase. The substantial peak broadening that occurred as a function of pressure did not permit us to carry out detailed refinement of the diffraction data. Instead we investigated the nature of the structural changes at high pressure using density functional theory (DFT) calculations, that enabled us to gain understanding of the pressure induced transformation by comparing the predicted diffraction profiles with experiment (Fig. 3).

DFT calculations

In a first series of DFT studies we geometry-optimized the structure of g-$C_6N_9H_3 \cdot HCl$ between P = 0–100 GPa in 10 GPa steps, starting each optimization from the ambient pressure lattice spacings obtained from experiment (a,b = 8.42 Å; c = 6.34 Å). We generated a proton ordered version of the 40-atom crystallographic unit cell within space group P1, removing all local symmetry constraints to allow the atoms to relax both within and away from the graphitic planes. This differs from our experimental data analysis that assumed a spatially averaged $P6_3/m$ unit cell8'18. The proton-ordered structure exhibits a small shear distortion between adjacent planes along (010) to maximize electrostatic interactions between the H^+ and Cl^- ions within layer voids. This leads to a small separation between a and b cell parameters that increases slightly with compression but that remains insignificant until above 20 GPa (Fig. 4). At P = 0 GPa the optimized lattice parameters were found to be a,b = 8.37 Å; c = 6.23 Å, slightly smaller than those found experimentally (8.42, 6.34Å) as is typical for LDA results. During compression there is a rapid decrease in the c-axis parameter consistent with experiment: between 0–40 GPa there is a 17% reduction in c, compared with 3.7% and 5.8% in the a and b lattice parameters (Fig. 4). Between 60–70 GPa a sudden collapse of the c-axis occurred in these initial compression experiments, and then this parameter remained approximately constant up to 100 GPa (Fig. 4).

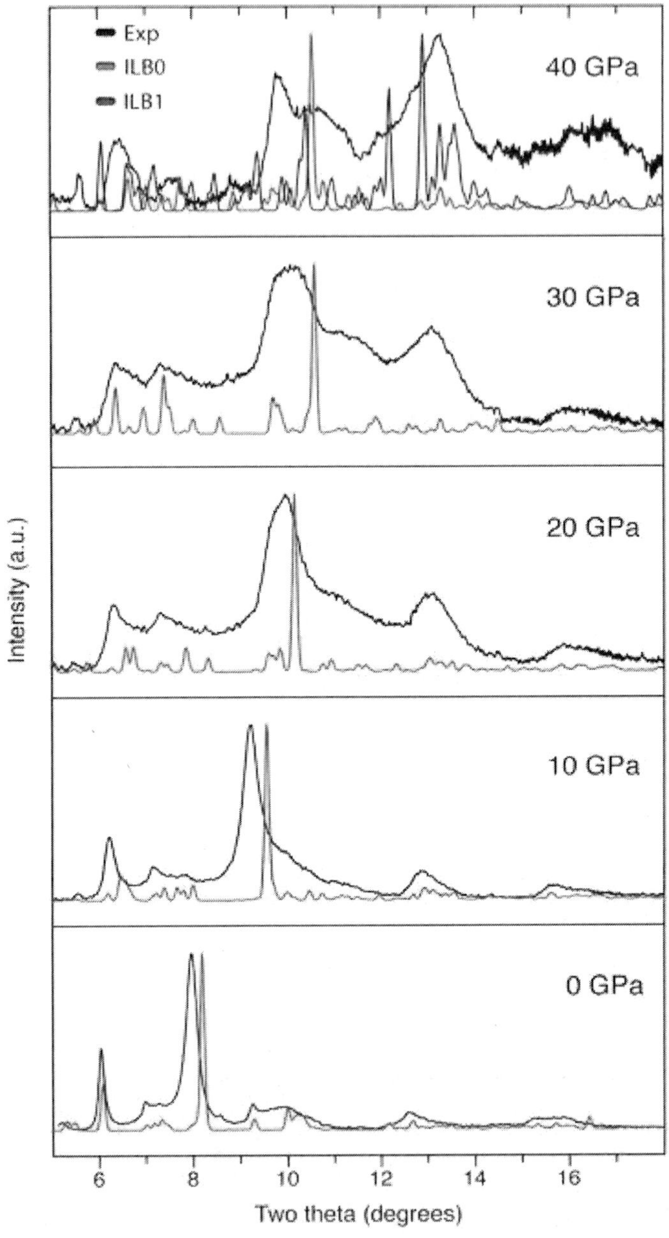

Figure 3: Comparison of experimental (black line) with the DFT predicted (ILB0 phase, red line) X-ray patterns for the graphitic phase at various pressures. At 40 GPa the blue pattern corresponds to dense forms with interlayer bonding (ILB1) as described in the text.

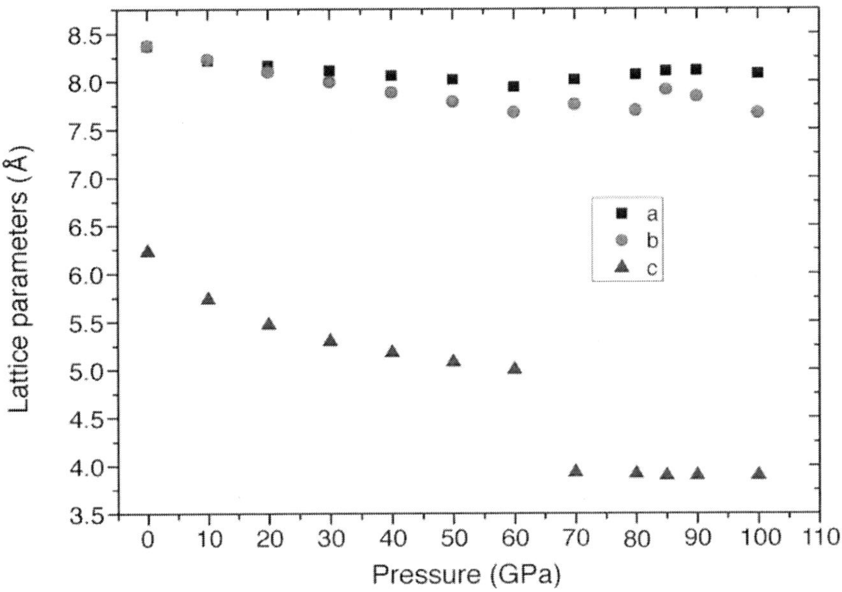

Figure 4: Changes in the DFT calculated lattice parameters for a proton ordered version of the g-$C_6N_9H_3$·HCl structure as a function of pressure.

These initial results showed that a dramatic change had occurred in the structure as the c axis dimension was reduced by compression, implying a separation between the initially planar layers of approximately 2.0 Å, without considering the reduction in distances due to layer buckling. Such a small interlayer separation could lead to the onset of C-N interlayer bonding and we investigated this possibility by analyzing the fully relaxed structures as a function of densification. The results led us to identify the structure optimised at high pressure as a new phase of the material, characterized by the occurrence of bonding between a subset of sp^3 hybridized C and N atoms in adjacent layers (Fig. 5). This new phase is termed the ILB1 structure to distinguish it from the densified graphite-like phase with buckled but independent layers obtained at lower compression (ILB0). The C-N bonding occurs between atoms associated with triazine rings that in the ambient pressure phase overlapped each other in adjacent layers; the others located above and below the $C_{12}N_{12}$ voids remain sp^2-hybridised giving rise to a unique new pillared-layered structure that contains a mixture of sp^2- and sp^3-bonded C and N atoms (Fig. 5).

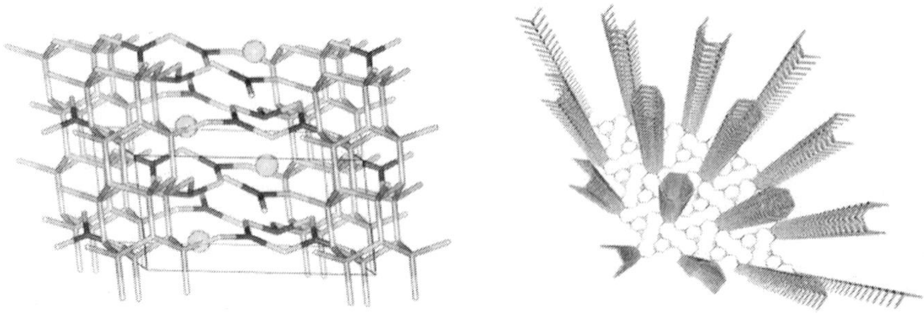

Figure 5: Left: The ILB1 phase of $C_6N_9H_3 \cdot HCl$ obtained at high pressure from DFT calculations.In the real structure studied experimentally the Cl^- displacements above and below the layers can occur in a random or locally ordered fashion. Right: picture highlighting the formation of pillars with three-dimensional connectivity between the carbon nitride layers produced by interlayer bonding within a mixed sp^2–sp^3 bonded state; the pillars are connected within the planes by relatively undistorted triazine rings.

Once we had identified the occurrence and connectivity of the new high-pressure phase, we examined the relative enthalpies of the ILB0 and ILB1 polymorphs over a wide pressure range, starting the studies using the compressed but still graphite-like (ILB0) structure obtained at 60 GPa and then gradually increasing P, compared with results during gradual decompression of the ILB1 phase obtained at 70 GPa (Fig. 6). Each calculation was initiated from the geometry-optimized structure at the immediately adjacent pressure value. The C-N connectivity was maintained throughout in both series of calculations, indicating that the ILB0 and ILB1 phases correspond to separate functions in the potential energy surface of the system and that their interconversion is an activated process that does involve interlayer but not intralayer C-N bond breaking or formation. Comparing the calculated enthalpies we identified 47 GPa as the equilibrium transition pressure between the ILB0 and ILB1 polymorphs at T = 0 K (Fig. 6). Following the compression studies we relaxed the ILB1 structure to ambient pressure, and found that the pillared connectivity was retained. Similar calculations repeated for the hypothetical Cl-free compound showed a similar structural transformation occurring at high pressure, but with the isoenthalpic point between ILB0 and ILB1 phases reduced to ~5 GPa. That result indicates that varying the Cl content of these gCNM phases should cause the transition pressures and recoverability of the pillared-layered material to be adjusted experimentally.

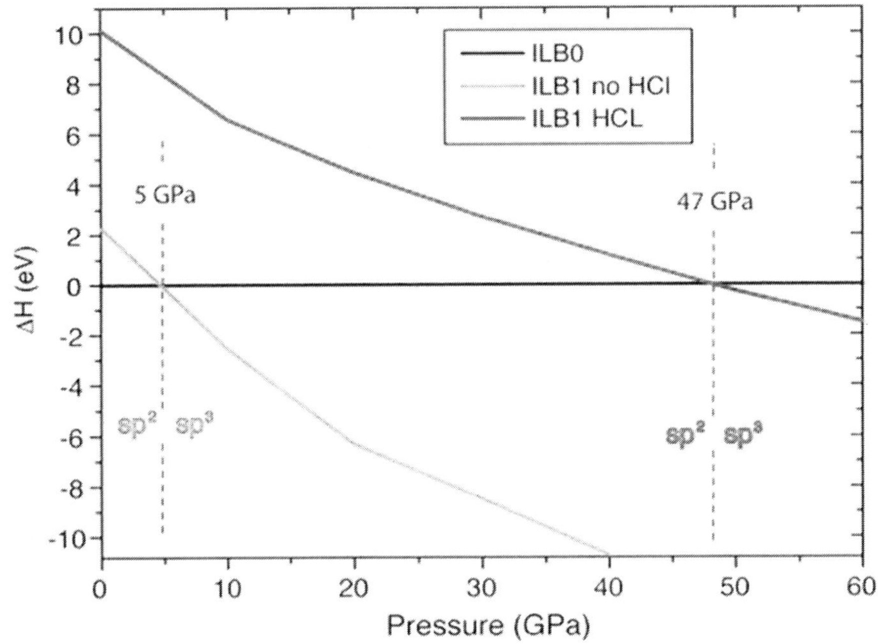

Figure 6: Calculated enthalpy as a function of P for the layered-pillared ILB1 phase relative to the layered graphitic (ILB0) polymorph of $C_6N_9H_3$·HCl and of the hypothetical Cl-free $C_6N_9H_3$system.

In order to fully characterize the structural changes associated with compression of $C_6N_9H_3$·HCl from its ambient pressure graphitic phase to the denser ILB1 polymorph at high pressure, we measured a set of dihedral angles in the minimum energy structures relaxed at each pressure. These angles (Fig. 7a) are chosen to describe the distortion of the initially planar graphitic layers: 1) the internal dihedral angle of a protonated C_3N_3 ring (ABCD); 2) the dihedral angle formed by the skeletal proton (i.e. the bridging NH group) with respect to the neighbouring triazine ring (ABFG); and 3) the internal dihedral angle of a non-protonated C_3N_3 ring (HIJK). All these values are identically zero for fully planar layers and any deviations thus record buckling of the gCNM framework. Achieving a value of 60° in any of these parameters indicates a change of hybridization at the central C and N atoms from sp^2 to sp^3. At each pressure reported in Fig. 7 we employ the thermodynamically stable phase from the DFT calculations, i.e. ILB0 below the transition pressure of 47 GPa, and ILB1 above.

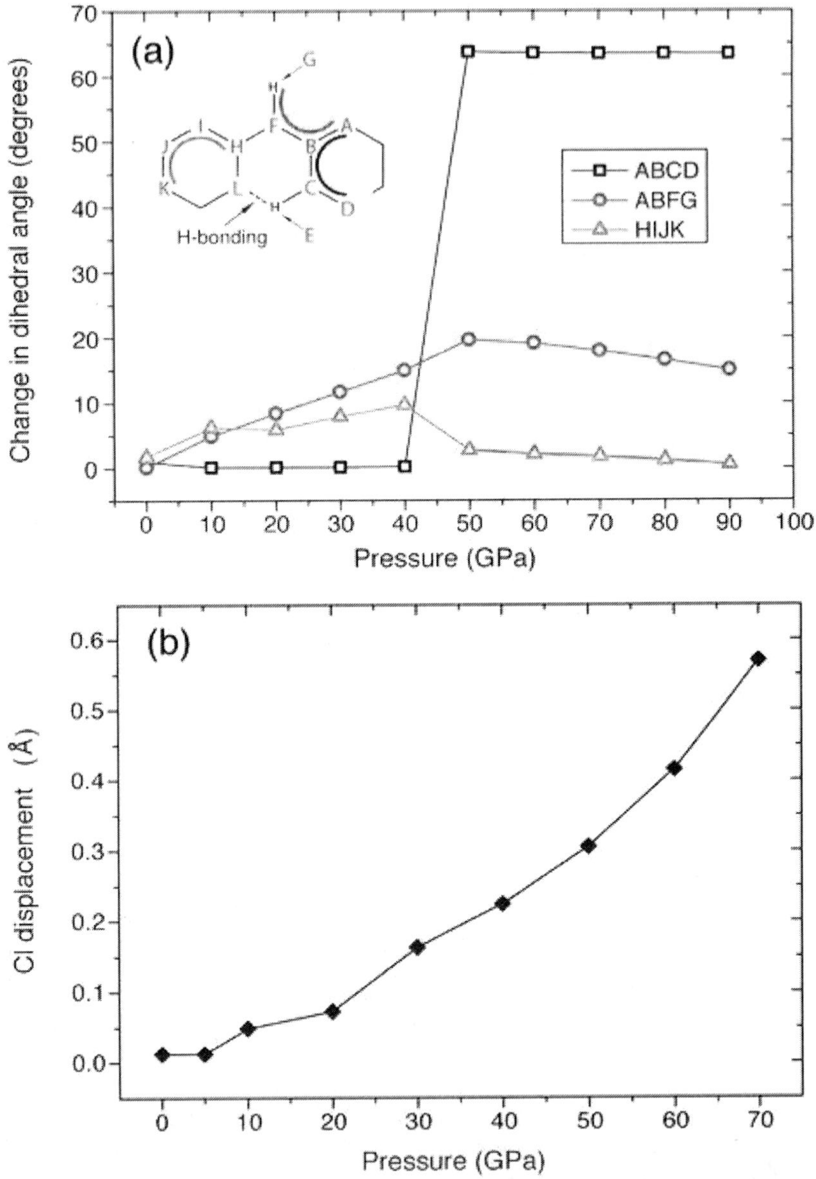

Figure 7: (a) Dihedral angles with reference to the initially planar g-$C_6N_9H_3$·HCl structure as a function of pressure.Here the dihedral angle changes are shown with reference to the fully relaxed and thermodynamically stable ILB0 or ILB1 states at each pressure. The structural phase transformation occurs at 47 GPa. (b) Displacement of the Cl^- ions along the *c* axis relative to the initial graphitic planes within the calculated ILB0 structure.

At ambient pressure the ILB0 structure is nearly planar with all dihedral angles <2°, as expected for a graphitic phase with delocalized sp^2 bonds. However, this condition is lost as soon as the structure is compressed even slightly. By 10 GPa the dihedral angles HIJK and ABFG differ noticeably from zero but those within the C_3N_3 rings (ABCD) remain constant. This reveals a buckling of the layers around the bridging -NH- groups (ABFG) beginning at low pressure that is accompanied by a displacement of the Cl^- ions out of the graphitic planes (Fig. 7b). Between 40–50 GPa the ABCD dihedral angle jumps from 2° to 65°, indicating the sp^2 to sp^3 hybridization change that has taken place around the C and N atoms associated with the triazine rings. At this point the reduction in the *c* axis is such that the interlayer C-N bonding has occurred to yield the ILB1 structure. The structural changes can also be correlated with the behavior of the Cl^- ions. While these remain within the layers, they are clamped tightly by strong H-bonding interactions with the NH groups around the $C_{12}N_{12}$ ring sites. On application of pressure, these voids are diminished in size and they become too small to host the Cl^- ions that are progressively squeezed out of the plane. This effect becomes more marked with increasing densification until it reaches nearly 0.6 Å by 70 GPa. These Cl^-displacements also lead to structural disorder in the real system, as the Cl^- ions can be located either above or below the planes and this effect can contribute to the peak broadening observed in the experimental X-ray diffraction patterns. Once removed from the graphitic layers and into the interlayer space the Cl^- ions also become free to be displaced along the (*a,b*) directions.

Comparison with experiment: ILB0 vs ILB1 formation at high pressure

Following our theoretical prediction of formation of a novel ILB1 structured phase at high pressure, we compared our calculated X-ray diffraction patterns for this material with experimental data (Fig. 8). The diffraction pattern recorded experimentally at 70 GPa showed broad features extending throughout the 5–18° 2θ range, that could not be refined to provide structural details. However comparison with the calculated results for the predicted ILB1 and ILB0 structures at the same pressure is revealing. The predicted X-ray pattern for the optimized ILB1 phase at 70 GPa matches the observed diffraction profile, with the main diffraction intensity concentrated near 13° 2θ, along with other features occurring near 10° and 6.5° 2θ. The pattern is significantly different from that expected for the ILB0 structure at the same pressure, leading us to conclude that the high pressure structural transformation observed experimentally corresponds to a transition between ILB0 and ILB1 phases.

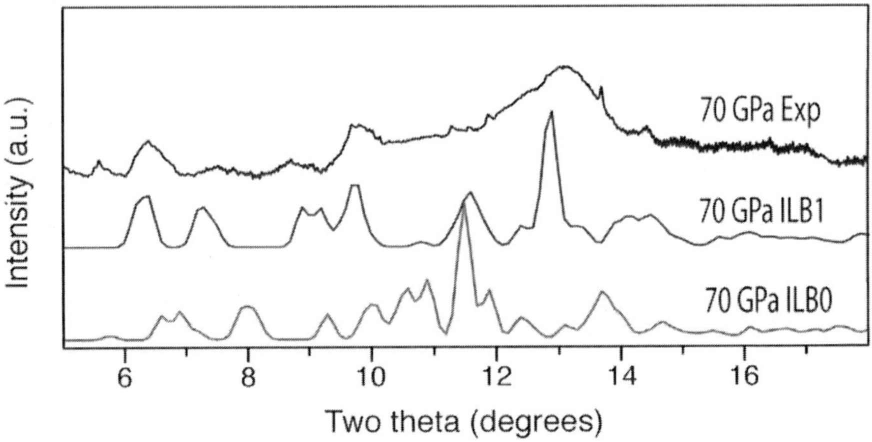

Figure 8: Comparison of the experimental XRD pattern obtained at 70 GPa (top) vs simulated patterns generated from DFT calculations (bottom), derived from geometry optimised ILB1 and ILB0 structures of $C_6N_9H_3{\cdot}HCl$, with or without interlayer bonding occurring at the same pressure.

Low pressure V(P) results

The computational results enabled us not only to identify the structural features of the new high-pressure ILB1 polymorph of the carbon nitride material under study, but also to further characterize the evolution of the graphitic ILB0 phase at intermediate pressures, highlighting the occurrence of buckling distortions and displacements of the Cl^- ions. We carried out a detailed analysis of our angle dispersive X-ray diffraction data below 36 GPa to study the compression behavior of the ILB0 structure and validate the computational results in relation to the experimental measurements. Refinements were performed within $P6_3/m$ using PowderCell software by first assigning the position of the main diffraction peak to the (002) reflection of the ABAB layer-ordered structure, using the DFT results to guide the structural analysis in the highest pressure ranges where substantial peak broadening was present (Fig. 2). The variations in refined *a,b,c* cell parameters led to the V(P) plot shown in Figure 9.

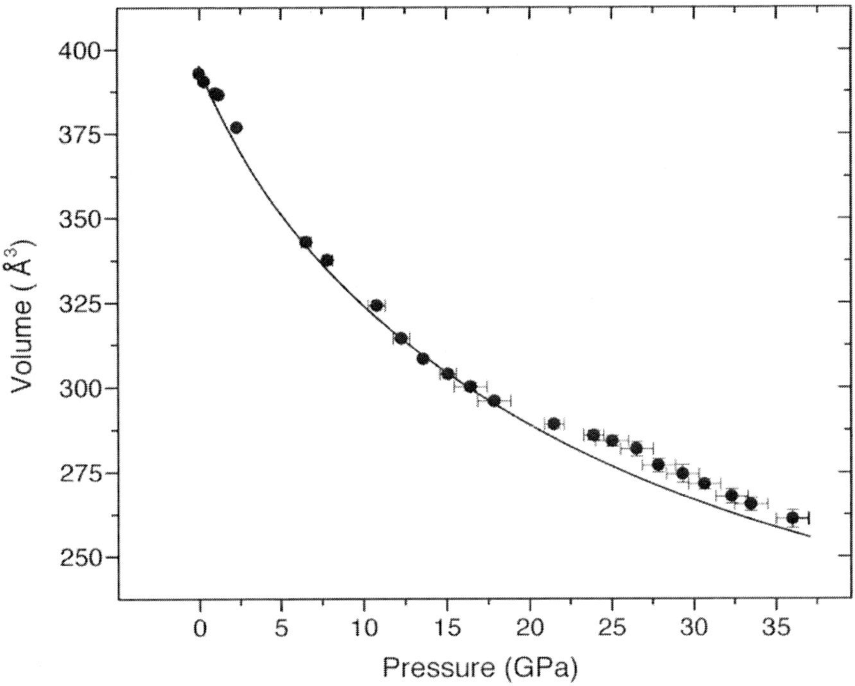

Figure 9: V(P) plot obtained from our experimental data (circles) for g-$C_6N_9H_3$.HCl analyzed within the spatially averaged space group *$P6_3/m$*. The data were fitted up to 21 GPa using a third order Birch-Murnaghan equation of state (K_0= 32.8 ± 7.3 GPa and K_0' = 4.8 ± 0.5) and extrapolated to 36 GPa (line).

Between 20–30 GPa we note a slight deviation of the V(P) relation upwards from the expected equation of state (EoS) curve obtained by analyzing the data using the spatially averaged space group $P6_3/m$. This can occur due to layer buckling along with displacements of the Cl^- ions away from the graphitic planes, that become significant above 20 GPa as shown by the DFT results (Fig. 7b). We attempted to analyze our data in this range using lower symmetry space group solutions $P6_3$, *P6 and* *P1*but these all gave similar R_{wp} and χ^2 values despite the increased structural degrees of freedom.

We fitted the *V(P)* data up to 21 GPa using a third order Birch-Murnaghan (BM-3) equation of state:

$$P(V)=3K_0f(1+2f)^{5/2}\left(1+\frac{3}{2}(K'_0-4)f\right)$$

with K_0 the zero pressure bulk modulus and K_0' its pressure derivative, and the volume strain variable expressed by

$$f=\frac{1}{2}\left[\left(\frac{V_0}{V}\right)^{2/3}-1\right]$$

This fit yielded K_0 = 32.8 ± 7.3 GPa and K_0' = 4.8 ± 0.5, in good agreement with our DFT results for the ILB0 polymorph (K_0 = 29.1 GPa;K_0' = 6.7). The compressibility is comparable with crystalline graphite (K_o = 34 GPa), although polycrystalline samples have been reported with K_o values extending down 10 GPa19.

Fitting our DFT *V(P)* results for the high-pressure ILB1 $C_6N_9H_3{\cdot}HCl$ structure to a BM-3 EoS yields a bulk modulus K_0 = 130 ± 5 GPa and K_0' = 3.57 ± 0.2, clearly showing that the ILB1 structure has much lower compressibility than the sp^2 bonded graphite-like ILB0 phase (K_0 = 29.1 ± 4.1 GPa). The new high pressure ILB1 phase, despite partly retaining the void intralayer structure of the low pressure ILB0 polymorph, has therefore substantially higher structural strength, a feature of potential interest for practical applications.

DISCUSSION

We conducted a high pressure study of graphitic $C_6N_9H_3{\cdot}HCl$ using complementary experimental and theoretical approaches. The graphitic layered compound $C_6N_9H_3{\cdot}HCl$ was subjected to pressures of up to 70 GPa in a diamond anvil cell, and its structural behaviour was examined using synchrotron X-ray diffraction. The structural model proposed by Zhang et al. with a layered hexagonal ($P6_3/m$) structure was employed for the experimental data analysis, for which it provides a useful description up to a pressure of ~40 GPa. Close examination of the V(P) relation indicates an anomaly near 20 GPa that can be explained by a deviation of the structure from $P6_3/m$ symmetry that can be interpreted from the computational results. On increasing pressure from ambient conditions, we observe three distinct structural rearrangements: at ambient pressure the carbon nitride layers are atomically flat and host the Cl^- ions within the layers; between 10–20 GPa the layers buckle and the Cl^- ions displace vertically but not horizontally with respect to the layers; above 20 GPa the Cl^- ions can displace significantly both horizontally and vertically, making the $P6_3/m$ model less and less appropriate for a detailed structural description.

The calculations identify the formation of a new pillared structure (referred to as ILB1), that is thermodynamically stable compared to the graphitic phase above 47 GPa. Simulated XRD patterns using graphitic and pillared phases from the computational work were compared with the experimental XRD data, and these indicate that the predicted pillared structure provides the most plausible match to the observed data above 40 GPa. This pillared structure is formed by buckled sp^2-hybridized platelets joined by sp^3-hybridised "tubes" running along the (001) direction (Fig. 5). The close match between the experimental data and that of the ILB1 structure suggests that the initially graphite-like system undergoes a change in hybridization and experiences interlayer C-N bonding as predicted by the calculations.

In conclusion, our experimental and DFT results indicate that compression of the layered compound g-$C_6N_9H_3$·HCl at ambient temperature results in a polymorphic transformation leading to a new interlayer bonded phase. This has mixed sp^2–sp^3 bonding between C and N atoms in adjacent layers and it gives rise to a new family of pillared-layered compounds that can combine the essential characteristics of the 2D material with buckled $C_xN_y(H_z)$ planes and the mechanical properties of a 3D-bonded compound.

METHODS

The starting compound g-$C_6N_9H_3$·HCl was formed from a 1:2 mixture of melamine ($C_3N_6H_6$) and cyanuric chloride ($C_3N_3Cl_3$) reacted under high-P,T conditions (1.5 GPa, 550°C) in a piston-cylinder device. The synthesis details and structural characterization of this material were described previously8′18.

Synchrotron X-ray diffraction studies were carried out at up to 70 GPa using angle dispersive techniques at ESRF ID27 (Grenoble, France: λ_o = 0.3738 Å: Figure 2). Additional data were obtained at Diamond Light Source I15, UK (λ_o = 0.4441 Å). Samples were loaded into mechanically (four post Mao-type symmetrical cells) or gas-driven He membrane diamond anvil cells (DACs). Diamonds with 200–300 μm culet sizes were used to collect data up to 70 GPa. Re gaskets pre-indented to ~30 μm with gasket holes ~80 μm in diameter were drilled by electro-erosion. Samples were packed into the cell chamber inside a glove box without a pressure-transmitting medium (PTM). Ruby fluorescence spectra indicated that the pressurisation environment remained reasonably "hydrostatic" throughout most of the runs: pressure gradients of only ~1–2 GPa across the cell were detected and the ruby signals showed little broadening. The samples were loaded without a PTM to avoid interference with the weak and often broadened diffraction signals from the low-Z carbon nitride material, and to avoid potential effects associated with intercalation of fluid species such as He, N_2 or H_2 into the layered structure. X-ray diffraction data were recorded using MAR image plate or CCD detectors and analysed using Fit2D20, FullProf21 and GSAS22′23 software.

Electronic structure calculations were carried out under periodic boundary conditions within the local density approximation (LDA) of density functional theory (DFT) using the planewave code CASTEP24. We used ultra-soft pseudopotentials and chose the cut-off energy (450 eV) and k-point grid size (3 × 3 × 4) to obtain best converged results25. Calculations were performed under hydrostatic pressure conditions: full geometry optimizations were carried out at each pressure within space group *P1* using a 40-atom unit crystallographic cell. Prior to each geometry optimization, the atoms were subjected to small random displacements to produce symmetry-free initial structures.

ACKNOWLEDGEMENTS

Work in PFM's group has been supported by EPSRC and a Wolfson-Royal Society research merit award. The authors thank Andrea Sella, Edward Bailey, Victoria Lees, and Mohamed Matar for assistance with sample preparation, synchrotron experiments and DFT calculations. We thank Diamond Light Source (UK) and ESRF (ID27) for access to synchrotron facilities.

CONTRIBUTIONS

A.S. carried out experiments and analysis assisted by R.Q.C. Calculations were carried out by F.C. and M.D., P.F.M., A.S. and F.C. conceived and wrote the paper.

COMPETING INTERESTS

The authors declare no competing financial interests.

REFERENCES

1. Liu, A. Y. & Cohen, M. L. Prediction of new low compressibility solids. *Science* **245**, 841–842 (1989).
2. Teter, D. M. & Hemley, R. J. Low-compressibility carbon nitrides.*Science* **271**, 53–55 (1996).
3. Horvath-Bordon, E. *et al.* High-pressure synthesis of crystalline carbon nitride imide, $C_2N_2(NH)$. *Angew. Chem. Int. Edit.* **46**, 1476–1480 (2007).
4. Salamat, A. *et al.* Tetrahedrally bonded dense C_2N_3H with a defective wurtzite structure: X-ray diffraction and Raman scattering results at high pressure and ambient conditions. *Phys. Rev. B* **80**, 1041066 (2009).

5. Bojdys, M. J., Muller, J. O., Antonietti, M. & Thomas, A.Ionothermal Synthesis of Crystalline, Condensed, Graphitic Carbon Nitride. *Chem. Eur. J.* **14**, 8177–8182 (2008).
6. Miller, D. R., Swenson, D. C. & Gillan, E. G. Synthesis and structure of 2,5,8-triazido-s-heptazine: An energetic and luminescent precursor to nitrogen-rich carbon nitrides. *J. Am. Chem. Soc.* **126**, 5372–5373 (2004).
7. Wang, X. C. *et al.* A metal-free polymeric photocatalyst for hydrogen production from water under visible light. *Nature Mater.* **8**, 76–80 (2009).
8. Zhang, Z. H. *et al.* High-pressure bulk synthesis of crystalline $C_6N_9H_3$·HCl: A novel C_3N_4 graphitic derivative. *J. Am. Chem. Soc.***123**, 7788–7796 (2001).
9. Ueno, M. *et al.* Room-temperature transition of rhombohedral-type boron nitride under high static pressure. *Phys. Rev. B* **45**, 10226–10230 (1992).
10. Yagi, T. *et al.* High-pressure *in situ* x-ray-diffraction study of the phase transformation from graphite to hexagonal diamond at room temperature. *Phys. Rev. B* **46**, 6031–6039 (1992).
11. Solozhenko, V. L. & Kurakevych, O. O. Raman scattering from turbostratic graphitelike BC_4 under pressure. *J. Appl. Phys.* **102**, 063509 (2007).
12. Solozhenko, V. L. & Kurakevych, O. O. Reversible pressure-induced structure changes in turbostratic BN-C solid solutions.*Acta Crystallogr. B* **61**, 498–503 (2005).
13. Mao, W. L. *et al.* Bonding changes in compressed superhard graphite. *Science* **302**, 425–427 (2003).
14. Meng, Y. *et al.* The formation of sp^3 bonding in compressed BN.*Nature Materials* **3**, 111–114 (2004).
15. Kroke, E. *et al.* Tri-s-triazine derivatives. Part I. From trichloro-tri-s-triazine to graphitic C_3N_4 structures. *New J. Chem.* **26**, 508–512 (2002).
16. Kouvetakis, J., Bandari, A., Todd, M., Wilkens, B. & Cave, N. Novel synthetic routes to carbon-nitrogen thin-films. *Chem. Mat.* **6**, 811–814 (1994).
17. Wirnhier, E. *et al.* Poly(triazine imide) with Intercalation of Lithium and Chloride Ions [$(C_3N_3)_2(NH_xLi_{1-x})_3$·LiCl]: A Crystalline 2D Carbon Nitride Network. *Chem. Eur. J.* **17**, 3213–3221 (2011).
18. McMillan, P. F. *et al.* Graphitic carbon nitride $C_6N_9H_3$·HCl: Characterization by UV and near-IR FT Raman spectroscopy. *J. Solid State Chem.* **182**, 2670–2677 (2009).
19. Krishnan, A., Dujardin, E., Ebbesen, T. W., Yianilos, P. N. & Treacy, M. M. J. Young's modulus of single-walled nanotubes. *Phys. Rev. B***58**, 14013–14019 (1998).

20. Hammersley, A. P., Svensson, S. O., Hanfland, M., Fitch, A. N. &Hausermann, D. Two-dimensional detector software: From real detector to idealised image or two-theta scan. *High Pressure Res.* **14**, 235–248 (1996).
21. Rodriguez-Carvajal, J. Recent advances in magnetic-structure determination by neutron powder diffraction. *Physica B* **192**, 55–69 (1993).
22. Larson, A. C. & Von Dreele, R. B. General Structure Analysis System (GSAS). *Los Alamos National Laboratory Report LAUR* 86–748 (2000).
23. Toby, B. H., EXPGUI, a graphical user interface for *GSAS*. *J. Appl. Cryst.* **34**, 210–213 (2001).
24. Clark, S. J. *et al.* First principles methods using CASTEP. *Z. Kristall.* **220**, 567–570 (2005).
25. Deifallah, M., McMillan, P. F. & Cora, F. Electronic and structural properties of two-dimensional carbon nitride graphenes. *J. Phys. Chem. C* **112**, 5447–5453 (2008).

Index